Communications
in Computer and Information Science 1572

More information about this series at https://link.springer.com/bookseries/7899

Kanubhai K. Patel · Gayatri Doctor · Atul Patel ·
Pawan Lingras (Eds.)

Soft Computing and its Engineering Applications

Third International Conference, icSoftComp 2021
Changa, Anand, India, December 10–11, 2021
Revised Selected Papers

 Springer

Editors
Kanubhai K. Patel 🆔
Charotar University of Science
and Technology
Changa, Anand, India

Atul Patel 🆔
Charotar University of Science
and Technology
Changa, Anand, India

Gayatri Doctor 🆔
CEPT University
Ahmedabad, India

Pawan Lingras
Saint Mary's University
Halifax, NS, Canada

ISSN 1865-0929   ISSN 1865-0937 (electronic)
Communications in Computer and Information Science
ISBN 978-3-031-05766-3   ISBN 978-3-031-05767-0 (eBook)
https://doi.org/10.1007/978-3-031-05767-0

This Springer imprint is published by the registered company Springer Nature Switzerland AG
The registered company address is: Gewerbestrasse 11, 6330 Cham, Switzerland

Preface

It is a matter of great privilege to have been tasked with the writing of this preface for the proceedings of the 3rd International Conference on Soft Computing and its Engineering Applications (icSoftComp 2021). The conference aimed to provide an excellent international forum to the researchers, academicians, students, and professionals in the areas of computer science and engineering to present their research, knowledge, new ideas, and innovations. The theme of the conference was "Soft Computing Techniques for Sustainable Development". The conference was held during December 10–11, 2021, at Charotar University of Science and Technology (CHARUSAT), Changa, India, and organized by the Faculty of Computer Science and Applications, CHARUSAT.

There are three pillars of soft computing, viz., i) fuzzy computing, ii) neuro computing, and iii) evolutionary computing. Research submissions in these three areas were received. The Program Committee of icSoftComp 2021 is extremely grateful to the authors from 17 different countries, including the USA, Poland, Russia, Turkey, United Arab Emirates, Denmark, Mauritius, Saudi Arabia, South Africa, Nigeria, Ecuador, Libya, Taiwan, Bangladesh, Ukraine, and UK, who showed an overwhelming response to the call for papers, submitting over 247 papers. The entire review team (the Technical Program Committee members along with three additional reviewers) expended tremendous effort to ensure fairness and consistency during the selection process resulting in the best quality papers being selected for presentation and publication. It was ensured that every paper received at least three, and in most cases four, reviews. Similarity checks were also performed based on the international norms and standards.

After a rigorous peer review process, 33 papers were accepted giving an acceptance rate of 13.36%. The papers are organized according to the following topics: Theory and Methods and Systems and Applications. The proceedings of the conference are published as one volume in the Communications in Computer and Information Science (CCIS) series by Springer, and are also indexed by WoS, dblp, Ulrich's, Ei Compendex, Scopus, zbMATH, Metapress, and SpringerLink. We, in our capacity as volume editors, convey our sincere gratitude to Springer for providing the opportunity to publish the proceedings of icSoftComp 2021 in their CCIS series.

icSoftComp 2021 provided an excellent international virtual forum to the conference delegates to present their research, knowledge, new ideas, and innovations. The conference exhibited an exciting technical program. It also featured high-quality workshops, a keynote, and six expert talks from prominent research and industry leaders. The keynote speech was given by V. Susheela Devi (Principal Research Scientist, Indian Institute of Science Bangalore, India). Experts talks were given by Ashish Ghosh (Indian Statistical Institute Kolkata, India), Shailesh Kumar (Chief Data Scientist, Center of Excellence in AI/ML, Reliance Jio, Hyderabad, India), Massimiliano Cannata (University of Applied Sciences and Arts of Southern Switzerland, Switzerland), Vishnu Pendyala (San José State University, USA), Kiran Trivedi (Vishwakarma Government Engineering College, India), and Pritpal Singh (Jagiellonian University, Poland). We are grateful to them for sharing their insights on their latest research with us.

The Organizing Committee of icSoftComp 2021 is indebted to R V Upadhyay, Provost of Charotar University of Science and Technology and Patron, for the confidence that he invested in us in organizing this international conference. We would also like to take this opportunity to extend our heartfelt thanks to the honorary chairs of this conference, Kalyanmoy Deb (Michigan State University, USA), Janusz Kacprzyk (Polish Academy of Sciences, Poland), and Leszek Rutkowski (Czestochowa University of Technology, Poland) for their active involvement from the very beginning until the end of the conference. The quality of a refereed volume primarily depends on the expertise and dedication of the reviewers who volunteer with a smiling face. The editors are further indebted to the Technical Program Committee members and external reviewers who not only produced excellent reviews but also did so in a short time frame, in spite of their very busy schedules. Because of their quality work it was possible to maintain the high academic standard of the proceedings. Without their support, this conference could never have assumed such a successful shape. Special words of appreciation are due to note the enthusiasm of all the staff and students of the Faculty of Computer Science and Applications of CHARUSAT, who organized the conference in a professional manner.

It is needless to say that the conference would not have been possible without the contributors. The editors would like to take this opportunity to thank the authors of all submitted papers for not only their hard work but also for considering the conference a viable platform to showcase some of their latest findings, not to mention their adherence to the deadlines and patience with the tedious review process. Special thanks to the team of OCS, whose paper submission platform was used to organize reviews and collate the files for these proceedings. We also wish to express our thanks to Amin Mobasheri (Springer Heidelberg) for his help and cooperation. We gratefully acknowledge the financial (partial) support received from the Department of Science and Technology, Government of India, and the Gujarat Council on Science and Technology (GUJCOST), Government of Gujarat, Gandhinagar, India, for organizing the conference. Last but not least, the editors profusely thank all who directly or indirectly helped us in making icSoftComp 2021 a grand success and allowed for the conference to achieve its goal, academic or otherwise.

December 2021

Kanubhai K. Patel
Gayatri Doctor
Atul Patel
Pawan Lingras

Organization

Patron

R. V. Upadhyay   Charotar University of Science and Technology, India

Honorary Chairs

Kalyanmoy Deb   Michigan State University, USA
Janusz Kacprzyk   Polish Academy of Sciences, Poland
Leszek Rutkowski   Czestochowa University of Technology, Poland

General Chairs

Atul Patel   Charotar University of Science and Technology, India

Pawan Lingras   Saint Mary's University, Canada

Technical Program Committee Chair

Kanubhai K. Patel   Charotar University of Science and Technology, India

Technical Program Committee Co-chairs

Deepak Garg   Bennett University, India
Gayatri Doctor   CEPT University, India

Advisory Committee

Arup Dasgupta   Geospatial Media and Communications, India
Ashish Ghosh   ISI Kolkata, India
Balas Valentina Emilia   University of Arad, Romania
Bhushan Trivedi   GLS University, India
Bhuvan Unhelkar   University of South Florida Sarasota-Manatee, USA
D. K. Pratihar   Indian Institute of Technology Kharagpur, India
J. C. Bansal   Soft Computing Research Society, India
Narendra S. Chaudhari   Indian Institute of Technology Indore, India
Rajendra Akerkar   Vestlandsforsking, Norway
Sudhir Kumar Barai   BITS Pilani, India

Suman Mitra	DAIICT, India
Devang Joshi	Charotar University of Science and Technology, India
S. P. Kosta	Charotar University of Science and Technology, India
Dharmendra T. Patel	Charotar University of Science and Technology, India

Technical Program Committee

Abdulla Omeer	Dr. Babasaheb Ambedkar Marathwada University, India
Abhilash Shukla	Charotar University of Science and Technology, India
Abhineet Anand	Chitkara University, India
Aditya Patel	Kamdhenu University, India
Adrijan Božinovski	University American College Skopje, Macedonia
Aji S.	University of Kerala, India
Akhil Meerja	Vardhaman College of Engineering, India
Aman Sharma	Jaypee University of Information Technology, India
Ami Choksi	C. K. Pithawala College of Engineering and Technology, India
Amit Joshi	Malaviya National Institute of Technology, India
Amit Thakkar	Charotar University of Science and Technology, India
Amol Vibhute	MIT World Peace University, India
Anand Nayyar	Duy Tan University, Vietnam
Angshuman Jana	IIIT Guwahati, India
Ansuman Bhattacharya	IIT (ISM) Dhanbad, India
Anurag Singh	IIIT Naya Raipur, India
Aravind Rajam	Washington State University, USA
Arjun Mane	Government Institute of Forensic Science, India
Arpankumar Raval	Charotar University of Science and Technology, India
Arti Jain	Jaypee Institute of Information Technology, India
Arunima Jaiswal	Indira Gandhi Delhi Technical University for Women, India
Asha Manek	RVITM Engineering College, India
Ashok Patel	Florida Polytechnic University, USA
Ashok Sharma	Lovely Professional University, India
Ashraf Elnagar	University of Sharjah, UAE
Ashutosh Kumar Dubey	Chitkara University, India

Ashwin Makwana	Charotar University of Science and Technology, India
Avimanyou Vatsa	Fairleigh Dickinson University, Teaneck, USA
Avinash Kadam	Dr. Babasaheb Ambedkar Marathwada University, India
Ayad Mousa	University of Kerbala, Iraq
Bhaskar Karn	BIT Mesra, India
Bhavik Pandya	Navgujarat College of Computer Applications, India
Bhogeswar Borah	Tezpur University, India
Bhuvaneswari Amma	IIIT Una, India
Chaman Sabharwal	Missouri University of Science and Technology, USA
Charu Gandhi	Jaypee University of Information Technology, India
Chirag Patel	Innovate Tax, UK
Chirag Paunwala	SCET, India
Costas Vassilakis	University of the Peloponnese, Greece
Darshana Patel	Navgujarat College of Computer Applications, India
Dattatraya Kodavade	DKTE Society's Textile and Engineering Institute, India
Dayashankar Singh	Madan Mohan Malaviya University of Technology, India
Deepa Thilak	SRM University, India
Deepak N. A.	RV Institute of Technology and Management, India
Deepak Singh	IIIT Lucknow, India
Delampady Narasimha	IIT Dharwad, India
Dharmendra Bhatti	Uka Tarsadia University, India
Digvijaysinh Rathod	National Forensic Sciences University, India
Dinesh Acharya	Manipal Institute of Technology, India
Divyansh Thakur	IIIT Una, India
Dushyantsinh Rathod	Alpha College of Engineering and Technology, India
E. Rajesh	Galgotias University, India
Gururaj Mukarambi	Central University of Karnataka, India
Gururaj H. L.	Vidyavardhaka College of Engineering, India
Hardik Joshi	Gujarat University, India
Harshal Arolkar	GLS University, India
Himanshu Jindal	Jaypee University of Information Technology, India
Hiren Joshi	Gujarat University, India

Hiren Mewada	Prince Mohammad Bin Fahd University, Saudi Arabia
Irene Govender	University of KwaZulu-Natal, South Africa
Jagadeesha Bhatt	IIIT Dharwad, India
Jaimin Undavia	Charotar University of Science and Technology, India
Jaishree Tailor	Uka Tarsadia University, India
Janmenjoy Nayak	AITAM, India
Jaspher Kathrine	Karunya Institute of Technology and Sciences, India
Jimitkumar Patel	Charotar University of Science and Technology, India
Joydip Dhar	ABV-IIITM, India
József Dombi	University of Szeged, Hungary
Kamlendu Pandey	VNSGU, India
Kamlesh Dutta	NIT Hamirpur, India
Kiran Trivedi	Vishwakarma Government Engineering College, India
Kiran Sree Pokkuluri	Shri Vishnu Engineering College for Women, India
Krishan Kumar	NIT Uttarakhand, India
Kuldip Singh Patel	IIIT Naya Raipur, India
Kuntal Patel	Ahmedabad University, India
Latika Singh	Ansal University, India
M. Srinivas	NIT Warangal, India
M. A. Jabbar	Vardhaman College of Engineering, India
Maciej Ławryńczuk	Warsaw University of Technology, Poland
Mahmoud Elish	Gulf University for Science and Technology, Kuwait
Mandeep Kaur	Sharda University, India
Manoj Majumder	IIIT Naya Raipur, India
Michał Chlebiej	Nicolaus Copernicus University, Poland
Mittal Desai	Charotar University of Science and Technology, India
Mohamad Ijab	National University of Malaysia, Malaysia
Mohini Agarwal	Amity University Noida, India
Monika Patel	NVP College of Pure Applied Sciences, India
Mukti Jadhav	Marathwada Institute of Technology, India
Neepa Shah	Gujarat Vidyapith, India
Neetu Sardana	Jaypee University of Information Technology, India
Nidhi Arora	Solusoft Technologies Pvt. Ltd., India

Nilay Ganatra	Charotar University of Science and Technology, India
Nilay Vaidya	Charotar University of Science and Technology, India
Nirali Honest	Charotar University of Science and Technology, India
Nitin Kumar	NIT Uttarakhand, India
Parag Rughani	GFSU, India
Parul Patel	VNSGU, India
Pranav Vyas	Charotar University of Science and Technology, India
Prashant Pittalia	Sardar Patel University, India
Priti Sajja	Sardar Patel University, India
Pritpal Singh	Jagiellonian University, Poland
Punya Paltani	IIIT Naya Raipur, India
Rajeev Kumar	NIT Hamirpur, India
Rajesh Thakker	Vishwakarma Government Engineering College, India
Ramesh Prajapati	LJ Institute of Engineering and Technology, India
Ramzi Guetari	University of Tunis El Manar, Tunisia
Rana Mukherji	ICFAI University, Jaipur, India
Rashmi Saini	GB Pant Institute of Engineering and Technology, India
Rathinaraja Jeyaraj	NIT Karnataka, India
Rekha A. G.	State Bank of India, India
Rohini Rao	Manipal Academy of Higher Education, India
S. Shanmugam	Concordia University Chicago, USA
S. Srinivasulu Raju	VR Siddhartha Engineering College, India
Sailesh Iyer	Rai University, India
Saman Chaeikar	Iranians University, Iran
Sameerchand Pudaruth	University of Mauritius, Mauritius
Samir Patel	PDPU, India
Sandeep Gaikwad	Charotar University of Science and Technology, India
Sandhya Dubey	Manipal Academy of Higher Education, India
Sanjay Moulik	IIIT Guwahati, India
Sannidhan M. S.	NMAM Institute of Technology, India
Sanskruti Patel	Charotar University of Science and Technology, India
Saurabh Das	University of Calcutta, India
S. B. Goyal	City University of Malaysia, Malaysia
Shachi Sharma	South Asian University, India

Shailesh Khant	Charotar University of Science and Technology, India
Shefali Naik	Ahmedabad University, India
Shilpa Gite	Symbiosis Institute of Technology, India
Shravan Kumar Garg	Swami Vivekanand Subharti University, India
Sohil Pandya	Charotar University of Science and Technology, India
Spiros Skiadopoulos	University of the Peloponnese, Greece
Srinibas Swain	IIIT Guwahati, India
Srinivasan Sriramulu	Galgotias University, India
Subhasish Dhal	IIIT Guwahati, India
Sudhanshu Maurya	Graphic Era Hill University, Malaysia
Sujit Das	NIT Warangal, India
Sumegh Tharewal	Dr. Babasaheb Ambedkar Marathwada University, India
Sunil Bajeja	Marwadi University, India
Swati Gupta	Jaypee University of Information Technology, India
Tanima Dutta	Indian Institute of Technology (BHU), India
Tanuja S. Dhope	Rajarshi Shahu College of Engineering, India
Thoudam Singh	NIT Silchar, India
Trushit Upadhyaya	Charotar University of Science and Technology, India
Tzung-Pei Hong	National University of Kaohsiung, Taiwan
Vana Kalogeraki	Athens University of Economics and Business, Greece
Vasudha M. P.	Jain University, India
Vatsal Shah	BVM Engineering, India
Veena Jokhakar	VNSGU, India
Vibhakar Pathak	Arya College of Engineering and IT, India
Vijaya Rajanala	SR Engineering College, India
Vinay Vachharajani	Ahmedabad University, India
Vinod Kumar	IIIT Lucknow, India
Vishnu Pendyala	San José State University, USA
Yogesh Rode	Jijamata Mahavidhyalaya Buldana, India
Zina Miled	Indiana University, USA

Additional Reviewers

Ankur Bist
Pooja Ajwani
Preeti Kathiria

Contents

Theory and Methods

Fuzzy C-Mean Clustering Based Soccer Result Analysis 3
   Richa and Jyotsna Yadav

Segregation of Areca Nuts Using Three Band Photometry and Deep
Neural Network 15
   Saurav Dosi, Bala Vamsi, Samarth S. Raut, and D. Narasimha

CT/MRI 3D Fusion for Cerebral System Analysis 28
   Michal Chlebiej, Anna Zurada, and Jerzy Gielecki

Soft Optimal Computing to Identify Surface Roughness in Manufacturing
Using a Gaussian and a Trigonometric Regressor 41
   Benedikt Haus, Paolo Mercorelli, Jin Siang Yap, and Lennart Schäfer

An Improved Crow Search Algorithm with Grey Wolf Optimizer
for High-Dimensional Optimization Problems 51
   Artee Abudayor and Özkan Ufuk Nalbantoğlu

A Fuzzy Rule Based Directional Approach for Salt and Pepper Noise
Removal with Edge Preservation 65
   Aritra Bandyopadhyay and Devadatta Das

Utility Distribution Based Measures of Probabilistic Single Valued
Neutrosophic Information, Hybrid Ambiguity and Information
Improvement 78
   Mahima Poonia and Rakesh Kumar Bajaj

Intelligent Friendship Graphs: A Theoretical Framework 90
   Indradeep Bhattacharya and Shibakali Gupta

Modeling of the Koch-Type Fractal Wire Dipole Antenna with the Random
Forest Algorithm 103
   Ilya Pershin and Dmitrii Tumakov

Comparative Analysis of Deep Learning Techniques for Facemask
Detection 116
   Ghazala Furqan, Najme Zehra Naqvi, and Arunima Jaiswal

Relating Machine Learning to the Real-World: Analogies to Enhance
Learning Comprehension 127
   Vishnu S. Pendyala

Sentiment Analysis of Twitter Data Using Machine Learning Approaches 140
   Vishal Gaba and Vijay Verma

Mining Spatio-Temporal Sequential Patterns Using MapReduce Approach 153
   Sumalatha Saleti, P. RadhaKrishna, and D. JaswanthReddy

A Split-Then-Join Lightweight Hybrid Majority Vote Classifier 167
   Moses L. Gadebe, Sunday O. Ojo, and Okuthe P. Kogeda

Identification of Barriers in Adoption of IoT: Commercial Complexes
in India 181
   Nishani Salvi and Gayatri Doctor

A Dynamically Adapting Framework for Stock Price Prediction 194
   Shruti Mittal and C. K. Nagpal

Evaluating Binary Classifiers with Word Embedding Techniques for Public
Grievances 209
   Khushboo Shah, Hardik Joshi, and Hiren Joshi

Database Concentration Method for Efficient Image Retrieval Using
Clustering and Image Tag Comparison 222
   Soorya Ram Shimgekar, Preetham Reddy Pathi, and V. Vijayarajan

Microstructure Image Classification of Metals Using Texture Features
and Machine Learning 235
   Hrishikesh Sabnis, J. Angel Arul Jothi, and A. M. Deva Prasad

Early Diagnosis of Alzheimer's Disease from MRI Images Using
Scattering Wavelet Transforms (SWT) 249
   Deepthi Oommen and J. Arunnehru

Constraint Pushing Multi-threshold Framework for High Utility Time
Interval Sequential Pattern Mining 264
   *Sumalatha Saleti, N. Naga Sahithya, K. Rasagna, K. Hemalatha,
   B. Sai Charan, and P. V. Karthik Upendra*

Systems and Applications

KTSVidRec: A Knowledge-Based Topic Centric Semantically Compliant
Approach for Video Recommendation on the Web 277
    Akhil S. Krishnan and Gerard Deepak

Intelligent Facial Expression Evaluation to Assess Mental Health Through
Deep Learning 290
    Prajwal Gaikwad, Sanskruti Pardeshi, Shreya Sawant,
    Shrushti Rudrawar, and Ketaki Upare

Intelligent Mobility: A Proposal for Modeling Traffic Lights Using Fuzzy
Logic and IoT for Smart Cities 302
    Gabriel Gomes de Oliveira, Yuzo Iano, Gabriel Caumo Vaz,
    Pablo David Minango Negrete, Juan Carlos Minango Negrete,
    and Euclides Lourenço Chuma

Crop Disease Prediction Using Multiple Linear Regression Modelling 312
    Hudaa Neetoo, Yasser Chuttur, Azina Nazurally, Sandhya Takooree,
    and Nooreen Mamode Ally

Radial Basis Function Network Based Intelligent Scheme for Software
Quality Prediction 327
    Ritu and O. P. Sangwan

Generating the Base Map of Regions Using an Efficient Object
Segmentation Technique in Satellite Images 341
    Kavitha Srinivasan, Sudhamsu Gurijala, V. Sai Chitti Subrahmanyam,
    and B. Swetha

KCEPS: Knowledge Centric Entity Population Scheme for Research
Document Recommendation 356
    N. Krishnan and Gerard Deepak

Weighted Hybrid Recommendation System Using Singular Value
Decomposition and Cosine Similarity 367
    Sanket Shah, Yogesh Raisinghani, and Nilay Gandhi

Text Analysis and Classification for Preprocessing Phase of Automatic
Text Summarization Systems 382
    Vaishali P. Kadam, Kalpana B. Khandale, and Namrata Mahender C.

Arabic Cyberbullying Detection from Imbalanced Dataset Using Machine
Learning 397
    Meshari Essa AlFarah, Ibrahim Kamel, Zaher Al Aghbari,
    and Djedjiga Mouheb

A CNN Based Air-Writing Recognition Framework for Linguistic
Characters 410
   Prabhat Kumar, Abhishek Chaudhary, and Abhishek Sharma

Intraday Stock Trading Performance of Traditional Machine Learning
Algorithms: Comparing Performance with and Without Consideration
of Trading Costs 421
   Kashyap D. Soni

Author Index 433

Theory and Methods

Fuzzy C-Mean Clustering Based Soccer Result Analysis

Richa[✉] and Jyotsna Yadav

Guru Gobind Singh Indraprastha University, New Delhi, India
richabhardwaj897@gmail.com

Abstract. European Soccer also known as soccer is considered to be the most popular team sport in the world. It is believed to be played by more than a whopping 150 million men as well as women of all the different age groups in probably more than 200 countries. Various techniques that can predict the outcome for professional soccer matches have used the count of goals which was scored by each team and use it as a base measure that evaluates performance of the team and also helps to estimate future results. In the sector of Machine Learning, various clustering algorithms are commonly used for creating several clusters. In this paper, an efficient clustering algorithm that is Fuzzy C-Mean clustering is proposed which would make different clusters that result to clustering of the home team corner with respect to away team corner on Soccer dataset. This clustering will help coaches to take right decision for improving their performance.

Keywords: Machine learning · Clustering algorithm · Cluster validation

1 Introduction

As the power of computer is increasing day by day, using computer for large amount of data became more conventional. Also, several machine learning techniques have been applied to sports data over the past decades. Clubs started using machine learning techniques to analyze their players to maximize the performance. This paper will show how we have implemented clustering technique of fuzzy c-mean for making clusters and that would help coaches to make strategies to improve their performance. The data taken is in.csv form. Firstly the data of different years is merged into single dataset and fed to the machine learning model and then number of clusters and cluster centers of these clusters are defined. A threshold value is also defined to let the model know where the iteration should be stopped. Finally, silhouette scores are used to validate the performance of the clustering algorithm.

This paper has been divided in various sections as follows: Sect. 2, the review for literature has been written, Sect. 3 will explain the clustering, Sect. 4 would describe the methodology that we have used, in the Sect. 5 we have explained the results and lastly Sect. 6 talks about conclusion which has been obtained and finally paper end with last section, having references.

© Springer Nature Switzerland AG 2022
K. K. Patel et al. (Eds.): icSoftComp 2021, CCIS 1572, pp. 3–14, 2022.
https://doi.org/10.1007/978-3-031-05767-0_1

2   Related Work

Saima Bano et al. [1] discussed about the importance of data clustering in various fields that include data mining, artificial intelligence, pattern recognition, etc. They also said that the main idea behind data clustering is to classify all the similar entities. They compared as well as analyzed every clustering technique and after that found their advantages along with disadvantages to give an improved approach for those future researches that would be done in clustering of data item. According to this paper, there are various techniques of clustering of data which is available for the various types of nature of applications. There are two types of data clustering techniques. One of these techniques is Partitioning Procedures and the other one is Hierarchical Procedures. They said that hierarchy of clusters is created in hierarchical procedures that look like a tree. Various partitions of objects are made in partitioning method clustering and those partitions are evaluated by some standard. They also introduced the strengths and weaknesses of different clustering techniques.

Md. Abu Bakr Siddique et al. [6] presented considerable two-dimensional clustering schemes. In those schemes, soft partitioning clustering algorithms i.e. Possibilistic C-means (PCM) and Fuzzy C-means (FCM) were applied. After that, to investigate the clustering tendency visually, VAT was used. After that indices were used which are of three types (e.g., PC, DI, and DBI) to the check cluster validation. In this paper, they concluded that each clustering algorithm has its limitations. FCM was less robust to noise than PCM. The reason for this less robustness is that a noise point has to be recognized as a part of any of the cluster in FCM.

R. Suganya et al. [7] represents a survey on fuzzy c-mean clustering algorithm. They said that clustering can be defined the process of referring a set of objects to be characterized in groups called clusters. According to this paper, the classification of clustering algorithms is achieved in two categories. One is soft clustering which is also called as fuzzy clustering; another one is hard clustering. In hard clustering, the data points are classified into groups called clusters in such a way that every specific data point corresponds to precisely a single cluster. Whereas, in soft clustering, the data points are divided into clusters in a suitable way that the data elements can correspond to multiple clusters, and associated a set of membership levels with each point. In this paper, they concluded the advantage and disadvantages of fuzzy c-mean clustering algorithm. Afterwards, they discussed about the fuzzy possibilistic c-mean algorithm and the possibilistic c-mean algorithm. After that they discussed about the pros and cons of these algorithm. They said that FCM uses the squared-norm to ascertain the correlation between the data points and the prototypes. The clustering spherical character is the only case in which it functions well. In the interim of the survey, they also found some points which can be improved further in the future by using clustering techniques which are advanced in order to accomplish more efficient accuracy in the result. It will also decrease the information retrieval and/or time taken for data from large dataset.

Pham DT et al. [8] introduced the clustering algorithms from the context of data mining. They concentrated more on the techniques for improving these algorithms so that they can handle large datasets. They also proposed clustering algorithm as an efficient device to handle challenging real-world problems by describing a count of engineering applications to which such techniques can be applied. According to this paper, cluster-ing can be described as an important technique for data exploration having numerous applications in diverse domains of engineering which includes design quality assur-ance, manufacturing system design, engineering design, modelling, process planning and production planning, monitoring and control.

3 Preliminary Description

3.1 Clustering

Clustering is a kind of unsupervised learning method with which we have tendency to analyse from the datasets that contain input data without the classified responses. Usually, to seek out meaningful structure, interpretive hidden generative features, groupings and processes, implicit in a group of examples, the techniques of clustering is used. Process collection is the process of dividing data points into multiple groups in such a way that each group has the same data points and those data points are dissimilar to the data points in other group. Clustering can be referred to as the set of objects based on correlation and non-correlation between them.

Clustering technique is essential because it finds the unlabelled data present among the elemental grouping. There exist no benchmarks for a quality clustering because the benchmark is decided by the user or criteria to be used to fulfill their need. For example, any user wants to find representative for homogeneous groups or he can be interested in searching the natural clusters and want to depict their own properties i.e. natural data types. The user can also want to search suitable and useful groupings also called as unusual data object or useful data classes that is also called as outlier detection. Such algorithms must describe some assumptions to have correlation between points and every specific assumption generate different clusters which are correspondingly valid [1]. The steps for clustering process are shown in Fig. 1 below:

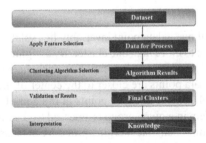

Fig. 1. Steps for clustering

3.1.1 Methods of Clustering

Density-Based Methods
In this method, the clusters are considered to be a dense region that involves some correlation and is differ from the region that has lower density of the space. These methods are very accurate and in line with that, it has the power to combine two groups.

Hierarchical Based Methods
In this method of clustering, the clusters generate a structure just like a tree on the basis of the hierarchy. The new clusters are generated on the basis of previously generated clusters. The hierarchical based clustering technique is divided into two categories:

- Agglomerative which is additionally called as bottom up approach.
- Divisive which is additionally called as top down approach.

Partitioning Methods
By partitioning the objects into k clusters, these methods form one cluster on each partition. An objective criterion similarity function can be optimized using partitioning method, where the distance is supposed to be a major parameter [13].

Grid-Based Methods
The data space is organized into a total number of cells in this method. Those cells form a grid-like structure. In these grids, all integration operations are performed and integration operations are independent and fast with a number of data objects [2].

Centroid-Based Clustering
Centroid-based clustering can be defined as the algorithm which organizes the data in such a way that, in contrast to hierarchical clustering, each data points form non-hierarchical clusters. K-mean clustering is the highly applied centroid-based clustering technique. Centroid-based algorithms are valuable. On the other-hand these algorithms are perceptive to outliers and initial conditions.

Fuzzy Clustering
In other clustering algorithms, a data point invariably lies solely within a cluster in such a way that it cannot belong to more than one cluster. By assigning a data point to multiple collections with a limited level of communication metrics, fuzzy merging methods alter this paradigm [10]. The data points are closer to the center of the collection, which can also be matched to a higher level than the points at the end of the collection [5]. The probability of being an item in a given group is assessed by a sufficiency of membership between 0 to 1. The incomprehensible merging process can be used with those dynamic data sets with a high degree of overlap [3].

3.2 Fuzzy C-Mean Clustering

The unsupervised k-means clustering algorithm descripts the values of any specific point which is residing in some appropriate cluster to have the values as 0 or 1 i.e. either true or false. On the other hand, fuzzy logic provides the fuzzy values of a specific data point that is residing in either of the clusters [4]. As far as the fuzzy c-means clustering algorithm is concerned, we calculate the distance of each data point from the given centroids (after searching for the centroid of the data points) until the clusters formed becomes constant. The assigned collection process is driven by this algorithm. The implementation of the FCM algorithm is almost identical to the k-means cluster assignment. The biggest difference is that according to this algorithm, the data point can be stored in multiple clusters and in some merger algorithms, the data point can only be stored in one cluster [11]. This method uses a method of collecting strange lumps. The FCM method works in the same way as the k-means cluster assignment. The main difference is that a data point can be stored in more than one cluster using this method, but with some merging algorithms, the data point can only be stored in one cluster [12].

Assume that, we have a finite dataset A with p clusters $C = \{c_1, c_2, c_3 \ldots \ldots \ldots \ldots c_p\}$ and a finite dataset A with $A = \{a_1, a_2, a_3 \ldots \ldots \ldots a_n\}$. Assume that $ai = \{a_{i1}, a_{i2}, a_{i3}, \ldots \ldots \ldots a_{io}\}$ is an object which is one dimensional and the ith object's j^{th} attribute is x_{ij}. Cluster centroids are represented by V, where $V = \{v_1, v_2, v_3 \ldots \ldots \ldots v_p\}$ and $v_i = \{v_{i1}, v_{i2}, v_{i3} \ldots \ldots \ldots v_{io}\}$ depicts cluster centroids in one dimension.

Now, assume that U is a matrix which is fuzzy partitioned (also called as fuzzy partition matrix) and defined as follows:

$$U = (u_{ik})_{(n*p)}$$

where, in k^{th} cluster, the degree of membership of ith object is denoted by u_{ik}. And also

$$\sum_{k=1}^{p} u_{ik} = 1 (\forall i = 1, 2 \ldots \ldots \ldots n)$$

Now objective function is defined as:

$$J_m(U, V) = \sum_{k=1}^{p} \sum_{i=1}^{n} (u_{ik})^m (d_{ik})^2 \qquad (1)$$

where, d_{ik} denotes the Euclidean distance of ith object from k^{th} cluster centroid and d_{ik} is represented as:

$$d_{ik} = \|a_i - v_k\|$$

And m is the fuzzification parameter, which determines how fuzzy a cluster is. It's a real number that ranges from 1.0 to infinity. When the value of m is close to 1, a cluster solution similar to the K-mean solution is produced, and when the value of m is close to, total fuzziness is produced.

According to Hathaway and Bezdek in 2001, the m value should be between 1.5 and 2.5, with m = 2 being a suitable choice.

Cluster centroid V and suitable fuzzy partition matrix U are now produced using the clustering criterion for minimising the objective function Jm. The Lagrange Multiplier technique is used to determine U and V.

$$u_{ik} = 1 \Big/ \sum_{j=1}^{p} (d_{ik}/d_{jk})^{2/(m-1)} \qquad (2)$$

$$V_k = \frac{\sum_{i=1}^{n}(u_{ik})^m . x_i}{\sum_{i=1}^{n}(u_{ik})^m} \tag{3}$$

Now, determine the objective function J_m with the help of U and V according to Eq. (1). If

$$\left| J_m^{(e)} - J_m^{(e-1)} \right| < \epsilon$$

or

$$e > I_{max}$$

Then stop here otherwise calculate $U^{(e+1)}$ according to Eq. (2). Here,

I_{max} = maximum number of iteration.
ϵ = Threshold.
m = fuzzification parameter or fuzziness index.

To figure out how many clusters of FCM are there, do the following steps:
Set the range for search the number of clusters as $[C_{min}, C_{max}]$ of number of clusters. Where $C_{min} = 2$ and $C_{max} = sqrt(n)$.
Choose a clustering validity index that is adequate for your needs.
Now, for each different point kn in that interval, run FCM. Where $kn \epsilon [C_{min}, C_{max}]$ Where $kn[C_{min}, C_{max}]$ is a number.
Evaluate the clustering validity index to assess the clustering results. Get the most clusters possible [4].

The steps for performing fuzzy c-mean clustering algorithm are:
Establish the number of clusters in which each data point is randomly grouped.
Define the fuzzification parameter, which is a greater-than-one constant, as well as the threshold value or halting condition.
The fuzzy partition matrix has now been initialized and set the loop counter to e.
After that, the cluster centroids are determined, as well as the goal function J.
In the matrix, compute the membership values for each object in each cluster.
Stop if the halting criterion is bigger than the objective function difference between two successive iterations. If not, set $e = e + 1$ and go to step 4.
Segmentation and defuzzification [6].

3.2.1 Advantages of Fuzzy C-Mean Clustering

It performs better than the k-means technique and produces the best results for data sets that overlap.
Data points in the k-means method must only belong to one cluster centre, however in the fuzzy c mean clustering technique, each data point is awarded membership to each cluster centre, allowing data points to belong to several cluster centres [7].

3.2.2 Disadvantages of Fuzzy C-Mean Clustering: Fuzzy C-Mean Clustering has the Following Drawbacks

- We must define the number of clusters.
- Hidden components can be unequally weighted using Euclidean distance metrics [7].

3.3 Clustering Application

- **Advertising:** Clustering can be used to characterise and define client categories for marketing objectives.
- **Biological Sciences:** For the categorization of diverse plant and animal species [14].
- **Libraries:** Clustering may be used to categorise distinct books based on subjects and metadata.
- **Obtaining insurance:** Clustering may be used to recognise consumers, their policies, and identify frauds [8].

4 Methodology Used

The dataset used is taken from github and is of 16 years. The dataset of each year contains 381 instances and 20 attributes. These attributes are- Div, Date, HomeTeam, AwayTeam, FTHG, FTAG, FTR, HTHG, HTAG, HTR, HS, AS, HST, AST, HF, AF, HC, AC, HY, AY, HR, AR. Where FTHG stands for Full Time Home Goal, FTAG stands for Half Time Away Goal, FTR stands for Full Time Result, HTHG stands for Half Time Home Goal, HTAG stands for Half Time Away Goal, HTR stands for Half Time Result, HS stands for Home Shots, AS stands for Away Shots, HST stands for Home Shots on Target, AST stands for Away Shots on Target, HF stands for Home Team Foul, AF stands for Away Team Foul, HC stands for Home Team Corner, AC stands for Away Team Corner, HY stands for Home Team Yellow Card, AY stands for Away Team Yellow Card, HR stands for Home Team Red Card, AR stands for Away Team Red Card. The data taken is in.csv form.

The above defined dataset is imported to google colaboratory for making use of device getting to know algorithm. Colaboratory is absolutely free Jupyter pocket book surroundings that don't require any setup. It runs totally withinside the cloud. Google Colaboratory comes with collaboration subsidized withinside the product. In fact, its miles a Jupyter pocket book that offers the benefit of Google Docs collaboration characteristic and runs on Google servers. The notebooks are stored on your Google Drive account.

After importing the dataset to Google colaboratory, Fuzzy C-Mean clustering is applied for making clusters of home team corner with respect to away team corner.Firstly the dataset of different years are merged into one dataset and is being loaded. Then feature selection is done using random forest technique to select feature properly. On the selected features, FCM is applied to encode as much information as possible concerning the task of our interest. After that, cluster validation is performed using silhouette scores for different number of clusters. Figure 2 describes all these steps followed for Fuzzy C-Mean Clustering:

Fig. 2. Steps for fuzzy c-means clustering

5   Experimental Results

This section is divided into three parts – First part contains the plot that was obtained after applying feature selection using random forest method on the dataset. Second part contains the cluster visualization that was obtained after applying fuzzy c-mean clustering on the dataset. Third part contains the plots of silhouette scores for different number of clusters and their visualization.

5.1   Plots Obtained After Feature Selection

The graph shown in Fig. 3 describes importance of all the features in the dataset. The x-axis contains all the features and y-axis contains the score for importance of those features. It is clear from that graph that the first four features are less important than other features in the dataset. So, we have discarded first four attributes i.e. Div, Date, HomeTeam and AwayTeam.

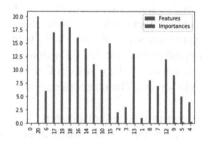

Fig. 3. Features importance after feature selection

5.2   Cluster Visualization After FCM

The plot shown in Fig. 4 contains initial plots of home team corner with respect to away team corner of the dataset.

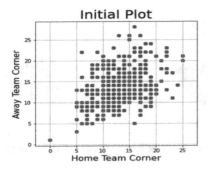

Fig. 4. Initial plot for HC with respect to AC

When variety of clusters is three and fuzzification parameter 'm' is taken as 1.7, the cluster visualization after making use of fuzzy c-approach clustering set of rules on the dataset is shown in Fig. 5 below:

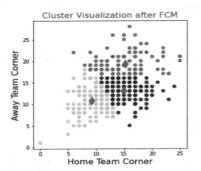

Fig. 5. Cluster visualization after applying FCM

5.3 Plots of Silhouette Analysis for FCM Clustering and Visualization of Corresponding Cluster

This part contains the Silhouette score plots for different clusters and the corresponding cluster display, if the number of clusters is two, the average Silhouette score is 0.3658918916004933. Figure 6 shows the Silhouette score plot for two total clusters and the grouped data visualization.

When number of clusters is three, the average silhouette score is 0.34260504867812985. Figure 7 shows the silhouette score plot for three total clusters and the corresponding cluster visualization.

When number of clusters is four, the average silhouette score is 0.36000557546607964. Figure 8 shows the silhouette score plot for four total clusters and the corresponding cluster visualization.

If the number of clusters is five, the average silhouette score is 0.3297407536588724. Figure 9 shows the silhouette scoring chart for five total clusters and grouped data display.

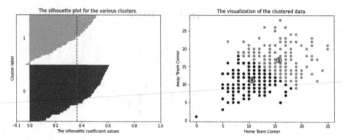

Fig. 6. Silhouette score plot for two clusters and the clusters visualization

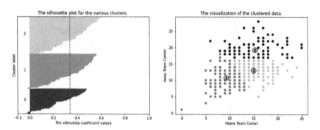

Fig. 7. Silhouette score plot for three clusters and the clusters visualization

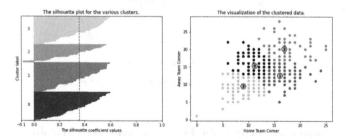

Fig. 8. Silhouette score plot for four clusters and the clusters visualization

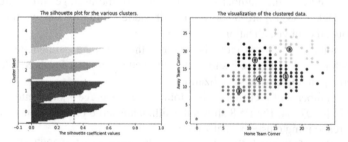

Fig. 9. Silhouette score plot for five clusters and the clusters visualization

If the number of clusters is six, the average silhouette score is 0.32079345770174683. Figure 10 shows the silhouette score plot for six total clusters and the clustered data visualization.

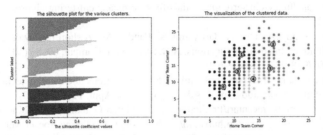

Fig. 10. Silhouette score plot for six clusters and the clusters visualization

6 Conclusion

Experimental result shows the various silhouette score plots for different number of clusters after applying fuzzy c-means clustering on the dataset. From all the plots shown above, it is clear that silhouette score for four cluster is the highest which is approaching to 1. So, optimal number of clusters should be four so that every data points correspond to its cluster.

References

1. Bano, S., Khan, M.: A survey of data clustering methods. Int. J. Adv. Sci. Technol. **113**, 133–142 (2018)
2. Blogs & Updates on Data Science, Business Analytics, AI Machine Learning. Types of Clustering Algorithms in Machine Learning With Examples (2021). https://www.analytixl abs.co.in/blog/types-of-clustering-algorithms/. Accessed 3 May 2021
3. Bezdek, J., Ehrlich, R., Full, W.: FCM: the fuzzy c-means clustering algorithm. Comput. Geosci. **10**(2–3), 191–203 (1984)
4. Ren, M., Liu, P., Wang, Z., Yi, J.: A self-adaptive fuzzy c-means algorithm for determining the optimal number of clusters. Comput. Intell. Neurosci. **2016**, 1–12 (2016)
5. Raman, R., Sharma, L., Akasapu, A.: Fuzzy clustering technique for numerical and categorical dataset. Int. J. Comput. Sci. Eng. (2010)
6. Siddique, M., Arif, R., Khan, M., Ashrafi, Z.: Implementation of Fuzzy C-Means and Possibilistic C-Means Clustering Algorithms, Cluster Tendency Analysis and Cluster Validation. arXiv e-Journal (2021). ISSN 2331-8422
7. Suganya, R., Shanthi, R.: Fuzzy C-means algorithm-a review. International Journal of Scientific and Research **2**(11), 1 (2012)
8. Pham, D., Afify, A.: Clustering techniques and their applications in engineering. Proc. Inst. Mech. Eng. Part C: J. Mech. Eng. Sci. **221**(11), 1445–1459.s (2007)
9. Halkidi, M., Batistakis, Y., Vazirgiannis, M.: On clustering validation techniques. J. Intell. Inform. Syst. **17**(2), 107–145 (2001)
10. En.wikipedia.org: Fuzzy clustering – Wikipedia (2021). https://en.wikipedia.org/wiki/Fuzzy_clustering. Accessed 6 May 2021

11. Sites.google.com: Data Clustering Algorithms - Fuzzy c-means clustering algorithm (2021). https://sites.google.com/site/dataclusteringalgorithms/fuzzy-c-means-clustering-algorithm. Accessed 6 May 2021

12. Datanovia: Fuzzy C-Means Clustering Algorithm – Datanovia (2021). https://www.datanovia.com/en/lessons/fuzzy-clustering-essentials/fuzzy-c-means-clustering-algorithm/. Accessed 6 May 2021

13. Ghosal, A., Nandy, A., Das, A., Goswami, S., Panday, M.: A short review on different clustering techniques and their applications. Adv. Intell. Syst. Comput. 69–83 (2019). https://doi.org/10.1007/978-981-13-7403-6_9

14. Faizan, M., Zuhairi, M.F., Ismail, S., Sultan, S.: Applications of clustering techniques in data mining: a comparative study. Int. J. Adv. Comput. Sci. Appl. **11**(12) (2020)

15. GitHub: prathameshtari/Predicting-Soccer-Match-Outcome-using-Machine-Learning dataset (2021). https://github.com/prathameshtari/Predicting-Soccer-Match-Outcome-using-Machine-Learning/tree/master/Predicting%20Soccer%20Match%20Outcome%20using%20Machine%20Learning/Datasets. Accessed 6 May 2021

Segregation of Areca Nuts Using Three Band Photometry and Deep Neural Network

Saurav Dosi, Bala Vamsi, Samarth S. Raut⬤, and D. Narasimha[✉]⬤

Indian Institute of Technology Dharwad, Dharwad, Karnataka, India
{sraut,d.narasimha}@iitdh.ac.in

Abstract. In the Malnadu region of Karnataka state in India, often tender Areca nuts of five to eight months old are harvested together, manually separated according to ripeness based on their colour, dehusked, sliced and boiled before marketing. This is a tedious process involving skilled labour and due to the remnant pesticides on the arecanut surface, the health of the worker is adversely affected. Hence to aid the farmers, a technique is developed to carry the task of segregation based on Three Band Photometry and Machine Learning. Light green to fully ripe orange Areca nuts from two different regions were used for Training as well as Testing, using state-of-the-art YOLOv3 Object Detection Deep Neural Network model. They were studied at various stages after plucking between fresh from harvest to one week old. The method developed was found to be effective for both samples at any of these stages.

Keywords: Machine learning · Deep Learning · Computer Vision · Artificial Intelligence · Fruit sorting · Agricultural informatics · Galaxy classification

1 Introduction

Areca nut (*Areca catechu*) is an important cash crop in some states of India, due to its commercial and social significance. India is the largest grower and consumer of Areca nut, with over 700000 tons of annual production having commercial value of over 3 billion US dollars. The major grower, Karnataka produces ripe sun dried (Chali) as well as tender red boiled (Red Supari) nuts. Processing Areca nut to produce Red Supari is time critical:

1. The tender green nut of less than 5 months since flowering to yellowish ~8 month old mature ones only are useful for this type of processing. The fully ripe orange or red ones have to be segregated from them, dried for about 45 days and dehusked to produce Chali.
2. Due to rain, labour scarcity and other operational reasons, plucking is feasible only limited number of times in the year.

Both first and second authors contributed equally.

© Springer Nature Switzerland AG 2022
K. K. Patel et al. (Eds.): icSoftComp 2021, CCIS 1572, pp. 15–27, 2022.
https://doi.org/10.1007/978-3-031-05767-0_2

3. Within three or four days of plucking from the palm, the very tender green ones, slightly grown greenish yellow and mature yellow nuts have to be segregated, peeled, sliced and suitably boiled.

It is a labour intensive skilled task and hence there is good scope for Artificial Intelligence and Machine Learning. There is acute labour shortage, partly because some residual fungicide Bordeax mixture, used to control the disease affecting the crop in rainy season, remains on the surface of the nut which has adverse health effect on the worker handling it. Consequently, automization of the process of segregation and peeling becomes a necessity. There is work describing segregation of the Areca nut after boiling and drying [1], but it does not solve the health issue, apart from the fact that the tender ones and the ripe ones, though might appear similar, have different taste and other intrinsic characteristics which the farmer is very conscious of. Also, for the least damage during peeling, it might be desirable to segregate before dehusking.

Our project to design a machine to segregate and dehusk arecanut started when the arecanut farmers' association of Sirsi requested Indian Institute of Technology Dharwad for help and one of the authors visited their office in 2018 for a detailed discussion followed by an examination of the arecanut processing. The present project aims on classifying freshly plucked Areca nuts based on their ripeness level which are traversing on a conveyor belt in an Areca Nut Processing equipment and sending them to various containers based on the ripeness level using the images of the fruits. Implementing an AI technology for this purpose by making use of a Deep Learning model to work in real-time would not only improve the performance but also reduce the human effort required and enable the farmer to finish this time critical process on–time. An accuracy close to 90% was achieved for the classification of the nuts by making use of Machine Learning Algorithms such as Logistic Regression, Support Vector Machine and YOLOv3. It is no surprise that these results are comparable [2] and our task remains to implement efficiently a feasible algorithm. In this paper, we would focus on the implementation of YOLOv3 model considering the deployment results.

The paper is arranged as follows: The essential idea of color determination from three band photometry is described in the second section. The data collection, sample preparation and data augmentation is described in the third section. The custom–trained YOLOv3 Deep Learning model based segregation is explained in the following section along with refinements, where the implementation of the scheme on a conveyor belt is also shown. The summary is given in the last section along with prospects.

We have to mention that this is not a work in Computer Science, but a project in response to request from farmers based on methods developed in Astronomy.

2   Separation Based on Three Band Photometry

The color of a perfect blackbody can be determined from two band photometry because the flux of radiation emitted from it obeys the Planck Spectrum, varies as the fourth power of its temperature and is proportional to the total emitting surface area. Consequently, the ratio of the flux received per unit solid angle per unit time by a detector in two fixed wavelength bands is a monotonic function of the blackbody temperature. Hence the three Principal Components of a color–color diagram of stars of specified chemical composition are the following:

- The color indices of a star are monotonic functions of the surface temperature.
- A deviation from the black body relation is caused by the surface gravity of the star, which depends on its mass.
- The observed color of a star is further modified due to absorption of the starlight by intervening medium in a characteristic way.

Consequently, the color-color diagram of stars in a star cluster tells us about the temperatures, an idea of their mass and possible extinction due to interstellar medium. cf. [3], Chap. 6. The arecanut color - color diagram is more similar to that of galaxies, the emission from which is composed of contribution from stars of a range of mass and age as well as the interstellar gas. Due to observational noise as well as model uncertainties in some of these processes, stellar type analysis is tough and consequently, Machine learning and Artificial Neural Network has been efficiently employed for classification of galaxies since 1980's. An illuminating discussion of the early methods is given by Lahav et al. [4]. We shall explain the essentials of their Principal Component Analysis using our figure for Arecanuts, later in this work. But the basic idea of our method is: The three band RGB (Red-Green-Blue) images taken with a Camera connected to a Raspberry Pi turns out to be a good indicator of the color, based on which we can get a measure of the age of an Areca nut. For example, Green indicates a tender nut of about five months old while a deep Yellow one is the mature one which is more than seven months old from the flowering time and an Orange or Red one, the fully ripe nut.

One hurdle we face for any such classification of a fruit is the abrupt change in the color with age and the skewness of the distribution of the fruits of a given class in the RGB space. Consequently, the Areca nuts of about 6 months age from flowering, which show abrupt change from Green to Yellow colour are hard to classify. To illustrate the problem, the classwise cluster distribution for our samples is plotted using median RGB values. It turns out that due to colour variation on the surface of the nut and occurrence of yellowish spots on the green nuts, the position of Greenish Yellow nuts in the RGB three color space will show significant separation between two instances of the same nut captured at different orientations. However, for a normal Green or a Yellow nut, the variation caused by orientation effects is almost negligible as the samples occupy the same region in the RGB space. Classification based on this approach requires imaging of the same nut at least twice.

The color defined by the Astronomer is not identical to that of a Computer Vision scientist. So, we use the Hue-Saturation-Value of Computer Vision researcher to get the RGB of Astronomer or directly use it for our Principal Component Analysis.

There have been rapid advances in object detection methods, both for static and moving objects, thanks to Computer Vision. The YOLO (You Only Look

Once) object detection model [6–8] is highly effective to detect an object on the conveyor belt and from the features of the image sample, it can determine the color of an object.

3   Data Collection

3.1   Sample Selection

Two samples of freshly plucked Areca nuts were obtained. A first sample of about 150 nuts of a range of ages and colour from green to red were procured locally from Dharwad region and another sample of about 260 nuts from Yellapur region one day after plucking. Using such varied sample, we can ensure that the result will not be restricted to specific type of Areca nuts grown in a certain region, having limited reflection properties.

The nuts are carefully manually classified into 5 distinct classes - namely light Green tender nuts, about one month older Greenish Yellow, fully grown Yellow nuts, ripe Orange (or Red) and Broken nuts. There was a small ambiguity between Green and Greenish Yellow or Greenish Yellow to Yellow. Due to age difference and consequent peeling issues, we continue with having them as separate classes despite the errors observed in the initial results. The Broken nuts could be dried ones, decomposed or those with one end split and hence any of these cannot be used for Red Supari (Fig. 3).

Prototype of the Design. A prototype of the set up for imaging and segregation of the Areca nuts is shown in Fig. 1. The nuts are placed in the container in the top, which is given a slow rotating motion. The Areca nut flow is controlled and when each nut enters the conveyor belt a light signal prompts the imaging action. The camera over the belt is shown and the exit for the segregated Areca nuts. But the software back-end including the Raspberry Pi and breadboard with trigger to rotate the segregating panel, is not shown here.

The set up is yet to be fine tuned and the design finalised. Due to abrupt termination of the project due to the pandemic, both in March 2020 and April 2021, the final touch to the segregation task awaits.

Multiple images of resolution 1280×720 pixels were taken for each Areca nut over a week to obtain a total of 1454 images, using a Raspberry Pi Camera and each image is annotated. 4 images having multiple Areca nuts were taken from the internet to

Fig. 1. Set up for arecanut segregation

take into account different lighting conditions as well as different colour distribution.

Here is the flowchart of the various steps we have taken. YOLOv3 allows imaging of the nut without stopping the conveyor belt (Fig. 2).

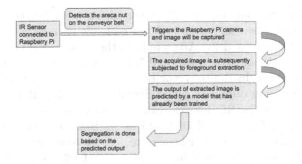

Fig. 2. Operation of our set up for Areca nut segregation

3.2 Data Preparation

We have in total 1458 images for the ~410 Areca nuts, taken over a period of a week. Each image is manually annotated with bounding box coordinates and the class labels to define the location of the object as well the class using a library called *labelImg* from GitHub repository. The annotations are required for training the Deep Learning Model.

Based on the data set analysis, the HSV (Hue-Saturation-Value) range of areca nuts in our sample is from [5, 80, 25] to [100, 255, 180]. From this, we have extracted the foreground or the Region of Interest (ROI), which is the Arecanut. The cropped images were later used for the RGB color space as well as the Principal Component Analysis (PCA) plots.

Some of the sample ROI images are shown in the Figs. 3 and 4 below. But the image of a given day is not exactly comparable to the one taken next day because of the lighting as well as orientation of the nut in the conveyor belt.

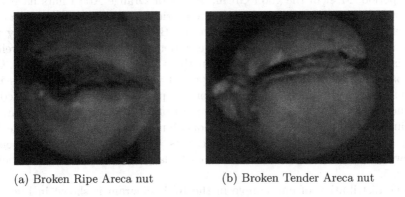

(a) Broken Ripe Areca nut   (b) Broken Tender Areca nut

Fig. 3. Sample of two broken Areca nuts, indicative of some problem with the crop which need remedial action.

The images of two typical undesired Areca nuts with split end and hence damaged or diseased kernel are displayed in Fig. 3. Evidently, they could have any mixed color, and more importantly, color might vary substantially depending on the orientation when the image is captured while in motion on the conveyor belt. There are also fully dried Areca nuts in the same category. Consequently, you do not expect them to occupy a well defined region in the RGB space of the Areca nuts. However, in the conveyor belt if you have two or multiple images captured of the same object at two different angles, they may occupy vastly different position in the RGB space. Consequently, this range of variation can be used to infer confidence level in classification. But the fraction of such Areca nuts in any sample is small and hence, we feel we have are able to segregate them reasonably well with single image using YOLOv3. This will become clear from the graphs later, where our 90% accuracy for the unusable Areca nuts is very similar to what we achieved for Green ones (Figs. 8 and 11).

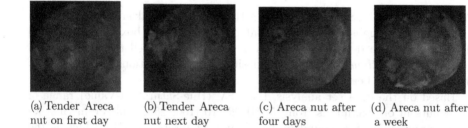

(a) Tender Areca nut on first day   (b) Tender Areca nut next day   (c) Areca nut after four days   (d) Areca nut after a week

Fig. 4. Tender Green Areca nut on four days. Orientation and lighting not identical. (Color figure online)

Displayed above is a sample of the images of a Green Areca nut over one week period. Most of the good Green, Yellow or Orange Areca nuts have very consistent image characteristics for different orientation. The images taken over a week period have a small shift in the RGB diagram but they occupy the same region in the diagram. The ambiguity between green and greenish yellow nuts is well known to farmers, but is sometimes important in classification. Our approach has been to use the irregular small yellow spots in Areca nuts just turning yellowish or the slight change of one part of the surface compared to another, and apply the deviation in the position in RGB diagram between two readings as a Data Augmentation technique. Since it is a continuous change due to age, segregation between tender arecanuts and those just turning greenish yellow is ambiguous. Typical sample of each class of Areca nut is shown below in Fig. 5.

The distribution of our sample in the RGB diagram is shown in Fig. 6. A tender Green nut and a ripe Yellow one are well separated in the RGB diagram and hence, Machine Learning can certainly help segregation and separate treatment for them as is evident from the diagram. The images taken over a week period have a small shift in the color, but still they all occupy the same color region of the RGB diagram.

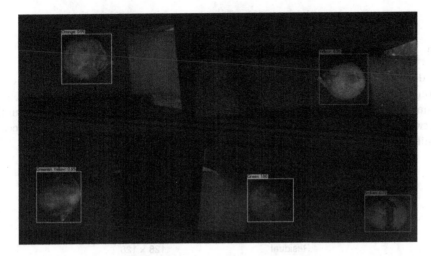

Fig. 5. Images of typical Areca nuts of various classes in our sample (Color figure online)

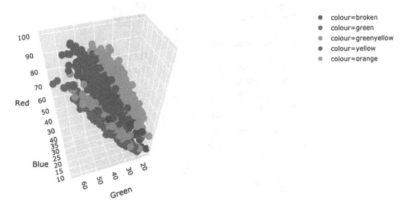

Fig. 6. Distribution of our sample of arecanuts in the RGB color diagram (Color figure online)

3.3   Data Augmentation

The Deep Learning approach requires considerable amount of data for each class for reliable training. In our sample, there were very few Areca nuts for the Broken set. There is a little bit of ambiguity in classifying as Green versus Greenish Yellow based on visual inspection only. So we tried Data Augmentation for these classes based on orientation of two images, the principle for which has been already explained.

4   Artificial Neural Network and Data Analysis

4.1   Detection and Classification

As described earlier, we shall discuss only the YOLOv3 based classification here, though the efficiency is comparable in the various methods we used. Custom–trained YOLOv3 is used to segregate Areca nuts. It is a Convolutional Neural Network and due to the heavy computational load, we adopt a two step process similar to Zheng at al. [9] (Fig. 7).

	Type	Filters	Size	Output
	Convolutional	32	3 × 3	256 × 256
	Convolutional	64	3 × 3 / 2	128 × 128
	Convolutional	32	1 × 1	
1×	Convolutional	64	3 × 3	
	Residual			128 × 128
	Convolutional	128	3 × 3 / 2	64 × 64
	Convolutional	64	1 × 1	
2×	Convolutional	128	3 × 3	
	Residual			64 × 64
	Convolutional	256	3 × 3 / 2	32 × 32
	Convolutional	128	1 × 1	
8×	Convolutional	256	3 × 3	
	Residual			32 × 32
	Convolutional	512	3 × 3 / 2	16 × 16
	Convolutional	256	1 × 1	
8×	Convolutional	512	3 × 3	
	Residual			16 × 16
	Convolutional	1024	3 × 3 / 2	8 × 8
	Convolutional	512	1 × 1	
4×	Convolutional	1024	3 × 3	
	Residual			8 × 8
	Avgpool		Global	
	Connected		1000	
	Softmax			

Table 1. **Darknet-53.**

Fig. 7. Architecture of YOLOv3 with Darknet as the backbone (ref. [7] and [8]).

We train the YOLOv3 model on a data set of 1458 images, and we apply a transfer learning approach. We use the pretrained convolutional weights and train the final dense layers of the model and the detection is nearly 95% accurate using the non-max suppression criterion but the classification accuracy is low. We implement the YOLOv3 model on the entire dataset (80% of our sample which is used for training).

The entire data set containing 1458 images was divided into two parts: 80% (1168 images) for training and the rest (290 images) for testing purposes. The images having multiple Areca nuts were retained in the training data.

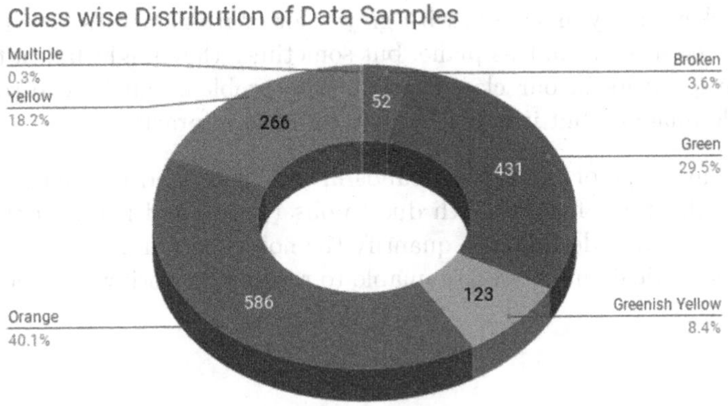

Class wise Distribution of Data Samples

Fig. 8. Class wise distribution of Areca nuts used for training and testing

4.2 Refining the Analysis

The two major reasons for refinement are

- The broken type of Areca nuts are a collection of a variety of nuts unusable for boiling and processing. They do not occupy a continuous region in the RGB diagram, and also their position in the diagram will depend on the orientation of the nut in the conveyor belt while imaging.
- Only a small number of nuts appear Greenish Yellow visually. But this is an important class as the nut ripens. Based on the slight orientation based change in the image characteristics or possible occurrence of yellow dots, they have to be segregated into a single class (Fig. 8).

We use the OpenCV function called *findContours* to extract the Region of Interest (ROI) and use it for the refinement. We compare the pixel values for the B, G, R layers and find their mean pixel values. We used two algorithms for the refinement:

- t-Distributed Stochastic Neighbour Embedding (**t-SNE**) [10]
- Principal Component Analysis (**PCA**) [11,12]

We used the algorithms in the *sklearn* library for the analyses. The result was qualitatively similar for the two algorithms. The distributions of the images in the transformed 3-d space for the PCA algorithm is shown in Fig. 9 next page. There are three Principal Components in this figure.

- All the good arecanuts fall in a plane in the 3-dimensional PCA diagram. The points along the various bands only indicate the size of the arecanut or amount of white light used for imaging.
- The mean central line of various bands make an angle with the 3rd direction axis in this figure. This angle increases monotonically for red (or orange), yellow to green arecanuts in our sample. This angle is a measure of the color (or the age of the arecanut). Ideally, a farmer should be able to change the angular width to control the segregation as a function of the age of the nut.

- The broken or dry arecanuts with grey color or irregularities are, in general, distributed away from this plane, but sometimes they might fall in the plane and cause errors in our classification. The problem can be overcome with multiple imaging, but it is not feasible to do so in practice.

The irregular shape of the various bands in this figure and the thickness of the band are indicators of noise, both due to observation and irregularities on the surface of nuts. It is desirable to quantify the noise spectrum, but obtaining the nuts is time critical and so we are unable to repeat the work with more sample.

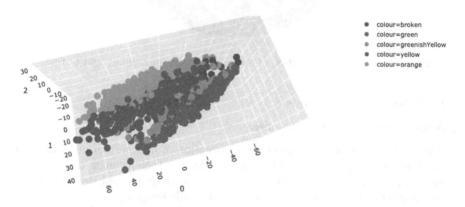

Fig. 9. Distribution of sample Areca nuts in the 3-color diagram after scaling based on the PCA analysis. (Color figure online)

The accuracy of the YOLOv3 neural network model as a function of the number of iterations is shown in Fig. 10. For our data set and refinements, 3500 appear to be the optimal number of iterations.

The final accuracy achieved for the various classes is shown in the Fig. 11. Generally this accuracy is ~90%. For the class of Greenish Yellow nuts the value of 46% corresponds to ~20 miss classifications which is mainly due to the large number in the neighboring classes. Refining the boundary between Green and Greenish Yellow class should be implemented after consultations of farmers as a subjectivity is involved for this class.

The final weights file was converted from float32 to float16 so that the file had manageable size and our model could be executable on the Raspberry Pi using TensorFlow Lite and camera interface for real–time detections.

The results obtained here are comparable to the other methods we tried as seen in Table 1: We have explained the reason for selecting YOLOv3.

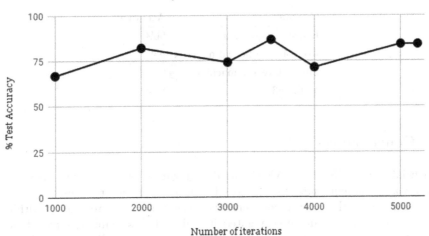

Fig. 10. Accuracy of the Machine Learning algorithm as a function of number of iterations

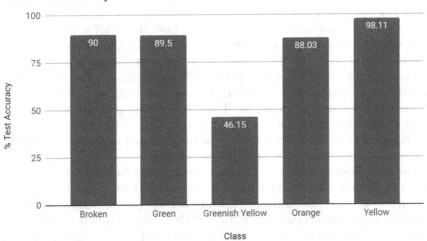

Fig. 11. Accuracy of classification of various classes of Areca nuts in our testing sample.

Table 1. Comparison of methods

Classifier	Accuracy
Logistic regression	96%
Convolution neural network	95%
Support vector machine	92%
YOLOv3	87%

5   Conclusions

Artificial Neural Network (ANN) based segregation of Areca nuts does more than what a human could do consistently, for a big set of Areca nuts.

Using the method developed here, the freshly plucked Areca nuts, within a week of harvesting, can be segregated into five classes, namely, Green tender ones, Greenish Yellow nuts about a month older, mature Yellow ones, fully ripe red or orange nuts which have to be separately processed and broken or diseased nuts which are to be discarded.

We used three band photometry based classification, incorporating statistical analysis of the skewness of the distribution as well as shift in the position of the three color diagram utilizing the powerful features of YOLOv3. But this requires substantial computational power.

The statistical tables of weights generated from this ANN were uploaded on a Raspberry Pi to drive the segregation mechanism on Areca nuts moving on a conveyor belt in real time. Some fine tuning of the mechanism is required.

The Data set we collected had specific lighting conditions. Possibly changing the lighting as well as background in the conveyor belt, the reliability of the scheme will have to be re-examined. Thus the robustness of our scheme can be tested and possible modifications can be worked out.

The size and shape information can be incorporated in the analysis because of the availability of the information of bounding box.

The method described here is fairly general. It can be used for classifying a variety of fruits or other objects based on Computer Vision. But the huge data set required in each class for reliable learning and the computational load implies that we can get the weight table from YOLOv3 with statistical additions, but it has to be uploaded in a device like Raspberry Pi for the actual classification in real time.

Acknowledgements. We thank Prof. Rajshekhar Bhat and Prof. Prabuchandran K. J. for their time and engaging discussions related to areca nuts are Machine Learning. Initial prototyping work for imaging of areca nuts done by Mr. Amit Kumar is acknowledged. We thank Mr. Aneesh, Mr. Manjunath, Mr. Rahul Maurya for offering their instrumentation expertise at a short notice. We are thankful to the referees for their constructive suggestions.

References

1. Pushparani, M.K.S., Vinod Kumar, D., Gubbi, A.: Areca nut grade analysis using image processing techniques. Int. J. Eng. Res. Technol. **7**(10), 1–6 (2019)
2. Jacot, A., Gabriel, F., Hongler, C: Neural tangent kernel: convergence and generalization in neural networks. In: 32nd Conference on Neural Information Processing Systems (NIPS 2018), Montreal (2020)
3. Unsold, A., Baschek, B.: The New Cosmos, An Introduction to Astronomy and Astrophysics, 5th edn. Springer, Heidelberg (2001). https://doi.org/10.1007/978-3-662-04356-1
4. Lahav, O., et al.: Neural computation as a tool for galaxy classification: methods and examples. MNRAS **283**, 207–221 (1996)
5. Szeliski, R.: Computer Vision: Algorithms and Applications. http://szeliski.org/Book/drafts/SzeliskiBook_20100903_draft.pdf
6. Redmon, J., Divvala, S., Girshick, R., Farhadi, A.: You only look once: unified, real-time object detection. In: Proceedings of the IEEE Conference on Computer Vision and Pattern Recognition. arXiv:1506.02640, July 2015. http://pjreddie.com/yolo/
7. Redmon, J., Farhadi, A.: YOLOv3: an incremental improvement. arXiv:1804.02767 (2018). cv-foundation.org/openaccess/content_cvpr_2016/papers/Redmon_You_Only_Look_CVPR_2016_paper.pdf
8. YOLOv3. http://www.pjreddie.com/media/files/papers/YOLOv3.pdf
9. Zheng, Y.-Y., et al.: CropDeep: the crop vision dataset for deep-learning-based classification and detection in precision agriculture. Sensors **7**(5), 1058 (2019). https://www.mdpi.com/1424-8220/19/5/1058. ISSN 1424–8220
10. van der Maarten, L., Hinton, G.: Visualizing data using t-SNE. J. Mach. Learn. Res. **9**, 2579–2605 (2008)
11. Pearson, K.: On lines and planes of closest fit to systems of points in space. Phil. Mag. **2**(11), 559–572 (1901)
12. Hotelling, H.: Analysis of a complex of statistical variables into principal components. J. Educ. Psychol. **24**, 417–441 (1933)

CT/MRI 3D Fusion for Cerebral System Analysis

Michal Chlebiej[1(✉)], Anna Zurada[2], and Jerzy Gielecki[2]

[1] Faculty of Mathematics and Computer Science, Nicolaus Copernicus University in Toruń, Chopina 12/18, 87-100 Torun, Poland
meow@mat.umk.pl

[2] Department of Anatomy, Collegium Medicum, School of Medicine, University of Warmia and Mazury, Olsztyn, Poland

Abstract. The main idea of our system is to create a fusion of CT & MRI data of the same patient. We care about the accuracy of fused data correlation, emphasising vascular structures. We propose volumetric correlation method, vascular system reconstruction using tubular structures, error measurement and analysis using surface distance measurements. The resulting data from the fusion system combines bone structures from CT imaging and the reconstructed vascular system from MRI. We ensure to match correctly cerebral vessels with corresponding data from CT. Such data enables extended analysis of vascular segments relative to bones. In this work, we also propose a new semi-circular coordinate system to unify the orientation of our fusion models. Such a system allows for a more convenient and precise numerical analysis of anatomical correspondence to the principal axes of the coordinate system. It is also possible to define precisely the left and right sides of the human head.

Keywords: Computer-aided detection and diagnosis · Magnetic resonance imaging (MRI) · Computed tomography (CT) · Cerebral vessels segmentation · 3D image fusion

1 Introduction

The primary purpose of the medical volume fusion method is to combine information delivered by different imaging modalities. Such solutions can improve diagnostic procedures, preoperative planning, interventional treatment and intraoperative guidance. Researchers successfully applied image fusion approaches in radiosurgery, neurosurgery and hypofractionated radiotherapy [1]. On the other hand, researchers are trying to analyze known data types in a novel way revealing new information. Many authors discussed this subject previously, but there are still many fields of investigation connected with this area that draws attention from modern researchers. When working with the theme of image fusion, we must always first correlate the data geometrically. Data-matching has been studied extensively in the literature [2–4]. Another fundamental question is what kind of information we want to match. According to authors in [5], the subject

K. K. Patel et al. (Eds.): icSoftComp 2021, CCIS 1572, pp. 28–40, 2022.
https://doi.org/10.1007/978-3-031-05767-0_3

of fusion can divide into three levels: pixel/signal level, object/feature level and symbolic/decision level. At intensity level, we combine data value information creating a new intensity value I_3 for two input values I_1, I_2 that correspond to the same point p in the matching sets. This approach has been widely discussed in much scientific literature [5–11]. Paper [12] describes the second level that refers to labels, property descriptors and objects obtained from various sources. The last category is high-level. It uses local decision markers derived from objective level fusion results to guide probabilistic decisions to extract information and perform fusion [13]. In this work, we use the object fusion approach. In the next steps of the method, we extract segmentation masks constituting the basis for reconstructing vascular objects and anatomical spatial points, which we provide in a combined form to the user. This work uses the development of the vascular reconstruction method proposed in work [14], allowing it to be customized and adapted to specific problems. We use fusion results to quantitatively analyse cerebral arteries and improve the proposed methodology [15].

2   Methodology

Many approaches to cerebral vessel segmentation have arisen, which we can divide into two categories - voxel-based methods and machine learning methods. Authors in [16] deliver a comprehensive review of such techniques. Several Open-source libraries such as VMTK [17] and TubeTK [18] provide API functions dedicated to vessel segmentation tasks. The tasks where the user should visualize the vessels hidden inside the bone structures are usually problematic. The primarily available reconstruction algorithms usually do not give satisfactory results in such situations. We have developed a fusion system that utilizes CT and MRI information enabling precise to deal with this problem. Our fusion method consists of several stages. Figure 1 presents the data flow. At first, we are dealing with MRI & CT datasets with contrast. Vascular structures are visible for both datasets, but we cannot separate them from high-density tissues for CT data. Especially immersed arteries in bone structures (ICA) cannot be segmented manually with sufficient precision. For this reason, we propose a method to extract vascular data from MRI and immerse them in bone marrow from tomography. In the first phase, we perform volume registration using normalised mutual information as a similarity function and Powell's [19, 20] method as an optimisation scheme. In the next step, we create a segmentation model based on MRI and use it to limit the segmentation volume from the CT data to separate the bone. The next step is to generate corresponding tubular vascular structures in both volumes. Such reconstructions we can use to compare corresponding segments to each other. We also create surface measurements to compare them quantitatively by analyzing their mutual distance. After models extraction, we can use both vascular versions - from CT and extended (with ICA segments) from the corresponding dataset. We can also extract bone structure data from CT data for adequate visualization, determining the location of selected anatomical points and making final measurements.

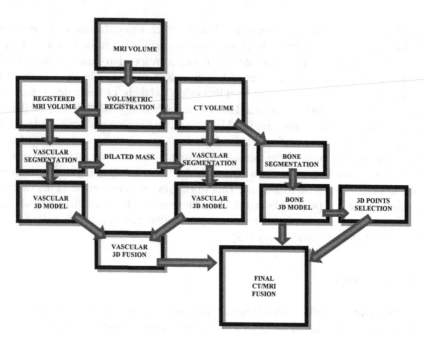

Fig. 1. Data flow diagram of our fusion system

2.1 3D Matching

In this step, we adjust the CT and MRI data to provide the combined information in a standard coordinate system. We are using rigid-body transformation model and normalized mutual information [2, 21, 22] as a similarity measure:

$$NMI(RI.FI) = \frac{h(RI) + h(FI)}{h(RI.FI)}$$

where $h(RI)$ is the entropy of reference images (CT), $h(FI)$ is the entropy of floating MRI image, and $h(RI.FI)$ is joint entropy.

$$h(I) = -\sum_{x=1}^{x_{max}} (p_I(x)log(p_I(x)))$$

$$h(RI.FI) = -\sum_{x=1}^{x_{max}} \sum_{y=1}^{y_{max}} (p_{RI.FI}(x.y)logp_{RI.FI}(x.y)))$$

We also use Powell's method as an optimization routine. We selected this deterministic routine because it is computationally efficient. It delivers the final result in seconds. As a result, we obtain a rigid body transformation and apply it to the MRI dataset to transform it into the CT coordinate system. We conclude this step by having two matched data sets with two corresponding values from the CT and MRI sets. After this step, we can use new data for visualization, processing, and measurements. Figure 2a presents the results of the registration process.

2.2 MRI and CT Segmentation Using Transformed MRI Mask

Segmentation of vascular structures can be performed based on MRI data with a contrast agent, using simple thresholding or our controlled front growth segmentation with probability. Our modified region growing algorithm utilizes probability maps for the homogeneity criterion proposed in [23]. Initial estimate segmentation model Φ_M (i.e. ellipsoid inside vascular region around seed points) is placed on an image I. Image region bounded by Φ_M is R_M with volume $V(R_M)$. We define probability of intensity i being consistent with model interior using Gaussian kernel:

$$P(i|\Phi_M) = \frac{1}{V(R_M)} \int \int \int_{R_M} \frac{1}{\sqrt{2\pi}\delta} e^{\frac{-i(i-I(p))^2}{2\delta^2}} dp$$

where δ is the standard deviation and $I(p)$ is the intensity at the position p in the dataset. We calculate $P(i|\Phi_M)$ for every voxel intensity value i to obtain a new intensity distribution, where new intensity values are probabilities of being consistent with initial segmentation. With this approach, the intensities initially separated in the histogram can become neighbours after remapping. A probability threshold parameter is only needed to parameterize the segmentation procedure.

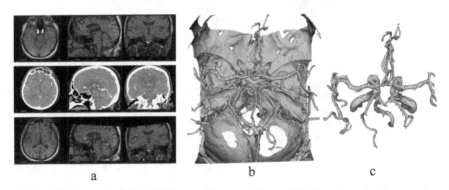

a   b   c

Fig. 2. (a) Top (MRI), middle (CT), bottom MRI matched with CT. Red rectangles represent a visual assessment of registration quality. (b) CT segmentation. (c) CT segmentation with MRI boundaries

For contrast-administered CT data, it is not apparent how to segment the vascular data. Intensities range covering contrasting systems overlapping intensity range for bone structures. Our approach uses dilation masks obtained from a matched MRI volume to create constraints for CT segmentation and solve this problem. A morphological filter [24] is to extend the primary segmentation by three voxels using a spherical structuring element. It is a sufficing operation to separate bone structures and perform rapid segmentation. The result of ICA segmentations shown in the image Fig. 2c was not correct. It was impossible to separate bone structures from ICA in the CT image. Therefore they cannot be included in the final reconstruction process. The resulting CT/MRI segmentations were pre-cut to cover only the circle of Willis region with the closest neighbouring structures. Figure 3c presented a visualisation of MRI/CT masks and superimposed images of both segmented regions.

2.3   Cerebral System Reconstruction and Fusion

At this stage, we apply our developed method for cerebral reconstruction using tubular structures. The method starts with region-based segmentation allowing for local growth control. After that, it traces skeletal lines and optimises them in terms of desired complexity and accuracy. The method can reduce the number of joints, remove short lines and simplify the final skeleton. In the last step, the algorithm reconstructs optimal tubular structures (in terms of minimum distance from a segmentation) using partitioned skeletons and defines joint volumes excluded from tubular structures. Figure 3d–f present the results of CT and MRI segmented models of vascular structures. They were a source for tubular 3D models with labelled Circle of Willis arteries. Every tubular segment has a smooth, skeletal line used for further processing and analysis. The details go far beyond the scope of this work, and we will not discuss them further.

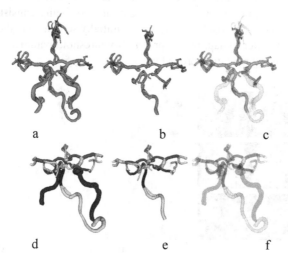

Fig. 3. (a) MRI vascular segmentation. (b) CT vascular segmentation. (c) CT segmentation with boundaries from corresponding MRI mask. Partitioned model into segments: (d) MRI tubular model. (e) matched CT tubular model (f) CT/MRI tubular fusion

2.4   Matching Quality Analysis

We need to ensure correct correlation when we want to use fusion results. We use spatial matching of the volumes of different modalities, where assessing the correlation quality is not a trivial task. We optimise the values of normalised mutual information or entropy measures during the matching process. This approach still does not give us precise information on corresponding tissues distance. In our work, we propose to analyse the spatial similarity between reconstructed objects. The implemented algorithm determines the Euclidean distance value of every surface point to the closest point of the corresponding dataset. Finally, we map all estimated values onto a geometrical tubular object using the colour lookup table (Fig. 4a). The maximum separation between corresponding points was above 4mm. However, such maxims only appeared in regions where we manually modified the segmentation, and we can ignore them. An average error was 0.35 mm (less than a voxel size - 0.5 mm), and the standard deviation was 0.39 mm. We created a distance histogram (see Fig. 4b) that shows the most likely occurring values. Estimations significantly above the mean value constitute a negligible fraction, and most of the analyzed points have values within the acceptable tolerance.

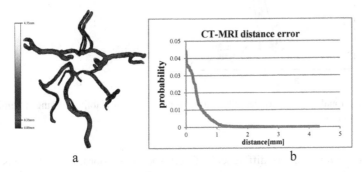

a

b

Fig. 4. Surface distance visualization. We coloured the surface reconstructed from the tomography was with the distance values to the nearest points belonging to the MRI data.

After such analysis, we can ensure that the fusion of CT and MRI is well correlated. We can use the result for further processing and and measurements (see Fig. 5).

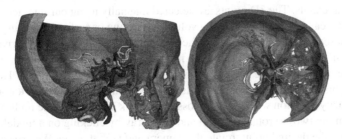

Fig. 5. Fusion of virtually sculpted CT bony structures and reconstructed vascular segments from MRI data.

2.5 Semi-circular Plane Definition

In [15], the authors proposed a method to analyse the spatial orientation of brain vessels using directional cosines/angles defined as angles between base vectors of the coordinate system and line segment representing the vascular segment. The proposed method assumes perfect levelling of the human head. Directional angles measured and analysed were an angular deviation from base vectors.

$$cos\alpha = \frac{\Delta x}{d} cos\beta = \frac{\Delta y}{d} cos\gamma = \frac{\Delta z}{d}$$

34   M. Chlebiej et al.

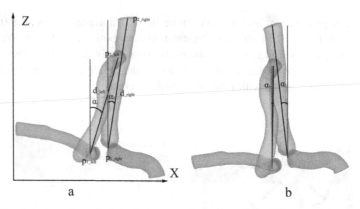

Fig. 6. Directional angles: (a) default dataset orientation. (b) stabilised using a semicircular coordinate system.

where $\Delta x.\Delta y$ *and* Δz are differences of corresponding coordinates of points p_1 and p_2 and d is the length of the $p1p2$ line segment. Figure 6a presents a slightly rotated head situation during the examination. Such measurement does not deliver any information - both angles have the same orientation, and values do not present any left-right correspondence. If we could stabilize the proper head and change the coordinate system, we can obtain the appropriate direction (Fig. 6b). It is necessary to define a new coordinate system to make correct angular global measurements. This work proposes a definition based on four anatomical points: the lowest points of both orbit margins and semi-circular canals. The 3D points are selected manually using our interactive system. Semi-circular canals are selected using volume rendering or surface renderings by picking points directly in 3D (see Fig. 7a). Picking points in 3D is not always trivial. When dealing with fully opaque surface mesh models, we can quickly estimate the closest point in the mesh using the projected line and the ray casting technique. However, it is more sophisticated when picking points in 3D volume rendering with a non-linearly defined opacity function where it is impossible to pick a point. We use multiple projections (minimum two) to rotate a model and cast a ray by clicking on a model. The user needs to pick the desired point, rotate the camera and pick it again. We use casted lines for calculation closest point to casted line-set in minimisation procedure where the sum of distances between the final point and its projection on lines is minimal. Semi-circular canals are selected using 2D slices with the guidance of a 3D pointer superimposed with the surface rendering of bony structures (using volume rendering or surface renderings by picking points directly in 3D (Fig. 7a). In the model presented in Fig. 7b, there are two lines visible - the first one connects orbit margin points while the second is parallel to the original X-axis. An angle between them shows how to rotate the coordinate system to obtain orbit margin points at the same level.

After selecting 4 points, we define the plane see Fig. 8a. In generally, 3 points in space define a plane. In our case. we use 4 points $P_1.P_2.P_3.P_4$ so we fit a plane in the least-squares sense minimizing the distance to plane error. In the next step, we define the new origin of our coordinate system. We calculate it as the centre of $P_1'.P_2'.P_3'.P_4'$ polygon resulting from where are the projection of $P_1.P_2.P_3.P_4$ points on a fitted plane.

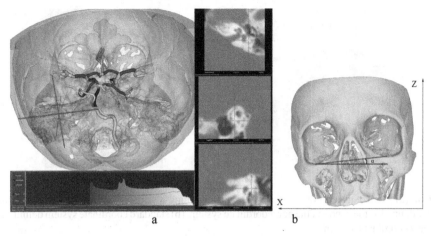

Fig. 7. (a) Semicircular canal points selection. (b) The lowest points of orbit margins were selected using picking points in the 3D interactive system. α angle shows the rotation that we apply in the new coordinate system to stabilize orbit margins at the same level.

As a result. we obtain a new origin point P_0 and $V_1.V_2.V_3$ normalized vectors defining a semicircular coordinate system (see Fig. 8b–c). The resulted perpendicular planes are visible in Fig. 9a. New coordinate allows for unified angular measurements, making it reliable and comparable. When we define new base vectors in our software, we perform all measurements using new coordinates. Additionally. P_{Origin} point and plane OY can be used for projection of symmetric left/right structures for further quantitative analysis.

2.6 Numerical Description

For all tubular objects, we deliver shape descriptors together for both CT and MRI reconstructions. In [15] authors proposed an angular analysis of selected arteries. The direction in the space of the tubular object was estimated by connecting its skeletal endpoints besides previously mentioned directional cosines describing global orientation in space authors defined shape descriptors using the skeletal line.

$$TI = \frac{l}{d} = \frac{\sum_{i=0}^{n-1} \sqrt{(x_{i+1} - x_1)^2 (y_{i+1} - y_1)^2 (z_{i+1} - z_1)^2}}{\sqrt{(x_2 - x_1)^2 (y_2 - y_1)^2 (z_2 - z_1)^2}}$$

where TI is the tortuosity index, l is the total length of the skeletal line, $p_1 = (x_1.y_1.z_1)$ and $p_2 = (x_2.y_2.z_2)$ are the endpoints of the vascular segment skeletal line, and d is the length of this line. Deviation index DI is a maximum distance skeletal point p_{max} from the line connecting points p_1 and p_2 and

$$CA = arccos(\alpha_1 \cdot \alpha_2 + \beta_1 \cdot \beta_2 + \gamma_1 \cdot \gamma_2)$$

where $\alpha.\beta.\gamma$ are directional angles of connected vascular segments. While we calculate TI and DI, taking into consideration all skeletal points. we consider CA as the folding angle between two lines $p_1 p_{max}$ and $p_{max} p_2$ (see Fig. 9b).

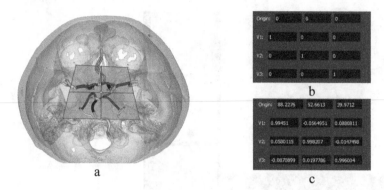

Fig. 8. (a) The connection of $P'_1.P'_2.P'_3.P'_4$ points defining a semicircular plane with P_Origin centre defined as the centre of the new coordinate system. (b) standard coordinate system definition (c) new coordinate system.

3 Results

We selected 18 pairs of CT/MRI datasets to evaluate the proposed methods. Using the proposed method, we applied the same procedure for each dataset for every sample data operator segmented vascular structures from MRI and CT datasets. Based on both segmentation models, software-generated automatically tubular models. We used Euclidean distance measure to quantify the accuracy of the reconstruction. Table 1 summarizes maximum, average and standard deviation values. CT/MR parts refer to the distance of a tubular model from the original segmentation. Most surface points (>95%) error was lower than voxel size in all cases. For less than 1% of points, the error was above 1mm. The highest error values resulted from cropped segmentation. That usually happened in regions with collisions of bones and vascular structures in CT images.

The fusion part of Table 1 presents the distance variability for matched CT and MRI tubular models. Error-values were slightly higher in these cases, but over 91% of corresponding surface points were lower than the voxel size. Such comparison ensures that our method is reliable and fusion is precise enough for further analysis.

For all 18 fused datasets, we selected left and right A2 segments (from MRI data) and calculated directional angles for every segment. We also used our semicircular coordinate system (using CT dataset as bone structure reference) to stabilise the dataset and unify the orientation of all datasets. Table 2 presents the results. We obtained lower standard deviation values for all angles when using stabilised data. For le left side the percentage decrease was $\Delta\alpha = 17.8\%$, $\Delta\beta = 19.8\%$, $\Delta\alpha = 20.2\%$ and for the right side respectively $\Delta\alpha = 5.9\%$, $\Delta\beta = 31.4\%$, $\Delta\alpha = 30.7\%$. As a result, we achieved a highly correlated orientation of A2 segments between datasets. We also compared differences between left and right side angles. Semicircular correlation decreased every angle difference: α - 0.2%, β - 4.1% and γ - 0.5. That fact also proves a better correlation of all 18 datasets and gives higher reliability of angular descriptors.

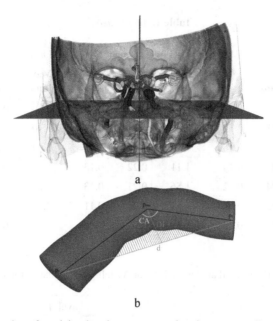

Fig. 9. (a) Visualisation of semicircular planes representing the new coordinate system. (b) Visualisation of the tubular segment with skeletal line l. lines resulting from connecting skeletal points and their projection onto the line p_1p_2 with length d. where CA is a curvature angle and DI is deviation index.

Table 1. Error analysis of tubular reconstructions and fusion models.

No	CT			MRI			Fusion		
	Max [mm]	Avg [mm]	δ [mm]	Max [mm]	Avg [mm]	δ [mm]	Max [mm]	Avg [mm]	δ [mm]
1	1.50	0.22	0.12	1.15	0.20	0.12	2.94	0.22	0.28
2	1.89	0.24	0.13	1.25	0.23	0.11	2.31	0.21	0.36
3	1.41	0.22	0.12	1.59	0.22	0.12	2.96	0.21	0.35
4	1.77	0.21	0.12	1.11	0.21	0.13	3.18	0.31	0.32
5	1.18	0.21	0.13	1.27	0.21	0.12	2.57	0.26	0.39
6	1.28	0.23	0.12	1.12	0.20	0.11	2.53	0.25	0.27
7	2.21	0.21	0.13	1.60	0.21	0.12	2.31	0.26	0.25
8	1.92	0.24	0.13	1.22	0.23	0.12	3.97	0.28	0.38
9	1.56	0.24	0.13	1.49	0.21	0.14	2.07	0.27	0.31
10	1.71	0.23	0.13	1.58	0.23	0.13	3.75	0.30	0.37
11	2.19	0.21	0.12	1.35	0.21	0.13	3.79	0.29	0.34

(*continued*)

Table 1. (*continued*)

No	CT			MRI			Fusion		
	Max [mm]	Avg [mm]	δ [mm]	Max [mm]	Avg [mm]	δ [mm]	Max [mm]	Avg [mm]	δ [mm]
12	1.92	0.23	0.13	1.54	0.22	0.13	2.20	0.25	0.28
13	1.39	0.21	0.12	1.31	0.21	0.11	3.90	0.25	0.26
14	2.08	0.27	0.16	1.53	0.21	0.11	3.03	0.26	0.39
15	2.42	0.21	0.14	1.11	0.22	0.13	3.67	0.24	0.31
16	1.77	0.24	0.12	1.56	0.21	0.13	3.60	0.29	0.38
17	2.72	0.22	0.13	1.31	0.21	0.11	2.37	0.28	0.32
18	1.31	0.22	0.12	1.19	0.22	0.12	2.10	0.30	0.28

Table 2. Directional angles analysis of A2 left/right segments for 18 cases.

			avg[°]	δ[°]
A2 left	Default	α	93.1	5.0
		β	41.6	13.4
		γ	49.0	13.3
	Semi-circular	α	92.2	4.1
		β	47.2	10.7
		γ	43.1	10.6
A2 right	Default	α	85.8	6.9
		β	35.7	13.7
		γ	55.6	12.7
	Semi-circular	α	84.9	6.5
		β	41.5	9.4
		γ	49.7	8.8

4 Conclusions

We have proposed a new CT & MRI fusion method dedicated to vascular systems and their correspondence with bone structures in this work. We have also proposed a method for assessing the quality of fusion using the Euclidean distance measure of the reconstructed surfaces. Then we presented a method for stabilising the human head using the semi-circular plane. The proposed method improved the calculation of directional angles proposed in [15], making it more reliable and resistant to the incorrect orientation of the patient's head during the examination. In publication [25], we used the fusion methodology shown in this work to build an interactive visualization system using an augmented reality system.

Acknowledgement. The research was partially supported by the Polish National Science Centre (NCN) (grant No. 2012/07/D/ST6/02479).

References

1. Inoue, H., Nakajima, A., Sato, H., Noda, S., Saitoh, J., Suzuki, Y.: Image fusion for radiosurgery, neurosurgery and hypofractionated radiotherapy. Cureus **7**, e252 (2015)
2. Maes, F., Vandermeulen, D., Suetens, P.: Medical image registration using mutual information. Proc. IEEE **91**, 1699–1722 (2003)
3. Van den Elsen, P., Pol, E., Viergever, M.: Medical image matching-a review with classification. IEEE Eng. Med. Biol. Mag. **12**, 26–39 (1993)
4. Rueckert, D.: Nonrigid Registration: Concepts, Algorithms, and Applications. Medical Image Registration. CRC Press (2001)
5. Bavirisetti, D., Kollu, V., Gang, X., Dhuli, R.: Fusion of MRI and CT images using guided image filter and image statistics. Int. J. Imaging Syst. Technol. **27**, 227–237 (2017)
6. Ben Hamza, A., He, Y., Krim, H., Willsky, A.: A multiscale approach to pixel-level image fusion. Integr. Comput.-Aided Eng. **12**, 135–146 (2005)
7. Li, H., Manjunath, B., Mitra, S.: Multisensor image fusion using the wavelet transform. Graph. Models Image Process. **57**, 235–245 (1995)
8. Petrovic, V.: Multisensor pixel-level image fusion. Ph.D. thesis, Department of Imaging Science and Biomedical Engineering Manchester School of Engineering, United Kingdom (2001)
9. James, A., Dasarathy, B.: Medical image fusion: a survey of the state of the art. Inf. Fusion **19**, 4–19 (2014)
10. Li, W., Lu, K., Xiao, B., Du, J.: An overview of multi-modal medical image fusion. Neurocomputing **215**, 3–20 (2016)
11. Perez, J., Mazo, C., Trujillo, M., Herrera, A.: MRI and CT fusion in stereotactic electroencephalography: a literature review. Appl. Sci. **11**, 5524 (2021)
12. Sasikala, M., Kumaravel, N.: A comparative analysis of feature based image fusion methods. Inf. Technol. J. **6**, 1224–1230 (2007)
13. Tao, Q., Veldhuis, R.: Threshold-optimized decision-level fusion and its application to biometrics. Pattern Recogn. **42**, 823–836 (2009)
14. Nowinski, W., Volkau, I., Marchenko, Y., Thirunavuukarasuu, A., Ng, T., Runge, V.: A 3D model of human cerebrovasculature derived from 3T magnetic resonance angiography. Neuroinformatics **7**, 23–36 (2008)
15. Żurada, A., Gajda, G., Nowak, D., Sienkiewicz-Zawilińska, Gielecki, J.: The description of vascular variations in three dimensional space: a novel method of spatial cerebral arteries evaluation. Med. Sci. Monitor **14**(9), 36–41 (2008)
16. Lesage, D., Angelini, E., Bloch, I., Funka-Lea, G.: A review of 3D vessel lumen segmentation techniques: models, features and extraction schemes. Med. Image Anal. **13**, 819–845 (2009)
17. Piccinelli, M., Veneziani, A., Steinman, D., Remuzzi, A., Antiga, L.: A framework for geometric analysis of vascular structures: application to cerebral aneurysms. IEEE Trans. Med. Imaging **28**, 1141–1155 (2009)
18. Pace, D., et al.: TubeTK, Segmentation, Registration, and Analysis of Tubular Structures in Images. Kitware Inc., Clifton Park (2012)

19. Powell, M.: An efficient method for finding the minimum of a function of several variables without calculating derivatives. Comput. J. **7**, 155–162 (1964)
20. Press, W., Teukolsky, S., Vetterling, W., Flannery, B.: Numerical Recipes in C. Cambridge University Press, Cambridge (1992)
21. Studholme, C., Hill, D., Hawkes, D.: An overlap invariant entropy measure of 3D medical image alignment. Pattern Recogn. **32**, 71–86 (1999)
22. Maes, F., Collignon, A., Vandermeulen, D., Marchal, G., Suetens, P.: Multimodality image registration by maximization of mutual information. IEEE Trans. Med. Imaging **16**, 187–198 (1997)
23. Huang, X., Metaxas, D., Chen, T.: MetaMorphs: deformable shape and texture models. In: IEEE Conference on Computer Vision and Pattern Recognition, vol. 1, pp. 496–503 (2004)
24. Serra, J.: Image analysis and mathematical morphology. Comput. Graphics Image Process. **20**, 96–97 (1982)
25. Chlebiej, M., Rutkowski, A., Zurada, A., Gielecki, J., Polak-Boron, K.: Interactive CT/MRI 3D Fusion for cerebral system analysis and as a preoperative surgical strategy and educational tool. Pol. Ann. Med. 1–7 (2021)

Soft Optimal Computing to Identify Surface Roughness in Manufacturing Using a Gaussian and a Trigonometric Regressor

Benedikt Haus, Paolo Mercorelli$^{(\boxtimes)}$, Jin Siang Yap, and Lennart Schäfer

Institute of Product and Process Innovation, Leuphana University of Lüneburg,
Universitätsallee 1, 21335 Lüneburg, Germany
mercorelli@leuphana.de
https://www.leuphana.de/en/institutes/ppi/staff/paolo-mercorelli.html

Abstract. This contribution deals with the identification of roughness as a function of gloss in manufacturing using Particle Swarm Optimization (PSO) methods. The proposed PSO method uses a Least Squares Method as a cost function to be optimized. The identification structure uses a Gaussian and a Trigonometric Regressor characterized by seven parameters to be estimated. In PSO algorithms, there is a delicate balance to maintain between exploration (global search) and exploitation (local search) and this is one of the most important issues of this optimization method. An analysis of an increment of the dimension of the search space of the PSO is proposed. This is realized through an increment of its exploitation dimension to improve the precision of the search phase of the PSO, at the cost of more computations in each iteration. Nevertheless, convergence time results to be shorter in the presented case. Thus, an optimal increment of the dimension exists which states a compromise between velocity of the convergence and precision. Measured results from a manufacturing system with and without enlargement of the search space are shown together with results obtained using a Genetic Algorithm (GA) for comparison. Advantages and drawbacks are pointed out.

Keywords: Particle Swarm Optimization · Curve fitting · Manufacturing applications

1 Introduction and Motivation

The maintaining of a constant surface quality in production and further processing of surface-treated materials, such as metal, plastic, wood, or paper surfaces, is more important than ever. In almost all industrial sectors, a uniform

Project "Optimierung der roboterbasierten, hybriden Fertigung (OPTIROB)" supported by Europäischer Fonds für regionale Entwicklung (EFRE) and Land Niedersachsen Programmgebiet Übergangsregion (ÜR); Förderperiode 2014-2020, grant number ZW 7-85049795.

and defined surface quality must be ensured. The mean roughness value is the arithmetical mean value of all deviations of the roughness profile from the middle line along the reference section. This means that the mean roughness value R_a theoretically corresponds to the distance between several lines that would arise if the mountains and valleys around the center line were converted into rectangles of the same size. There are different ways of evaluating the surface roughness and thus the surface quality. With the conventional tactile measuring method, information is collected at predefined points by scanning the surface using a mechanical probe element. This scattered data collected from several measurements provides information about the roughness, e.g. by principle of a stylus instrument profilometer, which is also applied in atomic force microscopy (AFM). In AFM, a cantilever is holding a small tip that is sliding along the horizontal direction over the object's surface. The cantilever is moving vertically following the profile. The vertical position is recorded as the measured profile. Another method for determining the surface quality is the gloss meter. The basic principle of the gloss meter is that electromagnetic waves stemming from a light source are directed at a predefined angle onto the surface to be characterized. A detector receives the reflected rays. The gloss can be determined by the ratio between transmitted and received light intensity. Depending on the nature of the surface, the light can be reflected differently. High-gloss surfaces reflect the light specularly, whereas the light is diffusely scattered with a matt surface. Also recent literature shows interest in this aspect. In [1], to control gloss defects, several methods have been suggested for enhancing the replication to avoid surface defects. Changes in roughness and gloss are indicators of changes in e.g. paint work due to various types of stress, such as the influence of weather and light or mechanical damage. For these reasons, in the more general context of the investigation of surface properties, in the literature particular emphasis is placed on the detection of roughness and gloss. A quite complex model is proposed in [2] in which gloss as a function of roughness is proposed. The proposed model essentially is a combination of Gaussian and trigonometric functions as a possible regressor. Once the regressor is chosen, the problem is to find an optimal approximation of the measured data. In this context, a possible approach to determine the coefficients of the regressor is the well-known Particle Swarm Optimization (PSO) which, in the context of random algorithms, is one of the leading optimizer structures [3]. In literature, many fitting techniques are known for different applications, for instance in [4] a polynomial fitting problem in the context of optimization of motions of an electromagnetic actuator is proposed. In [5] a trajectory optimization using Bernstein polynomials is proposed in the context of motions of a two-link robot. A fitting problem is formulated as an optimization problem.

1.1   General Applications of PSO and Main Contribution of the Paper

In general, PSO algorithms are widely used to search for optimal parameters of learning algorithms to improve the classification performance [6]. PSO are also used to adjust the weights of Back Propagation Neural Networks (BPNN), see [7] and in feature selection [8]. We can say that PSO algorithms are widely utilized in many non-covex optimization problems. In the contribution at hand, a PSO method is analyzed to calculate an optimal regressor in a fitting problem., as is done also in [9], where PSO is utilized in a fitting problem to improve the capability prediction of a preexisting method. In general, PSO can be applied in many fields, including optimal interpolation, with great success. An overview on the use of PSO in optimal interpretation is shown in [10] in which electromagnetic, magnetotelluric data, gravimetric, direct current and seismic data are optimally interpolated by PSO. More in general, the flexible character of PSO has been the reason for its success in many fields of application, such as structural design [11], solar photovoltaic systems [12], and more recently in epidemic modeling of Sars-Cov-2 [13, 14] and many other application fields. One of the most important issues in PSO is represented by finding an optimal balance to maintain between exploration (global search) and exploitation (local search) [15]. Moreover, it is known that PSO has the advantage of fast convergence speed, but it tends to converge prematurely on local optima. In this sense, to avoid stagnation in local optima with a low convergence accuracy, which, as explained, typically depends on the lack of proper balance between exploration and exploitation, in [16] acceleration coefficients are introduced to adaptively achieve this balance. In [17] the "social-learning" part of each particle is expanded from one exemplar to two exemplars. Secondly, different forgetting abilities are assigned to different particles. The main contribution of this paper is to show how, using a larger search space which is realized by increasing the dimension of the exploitation search space, improvements in terms of more accurate optimisation results in a shorter time are obtained. It is shown that there is an optimal increment of the dimension w.r.t. accuracy and convergence time. Results obtained from measured data are shown at the end of this analysis to compare the three different techniques. In particular, the proposed technique is compared to conventional PSO, as well as a Genetic Algorithm (GA). Advantages and drawbacks are pointed out. The paper is organized in the following way. Section 2 is dedicated to the introduction of the basic structure of the PSO. Section 3 shows the method which is used as a possible alternative to the basic PSO. The paper closes showing the results and their discussion together with the conclusions.

2   Particle Swarm Optimization Algorithm Background

The particles in conventional PSO are guided by three characteristic components, which are present movement, the personal experience, typically indicated as *pbest* in literature after optimization, and the overall experience of the swarm, typically indicated as *gbest*. Once the initial population (swarm) is fixed with the number of particles and the number of the parameters of the regressor, in accordance with the data to be approximated, the *size* and the *dimension* of the swarm is fixed. The following notation is adopted to describe the method.

The Main Nomenclature

$\mathbf{X} = [\mathbf{X}_1, \mathbf{X}_2, ..., \mathbf{X}_N]^T$: matrix $(N \times D)$ representing the swarm

$\mathbf{X}_i = [X_{i,1}, X_{i,2}, ..., X_{i,D}]$ with $i = 1, 2, ..., N$: individual particles

$\mathbf{V} = [\mathbf{V}_1, \mathbf{V}_2, ..., \mathbf{V}_N]^T$: swarm velocity

$\mathbf{V}_i = [V_{i,1}, V_{i,2}, ..., V_{i,D}]^T$ with $i = 1, 2, ..., N$: velocity of the particles

$w \in [w_{min}, w_{max}]$: inertia factor

k_1: acceleration factor of the personal best

k_2: acceleration factor of the (first) global best

$r_1 \in [0 \ \ 1]$: uniformly distributed random number

$r_2 \in [0 \ \ 1]$: uniformly distributed random number

$pbest_{i,j}^k$: j^{th} parameter of the personal best component of ith individual at iteration k

$^l gbest_j^k$: j^{th} parameter of the lth-best global component at iteration k

With the given nomenclature, conventional PSO can be described using the following particle- and parameter-wise update equations:

$$V_{i,j}^{k+1} = wV_{i,j}^k + k_1 r_1(pbest_{i,j}^k - X_{i,j}^k) + k_2 r_2(gbest_j^k - X_{i,j}^k) \tag{1}$$

$$X_{i,j}^{k+1} = X_{i,j}^k + V_{i,j}^{k+1}. \tag{2}$$

Remark 1. $k_{1,2}$ *are set to 2 if one wants to obtain* $\mathrm{E}[r_{1,2}k_{1,2}] = 1$ *for both components* **pbest**$_i^k$ *and* **gbest**^{k}, *given a uniform distribution of* $r_{1,2}$. *Concerning the nomenclature, it is to notice that in some literature the personal best component is also called* cognitive, individual, *or* exploration *component and the global best component is also called* team best, social, *or* exploitation *component.*

If we map this information into a matrix $\mathbf{X}$, then the individual particles are represented by the rows and the columns contain the parameters. In our case the dimension is $N \times D$, N individuals with D parameters are represented.

3   An Extended Search Space for PSO

In the presence of large dimension of the optimization problem (a large number of parameters D), the following variation of the PSO can be proposed, in which the global best component vector is renamed as ^{1}**gbest**, while the second-best is ^{2}**gbest**, and so on. The particle- and parameter-wise update equations are

$$V_{i,j}^{k+1} = wV_{i,j}^k + k_1 r_1(pbest_{i,j}^k - X_{i,j}^k) + k_2 r_2(^1 gbest_j^k - X_{i,j}^k)$$
$$+ \sum_{l=2}^{D-3} \underbrace{k_{l+1} r_{l+1}(^l gbest_j^k - X_{i,j}^k)}_{c_l}, \tag{3}$$

$$X_{i,j}^{k+1} = X_{i,j}^k + V_{i,j}^{k+1}. \tag{4}$$

Remark 2. *In (3), all terms are assumed to be linearly independent. This assumption represents a worst case analysis, since, ideally, the inertia component already points in a similar direction as the global or personal best components. In such a case of independency, a maximum number of $D - 3$ terms can be added in order to achieve a dimension of the search space that is equal to D.*

Remark 3. *The idea behind the structure represented in (3) is to move the search of the swarm throughout the full D-dimensional search space or at least to enlarge the dimension of that space. This technique results to require a larger number of particles, which is intuitively understandable considering that increasing the* volume *(in a D-dimensional sense) of the search phase decreases the D-volumetric particle density if the number of particles is not adjusted accordingly.*

Considering an explanatory variable $x > 0$ with the following regressor which represents a simplification of that proposed in [2]:

$$y = A_1 \exp\left(-(xA_2)^2\right) + A_3 + A_4 \cos(2\pi A_5 x) + A_6 \sin(2\pi A_7 x) \qquad (5)$$

in which parameter A_{1-7} are to be calculated. A possible strategy to achieve this is a space-extended PSO to estimate the parameters A_{1-7} of the defined regressor in which additional components ${}^2\mathbf{gbest}^k$, ${}^3\mathbf{gbest}^k$ are considered. As mentioned, these represent the second-best and third-best global components of the swarm at each time k. In the flowchart in Fig. 1 it is possible to see a sketch of this PSO variation. In the presented case, the algorithm uses parameters $k_2 = 2$, $k_3 = 1$ and parameter $k_3 = \frac{1}{2}$ as tuning parameters. It is straightforward to weight these components progressively less, as their number has increased w.r.t. conventional PSO. The pseudo code of Algorithm 1 sketches the proposed PSO structure. In the proposed algorithm, values of l from 2 to 3 are considered because when increasing the dimension of the search space, no improvements are noticed in the experiment. Moreover, increasing the dimension of the searching space, an increment of the calculation load is noticed.

4   Results and Discussion

The experiment is done using an Intel Core i7-6700K CPU (2016). In Table 1 as well Figs. 2, 3 and 4, possible interpretations of multiple different variants of the extended-space PSO are included. In fact, the algorithm with the search space without any social components (setting $k_2 = k_3 = k_4 = 0$) is called *PSO Anarchy*, in which just the exploration components $\mathbf{pbest}_i^k$ are considered. This metaphorical designation is related to an analogy where the particles are considered as workers, who all do their own exploration independently, while the chief is absent. The other variations which include different exploitation

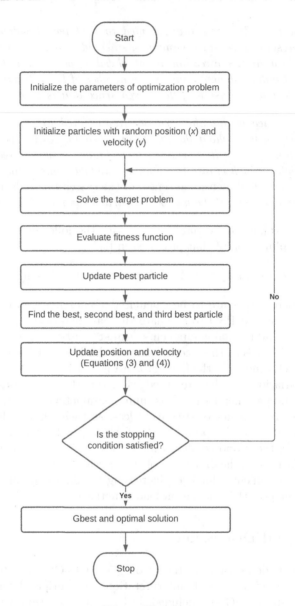

Fig. 1. Flowchart

components are called *PSO 1 Chief* ($k_2 \neq 0$, $k_3 = 0$, $k_4 = 0$), *PSO 2 Chiefs* ($k_2 \neq 0$, $k_3 \neq 0$, $k_4 = 0$) and *PSO 3 Chiefs* ($k_2 \neq 0$, $k_3 \neq 0$, $k_4 \neq 0$). From Table 1 as well from Fig. 2 it is possible to observe that the shortest time for the cost function (sum of squared errors, SSE) to fall below the given tolerance of 4.13×10^{-11} is reached with the variant of *PSO 2 Chiefs*, also the lowest final value of the SSE is obtained with this variant. However, when including another component (*PSO 3 Chiefs*), the convergence time increases again, with worse SSE. This implies that choosing the number of additional components is, in itself, a convex optimization problem, at least in the presented case.

Nevertheless, in terms of precision, the Genetic Algorithm (GA) is the best one, yet slower than all the tested PSO variants. It is to observe that GA variation and *PSO Anarchy* have almost the same performances in terms of time. In fact, the GA algorithm is looking inside the search space without identifying any chiefs. The GA was programmed with the same single objective optimization consisting of 7 variables.

Table 1. Algorithm comparison

Variant	Efficiency index $1/(Time \times SSE)$	Time (s)	SSE value
GA	1.7080×10^8	1.4214×10^2	4.1192×10^{-11}
PSO anarchy	1.7455×10^8	1.3426×10^2	4.2670×10^{-11}
PSO 1 chief	3.0794×10^8	0.7768×10^2	4.1803×10^{-11}
PSO 2 chief	3.8594×10^8	0.6281×10^2	4.1255×10^{-11}
PSO 3 chief	2.5707×10^8	0.9299×10^2	4.1831×10^{-11}

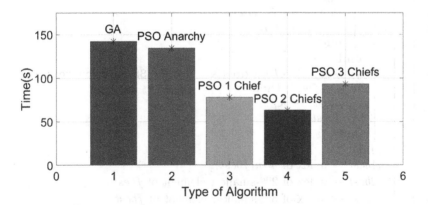

Fig. 2. Time vs. algorithm variant

Algorithm 1: PSO algorithm structure with extended search space

---

1 Initialize *pbest* $(N \times D)$ to store each particle's best parameters and f,
   fbest $(N \times 1)$ to store current and best SSE of each particle and *g1best*,
   g2best, *g3best* $(1 \times D)$ to store the best, 2nd-best and 3rd-best particle.
2 **for** *run from 1 to* run_{max} **do**
3   Initialize swarm positions **X** and velocities **V** randomly (uniform
        distribution within given bounds LB_j, UB_j for each parameter j).
4   **for** *i from 1 to N* **do**
5           $f_i \leftarrow objectiveFunction(\mathbf{X}_i)$ // SSE for initial particles
6   **if** $f_i \leq tolerance$ **then**
7   **return** $\mathbf{X}_i$ as best solution and exit
8   **end**
9   **end**
10      $fbest \leftarrow f$ // initialize
11      $i1best \leftarrow$ index of $min(fbest)$
12      $g1best \leftarrow \mathbf{X}_{i1best}$
13      $g2best \leftarrow \mathbf{X}_{i1best}$ // initialize
14      $g3best \leftarrow \mathbf{X}_{i1best}$ // initialize
15   **for** *iteration from 1 to* $iteration_{max}$ **do**
16   **for** *i from 1 to N* **do**
17   **for** *j from 1 to D* **do**
18   Update $V_{i,j}$ // according to eq. (3)
19   Update $X_{i,j}$ // according to eq. (4)
20   **if** $X_{i,j} < LB_j$ **then**
21                      $X_{i,j} \leftarrow LB_j$
22   **end**
23   **if** $X_{i,j} > UB_j$ **then**
24                      $X_{i,j} \leftarrow UB_j$
25   **end**
26   **end**
27              $f_i \leftarrow objectiveFunction(\mathbf{X}_i)$ // SSE for current part.
28   **if** $f_i \leq fbest_i$ **then**
29                  $pbest_i \leftarrow \mathbf{X}_i$
30                  $fbest_i \leftarrow f_i$
31   **end**
32   **end**
33          $i1best \leftarrow$ index of $min(fbest)$
34          $i2best \leftarrow$ index of 2nd-smallest element of $fbest$
35          $i3best \leftarrow$ index of 3rd-smallest element of $fbest$
36          $g1best \leftarrow pbest_{i1best}$
37          $g2best \leftarrow pbest_{i2best}$
38          $g3best \leftarrow pbest_{i3best}$
39   **if** $objectiveFunction(gbest) \leq tolerance$ **then**
40   **return** *gbest* as best solution and exit
41   **end**
42   **end**
43 **end**

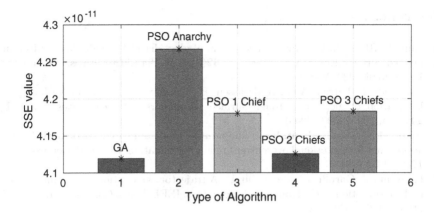

Fig. 3. Least squares cost function values (SSE) vs. algorithm type

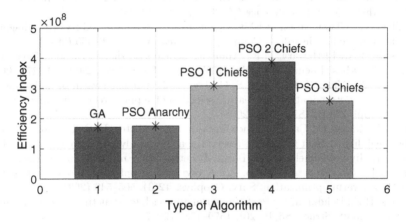

Fig. 4. Efficiency index $1/(Time \times SSE)$ vs. algorithm type

5   Conclusion

This contribution deals with the identification of roughness as a function of gloss in manufacturing using Particle Swarm Optimization methods. An identification structure is proposed consisting of a Gaussian and Trigonometric Regressor characterized by seven parameters to be estimated. An analysis of an increment of the dimension of the search space of the PSO is proposed. This is realized through an increment of its exploitation dimension to improve the precision of the search phase of the PSO. Nevertheless, an optimal increment of the dimension exists. Experimental results using measured data from a manufacturing system are shown with and without enlargement of the search space together with the results obtained by using a Genetic Algorithm. Advantages and drawbacks are pointed out.

References

1. Gim, J., Rhee, B.: Generation mechanism of gloss defect for high-glossy injection-molded surface. Korea Aust. Rheol. J. **32**(3), 183–194 (2020). https://doi.org/10.1007/s13367-020-0018-1
2. Simonsen, I., Larsen, Å.G., Andreassen, E., Ommundsen, E., Nord-Varhaug, K.: Estimation of gloss from rough surface parameters. Physica Status Solidi (b) **242**(15), 2995–3000 (2005)
3. Alam, M.N., Das, B., Pant, V.: A comparative study of metaheuristic optimization approaches for directional overcurrent relays coordination. Electr. Power Syst. Res. **128**, 39–52 (2015)
4. Fabbrini, A., Garulli, A., Mercorelli, P.: A trajectory generation algorithm for optimal consumption in electromagnetic actuators. IEEE Trans. Control Syst. Technol. **20**(4), 1025–1032 (2012)
5. Stephan, R., Mercorelli, P., Belda, K.: Energy optimization in motion planning of a two-link manipulator using Bernstein polynomials. In: 2021 22nd International Carpathian Control Conference (ICCC), pp. 1–6 (2021)
6. Subasi, A.: Classification of EMG signals using PSO optimized SVM for diagnosis of neuromuscular disorders. Comput. Biol. Med. **43**(5), 576–586 (2013)
7. Bashir, Z.A., El-Hawary, M.E.: Applying wavelets to short-term load forecasting using PSO-based neural networks. IEEE Trans. Power Syst. **24**(1), 20–27 (2009)
8. Lin, S.-W., Ying, K.-C., Chen, S.-C., Lee, Z.-J.: Particle swarm optimization for parameter determination and feature selection of support vector machines. Expert Syst. Appl. **35**(4), 1817–1824 (2008)
9. Gu, Y., Zhang, D., Bao, Z.: A new data-driven predictor, PSO-XGBoost, used for permeability of tight sandstone reservoirs: a case study of member of chang 4+5, western Jiyuan Oilfield, Ordos Basin. J. Petrol. Sci. Eng. **199**, 108350 (2021)
10. Pace, F., Santilano, A., Godio, A.: A review of geophysical modeling based on particle swarm optimization. Surv. Geophys. **42**(3), 505–549 (2021)
11. Perez, R.E., Behdinan, K.: Particle swarm approach for structural design optimization. Comput. Struct. **85**(19–20), 1579–1588 (2007)
12. Khare, A., Rangnekar, S.: A review of particle swarm optimization and its applications in solar photovoltaic system. Appl. Soft Comput. **13**(5), 2997–3006 (2013)
13. Godio, A., Pace, F., Vergnano, A.: SEIR modeling of the Italian epidemic of SARS-CoV-2 using computational swarm intelligence. Int. J. Environ. Res. Public Health **17**(10), 3535 (2020)
14. Al-qaness, M.A.A., Ewees, A.A., Fan, H., Abd El Aziz, M.: Optimization method for forecasting confirmed cases of COVID-19 in china. J. Clin. Med. **9**(3), 674 (2020)
15. Barshandeh, S., Haghzadeh, M.: A new hybrid chaotic atom search optimization based on tree-seed algorithm and Levy flight for solving optimization problems. Eng. Comput. **37**(4), 3079–3122 (2020). https://doi.org/10.1007/s00366-020-00994-0
16. Zhao, W., Shi, T., Wang, L., Cao, Q., Zhang, H.: An adaptive hybrid atom search optimization with particle swarm optimization and its application to optimal no-load PID design of hydro-turbine governor. J. Comput. Des. Eng. **8**(5), 1204–1233 (2021)
17. Xia, X., et al.: An expanded particle swarm optimization based on multi-exemplar and forgetting ability. Inf. Sci. **508**, 105–120 (2020)

An Improved Crow Search Algorithm with Grey Wolf Optimizer for High-Dimensional Optimization Problems

Artee Abudayor[✉] and Özkan Ufuk Nalbantoğlu[✉]

Department of Computer Engineering, Erciyes University, Talas, Kayseri, Turkey
jamendat@gmail.com, nalbantoglu@erciyes.edu.tr

Abstract. Crow search algorithm (CSA) mainly solves optimization problems. In high-dimensional optimization problems, CSA searches with moves toward the wrong crows' hiding position. Solving the problems of the CSA algorithm, this paper proposes an improved CSA with Grey Wolf Optimization (GWO) algorithms is called ICSAGWO for manipulating the high-dimensional optimization problem. The main idea is to hybrid both algorithms' strengths that utilize the efficient exploitation ability of CSA with good performance in the exploration ability and convergence speed of GWO. By hybridizing, the authors employ an adaptive inertia weight to control exploitation and exploration capacities. ICSAGWO algorithm is tested on twenty-three benchmark functions with 30 to 500 dimensions and compared among other algorithms, such as GSA, WOA, GWO, CSA, etc. Experimental results of the proposed algorithm ICSAGWO obtain high performance in both unimodal and multimodal and not affecting the search performance even in high dimension data over other algorithms.

Keywords: Crow search algorithm · Optimization · Hybrid approach · Grey-wolf optimizer · High-dimensional optimization problem

1 Introduction

Researchers are interested in challenging optimization problems in various science and engineering areas for solving real-world problems [1]. Moreover, simple and ideal mathematical models are defined to solve a small subset of these problems, the most popular usually be addressed by these approaches [2]. An example of the related problems is the non-convex, non-linear, multimodal, and high-dimensional optimization problems [3]. The metaheuristic algorithms (MAs) are stochastic population-based on global optimization methods, which successfully solve the various complexes and real optimization problems, such as travel salesman problems [4], image processing [5], etc. Because they obtain the optimal solutions or results in minimizing the execution time even problems are complex. MAs can be divided into three categories [6], such as evolutionary [7], swarm intelligence [8], and physics-based [9] algorithms. Examples of widely well-known algorithms in this field, namely, Ant Colony Optimization (ACO) [10], Marine Predators Algorithm [11], Grey Wolf Optimizer (GWO) [12], Genetic Algorithms (GA)

© Springer Nature Switzerland AG 2022
K. K. Patel et al. (Eds.): icSoftComp 2021, CCIS 1572, pp. 51–64, 2022.
https://doi.org/10.1007/978-3-031-05767-0_5

[13], Crow Search Algorithm (CSA) [14], Gravitational Search Algorithm (GSA) [15], Gradient-based optimizer (GBO) [3], Quantum Beetle Antennae Search (QBAS) [16], and Whale Optimization Algorithm (WOA) [17].

In addition, solving high-dimensional optimization problems in MAs also requires an algorithm that balances between two crucial search capacities: exploration and exploitation for exploring alternatives that are not discovered from the current in all entire search space and exploring near-optimal solutions, respectively. However, Many MAs algorithms have limited the decision variables of the scale problem with 30-dimensional variables like CSA. Enhancing the performance of CSA both hybrid and modified versions, such as CSA with niching technique (Niching CSA) [18] proposed to improve the exploration capacity because of the weakness of CSA, which is not good in the global search and is slow in convergence speed in multi-modal optimization problems. A modified CSA (MCSA) [19] gained improving the exploitation capacity of CSA by adaptive flight length (fl) strategy. Another group of CSA focuses on balancing between exploration and exploitation capacities: An improved CSA to balance exploration and exploitation by adaptive inertia weight strategy and another selection mechanism like roulette wheel selection is called ICSA [20]. CSA with neighborhood search (NICSA) can balance the exploration and exploration capacities by improving the search mechanism of CSA [21]. Lastly, there are modifications for the group of high-dimensional problems: An improved CSA with other extensions, as the equation of WOA that is a spiral search, and a gaussian variation strategy (ISCSA) [22]. This study aims to modify CSA with GWO by hybridization techniques because the weakness of CSA in exploration capacity can be overcome by GWO.

GWO was developed by Mirjalili in 2014 [12]. The basis of the algorithm mimics a hierarchy of the grey-wolf behavior in both social leadership and hunting in nature. Moreover, GWO has a strong exploration ability and it also converges quickly towards the optimum. The experimental results of GWO show that it has superior performance compared to PSO and GSA in finding the optimum with a decent convergence. An example of GWO for enhancing the exploitation or/and exploration capacities by hybridized or improved original algorithms, such as GWO with the Fireworks algorithm (FWA), is called FWGWO [23]. This is an algorithm proposed with the hybrid approach between FWA for the exploration ability and GWO for the exploitation ability for improving the convergence speed. Mean Grey Wolf Optimizer (MGWO) with WOA is called HAGWO [24] that balances two crucial searches in both exploration and exploitation and also avoids premature convergence trapping in local minima. A hybrid GWO with SCA is called GWOSCA [25], enhancing GWO with the exploration capacity and convergence speed of SCA to solve the performance of GWO in an alpha equation.

Moreover, there are hybridization techniques helping improve the performance and getting results efficiently at the same time, in optimizing problems of MAs [26]. Talbi in [27] defined the hybridization technique that combines various algorithms, which can divide into two hybridized methods in high-level or low-level. The high-level hybrid method consists of two algorithms in combination by parallel as heterogeneous. On the other hand, the low-level hybrid method embeds the main algorithm as homogeneous. Consequently, the hybridized algorithms are famous for improving their performance, such as PSO-GWO [28] for solving stability, convergence speed, and exploration ability;

GA-PSO with Symbiotic Organisms Search (SOS) for reducing the complexity and minimizing the execution time [26], and GSA-DE for improving a convergence speed in global optimization problems [29].

This paper aims to propose an improved CSA with GWO for improving the gap of the original CSA on high-dimensional problems. In [30] an algorithm that improves the weakness of CSA in the exploration ability with GWO algorithm is also proposed. However, while some meta-heuristic algorithms may achieve solving some problems successfully, the same algorithms may be unsuccessful in solving other problems, which means no algorithms can solve all of them according to the a rule of No Free Lunch (NFL) [31]. Consequently, this study attempts to propose a new hybrid algorithm that combines two algorithms based on stochastic population-based metaheuristic algorithms, which aid to fulfill a research gap.

An expectation from the proposed algorithm ICSAGWO is achieving high performance when solving high-dimensional problems. The performance of ICSAWOA is investigated in 23 benchmark functions for evaluating both efficiency and robustness. In addition, The ICSAGWO algorithm had demonstrated to be effective in finding global optima of high-dimensional optimization problems. The ICSAGWO can quickly identify the region in which the global optimum is located. Moreover, the experiment results on all function tests are found to be competitive among other well-known algorithms.

Inspired and mathematical models of CSA and GWO algorithms are explained in Sect. 2. A meticulously description of the proposed algorithm ICSAGWO is represented in Sect. 3. The experimental results and discussion on twenty-three commonly used benchmark problems of high dimensional data have shown in Sect. 4. Lastly, Sect. 5 includes the conclusions and future work.

2 Methods

2.1 Crow Search Algorithm (CSA)

CSA is mimicking crows' behavior, which was developed by Askazadeh et al. in 2016 [14]. The concept of CSA is to imitate a crow individual that attempts to hide a place to store food. Moreover, the crow takes precautions to protect their location from other crows, who could follow to steal the stored food.

Thus, the motion of each crow individual for storing food is induced by two main situations: Firstly, finding the place for hiding food of other crows; and secondly, protecting its hiding place.

Firstly, the crow q may be unaware that they are being followed by other crows as h. Therefore, the other crows h will know the place to hide food belonging to crow q. Then, the crow h will update the position, as shown in Eq. (1).

$$\vec{x}_h(ite + 1) = \vec{x}_h(ite) + fl_h(ite) \times rand() \times \left[\vec{M}_q(ite) - \vec{x}_h(ite) \right] \tag{1}$$

where h = 1, 2, ..., NF_C; NF_C is the number of all crows. The iteration $ite = 1, 2..., ite_m$; i te and ite_m represent the current iteration and the maximum number of iterations, respectively. $\vec{x}_h(ite)$ dedicates the current position of the h-th crow at ite-th iteration. $fl_h(ite)$ is flight length of the h-th crow. Additionally, the hiding food place of crow q at ite-th iteration is expressed by $\vec{M}_q(ite)$. The parameter rand() is a number randomly between 0 and 1.

Secondly, the crow q may be aware that it is being followed by other crows such as h. Therefore, the other crows q deceives crow h, and the crow h choose a position randomly, as shown in Eq. (2)

$$\vec{x}_h(ite + 1) = a \; random \; posiotion \; otherwise \qquad (2)$$

According to the motion of each crow individuals that mention as above two situations in which can be expressed in Eq. (3):

$$\vec{x}_h(ite + 1) = \begin{cases} \vec{x}_h(ite) + fl_h(ite) \times rand() \times \left[\vec{M}_q(ite) - \vec{x}_h(ite)\right] & r \geq AP_q(ite) \\ a \; random \; posiotion \; otherwise \end{cases} \qquad (3)$$

where, $AP_q(ite)$ dedicates the awareness probability of the q-th crow individual. The parameter r is a number randomly between 0 and 1.

2.2 Grey Wolf Optimizer Algorithm (GWO)

GWO mimics grey-wolf behaviors in which was developed by Mirjalili in 2014 [12]. The concept of GWO is to imitate both social leadership and hunting of the grey-wolf behavior. In addition, the hierarchy of the grey wolf consists of 4 classes, namely, alpha (α), beta (β), delta (δ), and omega (Ω). However, three leader wolves are considered in the GWO system, whereas the omega group is the rest of the candidate solutions. Therefore, the GWO system can be divided into three steps:

- **Encircling:** The mathematical model of grey wolves for encircling the prey, as shown in Eq. (4) and (5).

$$\vec{D} = \left| \vec{C}.\vec{X}_p(ite) - \vec{X}(ite) \right| \qquad (4)$$

$$\vec{X}(ite + 1) = \vec{X}_p(ite) - \vec{A}.\vec{D} \qquad (5)$$

where $\vec{X}(ite)$ and, $\vec{X}_p(ite)$ are the current position and the prey position at the ite-th iteration, respectively. $\vec{A}, \vec{D}$ are the coefficient variables; these parameters can be calculated by Eq. (6) and (7) as follow:

$$\vec{A} = 2\vec{\alpha}\,\vec{r_1} \qquad (6)$$

$$\vec{C} = 2\vec{r_2} \qquad (7)$$

where $\vec{r_1}, \vec{r_2}$ are random numbers in the range between 0 and 1. $\vec{\alpha}$ is a linear parameter that decreased from 2 to 0 over each iteration, as calculated by Eq. (8).

$$\vec{\alpha} = 2 - \left(2 \times \frac{ite}{ite_{max}}\right) \qquad (8)$$

- **Hunting:** The mathematical model for the hunting behavior, which the grey wolves have better knowledge about the preys' location by guided α. Hence, the updates of the location of α, β, and δ can be represented in Eq. (9)–(11), whereas the other wolves are obliged to update their positions by following them.

$$\vec{D_\alpha} = \vec{C_1}\vec{X_\alpha} - \vec{X}, \vec{D_\beta} = \vec{C_2}\vec{X_\beta} - \vec{X}, \vec{D_\delta} = \vec{C_3}\vec{X_\delta} - \vec{X} \tag{9}$$

where $\vec{C_1}$, $\vec{C_2}$, and $\vec{C_3}$ are calculated by Eq. (7).

$$\vec{X_1} = \vec{X_\alpha} - \vec{A_1}\vec{D_\alpha}, \vec{X_2} = \vec{X_\beta} - \vec{A_2}\vec{D_\beta}, \vec{X_3} = \vec{X_\delta} - \vec{A_3}\vec{D_\delta} \tag{10}$$

where the vectors $\vec{X}_1$, $\vec{X}_2$, and $\vec{X}_3$ are the first three best solutions at ite-th iteration. The parameters A_1, A_2, and A_3 can be calculated by utilizing Eq. (6). The parameters D_α, D_β, and D_δ are defined in Eq. (9).

$$\vec{X}(ite + 1) = \frac{\vec{X}_1 + \vec{X}_2 + \vec{X}_3}{3} \tag{11}$$

Attacking: The grey wolf terminates a hunt and starts an attack on the prey when they stop moving. To form a mathematical model of attacking a prey the parameter $\vec{a}$ is used. $\vec{a}$ is a linear parameter that decreases from 2 to 0 at each iteration gradually. Moreover, the parameter also controls the exploration and exploitation capacities. Furthermore, the fluctuation of $\vec{A}$ depends on $\vec{a}$, and the $\vec{A}$ value decreases by following $\vec{a}$.

3 The Proposed Algorithm

In this paper, we present an improved CSA with GWO by using the idea of the low-level hybrid algorithm by integrating the functionality of both algorithms to manipulate high-dimensional problems. Therefore, the process of ICSAGWO consists of two steps: adaptive inertia weight (in_w) strategy and modified position update mechanism as follow:

3.1 Adaptative Inertia Weight (W) Strategy

The original CSA algorithm searches the global optimum by mimicking the crows' behavior. A crow individual attempts its hiding place for storing their food, whereas the other crows could follow them to steal their food. Moreover, the proposed algorithm attempts to find a suitable balance of two search capacities to manipulate the high-dimension optimization problems.

To enhance both crucial search capacities of the original CSA algorithm, we propose to weight the position update equation in every iteration as shown in the following equations. The inertia wights in_w were proposed by Shi and Eberhart in 1998 [32] to control exploration and exploitation capacities, which are calculated as shown in Eq. (12).

$$in\_w = (in_{wmax} - in_{wmin}) \times \left(\frac{(ite_{max} + ite)}{(2 \times ite_{max})} \right) \tag{12}$$

where: in_w_{max}, in_w_{min} dedicate 0.9 and 0.4, respectively.

3.2 Modified Position Update Mechanism

The position updated mechanism in standard CSA utilizes Eq. (1) in the local optimal region with the boundaries to provide better exploitation. On the other hand, using Eq. (2) a search for global optimal region is introduced, which finds the other best solutions using a random walk. The concept of the proposed algorithm is the hybridization of two algorithms between CSA and GWO. Moreover, to enhance the efficiency exploitation phase of CSA with the updated position equation of $\vec{x}_h(ite + 1)$ can be calculated based on the inertia weight in_w (in Eq. (12)) obtained so far as calculated by Eq. (13).

$$\vec{x}_h(ite + 1) = (\vec{x}_h(ite) \times in\_w) + fl_h(ite) \times rand() \times \left[\vec{M}_q(ite) - \vec{x}_h(ite)\right] \quad (13)$$

In the exploration phase of ICSAGWO, the update position of GWO with inertia weighted in every iteration is calculated as follow: Firstly, the inertia weight (in_w) is utilized to control the movement as shown in Eq. (14) using in GWO are calculated based on coefficient vectors calculated as in Eq. (6), (7), and (12), respectively.

$$\begin{aligned} in\_w_1 &= in\_w \times \vec{A}_1 \times \vec{C}_1 \\ in\_w_2 &= in\_w \times \vec{A}_2 \times \vec{C}_2 \\ in\_w_3 &= in\_w \times \vec{A}_3 \times \vec{C}_3 \end{aligned} \quad (14)$$

Then, the update location equation is modified as per the calculated inertia weights, as shown in Eq. (15) based on considering $\vec{X}_1, \vec{X}_2, \vec{X}_3, in\_w_1, in\_w_2,$ and in_w_3. These are shown in Eq. (10), and (14), respectively.

$$\vec{x}_h(ite + 1) = \frac{\left(in\_w_1 \times \vec{X}_1\right) + \left(in\_w_2 \times \vec{X}_2\right) + \left(in\_w_3 \times \vec{X}_3\right)}{(in\_w_1 + in\_w_2 + in\_w_3)} \quad (15)$$

In summary, the updated position equation of the ICSAGWO that manifests the aforementioned two situations can be concluded in Eq. (16). Therefore, the detail of ICSAGWO is represented in Algorithm 1.

$$\vec{x}_h(ite + 1) = \begin{cases} (\vec{x}_h(ite) \times in\_w) + fl_h(ite) \times rand() \times \left[\vec{M}_q(ite) - \vec{x}_h(ite)\right] & r \geq AP_q(ite) \\ \frac{\left(in\_w_1 \times \vec{X}_1\right) + \left(in\_w_2 \times \vec{X}_2\right) + \left(in\_w_3 \times \vec{X}_3\right)}{(in\_w_1 + in\_w_2 + in\_w_3)} & otherwise \end{cases} \quad (16)$$

where, $AP_q(ite)$ dedicates the awareness probability of the q-th crow individual. The parameter r is a number randomly between 0 and 1.

Algorithm 1 Pseudocode of ICSAGWO algorithm

Set the initial parameters of NF_C, AP_q, fl_h, ite, and ite_{max}

Initialize the crow position h randomly $\vec{X}_h$; $h = 1, .., NF_C$

Evaluate the fitness of crows $Fn(\vec{X})$.

Evaluate the memory of crows' hiding food location $\vec{M}$

Repeat

    Find the three best fitness from $Fn(\vec{X})$ as $\vec{X}_\alpha$, $\vec{X}_\beta$ and $\vec{X}_\delta$, respectively.

    Update $in\_w, in\_w_1, in\_w_2, in\_w_3$, a, A and C

    for $(h = 1 : h \le NF_C)$ **do**

        One of crows randomly choose to follow crow q

        if rand $\ge AP_q$ **then**

$$\vec{x}_h(ite + 1) = (\vec{x}_h(ite) \times in\_w) + fl_h(ite) \times rand() \times [\vec{M}_q(ite) - \vec{x}_h(ite)], \text{ as Eq. (13)}$$

        else

$$\vec{X}_h(ite + 1) = \frac{(in\_w_1 \times \vec{X}_1) + (in\_w_2 \times \vec{X}_2) + (in\_w_3 \times \vec{X}_3)}{(in\_w_1 + in\_w_2 + in\_w_3)} \text{ , as Eq. (15)}$$

    End If

    End for

    Recheck a boundary of the possibility of $\vec{X}(ite + 1)$

    Evaluate the new fitness of $\vec{X}(ite + 1)$

    Update the memory of crows' hiding food location $\vec{M}$

    Set ite = ite+1.

Until $(ite < ite_{max})$.

Return the best memory of crows' hiding food location $\vec{M}$.

4 Experimental Results

The experiment in this study aims to show how consistent are the solutions of the proposed algorithm ICSAGWO by comparing it against well-known meta-heuristic algorithms. This study chooses the set of twenty-three benchmark functions. Moreover, the optimal results are presented in average values of over 30 independent runs.

4.1 Benchmark Function

To validate the exploration and exploitation abilities of ICSAGWO, we utilize three groups of the benchmark functions: UF-1 to UF-7 Functions for unimodal benchmark, MF-1 to MF-6 Functions for multimodal benchmark, and FMF-1 to FMF-10 for fixed-dimension multimodal functions. Tables 1, 2 and 3 are shown the details of these benchmark functions.

4.2 Performance Measure and Common Parameter Settings

The ICSAGWO algorithm was implemented in MATLAB R2018a and ran on a computer with 8 GB 1600 MHz DDR3 memory, 1.6 GHz Dual-Core Intel Core i5 CPU, macOS Big Sur operating system, and 128 GB HDD.

Table 1. UF-1 to UF-7 unimodal functions.

Name	Function	Range	F_{min}
UF-1	$f_{UF\_1}(x) = \sum_{i=1}^{n} x_i^2$	$[-100, 100]$	0
UF-2	$f_{UF\_2}(x) = \sum_{i=1}^{n} |x_i| + \prod_{i=1}^{n} |x_i|$	$[-10, 10]$	0
UF-3	$f_{UF\_3}(x) = \sum_{i=1}^{n} \left(\sum_{j-2}^{i} x_j\right)^2$	$[-100, 100]$	0
UF-4	$f_{UF\_4}(x) = \max_i\{|x_i|, 1 \leq i \leq n\}$	$[-100, 100]$	0
UF-5	$f_{UF\_5}(x) = \sum_{i=1}^{n-1}[100\left(x_{i+1} + x_i^2\right)^2 + (x_i - 1)^2]$	$[-30, 30]$	0
UF-6	$f_{UF\_6}(x) = \sum_{i=1}^{n}([x_i - 0.5])^2$	$[-100, 100]$	0
UF-7	$f_{UF\_7}(x) = \sum_{i=1}^{n} i x_i^2 + \text{random}[0, 1)$	$[-1.28, 1.28]$	0

Table 2. MF-1 to MF-6 multimodal functions.

Name	Function	Range	F_{min}
MF-1	$f_{MF\_1}(x) = \sum_{i=1}^{n} - x_i \sin(\sqrt{|x_i|})$	$[-500, 500]$	-418.982
MF-2	$f_{MF\_2}(x) = \sum_{i=1}^{n}[x_i^2 - 10\cos(2\pi x_i) + 10]$	$[-5.12, 5.12]$	0
MF-3	$f_{MF\_3}(x) = -20\exp(-20\sqrt{\frac{1}{n}\sum_{i=1}^{n} x_i^2}) - \exp\left(\frac{1}{n}\sum_{i+1}^{n}\cos(2\pi x_i)\right) + 20 + e$	$[-32, 32]$	0
MF-4	$f_{MF\_4}(x) = \frac{1}{4000}\sum_{i=1}^{n} x_1^2 - \prod_{i=1}^{n}\cos\left(\frac{x_i}{\sqrt{i}}\right) + 1$	$[-600, 600]$	0
MF-5	$f_{MF\_5}(x) = \frac{\pi}{n}\left\{100\sin(\pi y_1) + \sum_{i=1}^{n-1}(y_i - 1)^2\left[1 + 10\sin^2(\pi y_{i+1})\right] + (y_n - 1)^2\right\}$ $+ \sum_{i=1}^{n} u(x_i, 10, 100, 4)\}$ $y_i = 1 + \frac{x_i+1}{4}$ $u(x_i, \alpha, k, m = f(x) = \begin{cases} k(x_i - \alpha)^m, x_i > \alpha \\ 0, -\alpha < x_i < a \\ -k(x_i - \alpha)^m, x_i > \alpha \end{cases}$	$[-50, 50]$	0
MF-6	$f_{MF\_6}(x) = 0.1\{\sin^2(\pi x_1) + \sum_{i=1}^{n}(x_i - 1)^2\left[1 + \sin^2(3\pi x_i + 1)\right] +$ $(x_n - 1)^2\left[1 + \sin^2(2\pi x_n)\right] + \sum_{i=1}^{n} u(x_i, 5, 100, 4)\}$	$[-50, 50]$	0

Table 4 shows the list of all algorithms' parameter settings used for comparing their performance in this study. The experiments analyze the performance of ICSAGWO by considering the following two aspects: (1) ICSAGWO is compared with other meta-heuristic algorithms, namely, CSA [14], GSA [15], WOA [17], GA [33], DE [34], GWO [12], ICSA [35], and SCCSA [36]. These results were obtained by [12, 22], and [36] for solving the optimization problems with 30-dimensional data. (2) A comparative analysis of data in different dimensions test the search performance of ICSAGWO, CSA, and ICSA algorithms with 50, 100, 200, and 500-dimensional data. The results of CSA and ICSA were obtained by [22] and [35], respectively.

Table 3. FMF-1 to FMF-10 fixed-dimension multimodal functions.

Name	Function	Dim	Range	F_{min}
FMF-1	$f_{FMF\_1}(x) = \left(\frac{1}{500} + \sum_{l=1}^{25} \frac{1}{j + \sum_{i=1}^{2}(x_i - a_{ij})^6} \right)^{-1}$	2	[−65, 65]	1
FMF-2	$f_{FMF\_2}(x) = \sum_{i=1}^{n} \left[a_i - \frac{x_1(b_i^2 + b_i x_2)}{b_i^2 + b_i x_3 + x_4} \right]^2$	4	[−5, 5]	0.0003
FMF-3	$f_{FMF\_3}(x) = 4x_1^2 - 2.1x_1^4 + \frac{1}{3}x_1^6 + x_1 x_2 - 4x_2^2 + 4x_2^4$	2	[−5, 5]	−1.0316
FMF-4	$f_{FMF\_4}(x) = \left(x_2 - \frac{5.1}{4\pi^2}x_1^2 + \frac{5}{\pi}x_1 - 6 \right)^2 + 10(1 - \frac{1}{8\pi})\cos x_1 + 10$	2	[−5, 5]	0.398
FMF-5	$f_{FMF\_5}(x) = [1 + (x_1 + x_2 + 1)^2(19 - 14x_1 + 3x_1^2 - 14x_2 + 6x_1 x_2 + 3x_2^2)]$ $\times [30 + (2x_1 + 3x_2)^2(18 - 32x_1 + 12x_1^2 + 48x_2 + 36x_1 x_2 + 27x_2^2)]$	2	[−2, 2]	3
FMF-6	$f_{FMF\_6}(x) = -\sum_{i=1}^{4} c_i \exp\left(-\sum_{j=1}^{3} a_{ij}(x_j - p_{ij})^2 \right)$	3	[1, 3]	−3.86
FMF-7	$f_{FMF\_7}(x) = -\sum_{i=1}^{4} c_i \exp\left(-\sum_{j=1}^{6} a_{ij}(x_j - p_{ij})^2 \right)$	6	[0, 1]	−3.32
FMF-8	$f_{FMF\_8}(x) = -\sum_{i=1}^{5} \left[(X - a_i)(X - a_i)^T + c_i \right]^{-1}$	4	[0, 10]	−10.1532
FMF-9	$f_{FMF\_9}(x) = -\sum_{i=1}^{7} \left[(X - a_i)(X - a_i)^T + c_i \right]^{-1}$	4	[0, 10]	−10.4028
FMF-10	$f_{FMF\_10}(x) = -\sum_{i=1}^{10} \left[(X - a_i)(X - a_i)^T + c_i \right]^{-1}$	4	[0, 10]	−10.5363

Table 4. The list of all algorithms' parameter settings used for comparison in this study

Algorithms	No. of population	No. of iteration	Parameters
CSA	20	2000	AP = 0.1, fl = 2
WOA	20	2000	a ∈ [2, 0]
GWO	20	2000	a ∈ [2, 0]
GSA	20	2000	$G_0 = 100$ and $\alpha = 20$ and final_per = 2
ICSA	20 and 30	2,000	ef = 0.5, fl = 2 and AP = 0.1
SCCSA	30	100,000 function evaluations	fl = 2, $r_1 = r_2 = r_3 = r_4 \in$ [1, 0]
Our proposed	20 and 30	300 and 2,000	AP = 0.2, fl = 2, $in\_w_{max} = 0.9$, $in\_w_{min} = 0.4$, and a ∈ [2, 0]

4.3 Experimental Series1: Comparison with Other Meta-heuristic Algorithms

This sub-section aims to compare our proposed algorithm ICSAGWO with well-known meta-heuristic algorithms, namely, GA, DE, GSA, WOA, GWO, CSA, WOA, PSOGSA ICSA, and SCCSA; to find the best optimal solution on 30-dimensional data. These optimizations were run 30 times, with 20 population sizes, and for 2000 iteration cycles on each function, as shown in Table 5.

UF-1 to UF-7 unimodal benchmark functions test for the exploitation ability. The results of ICSAGWO were statistically significantly found to be the best in UF-1 to

60   A. Abudayor and Ö. U. Nalbantoğlu

UF-4, and UF-7 functions. Therefore, according to the proposed algorithm ICSAGWO, exhibited the most successful performance among others in terms of exploiting the optimum.

The multimodal benchmark functions MF-1 to MF-6 tested exploration ability. The experimental results of ICSAGWO were found to attain the best optimums in MF-2 to MF-4 functions. Fixed dimensional benchmark functions FMF-1 to FMF-10 also tested exploration capacity in which the results of our proposed algorithm can perform superior to ICSA and GA. Therefore, the performance of ICSAGWO that was evaluated on 30 dimensions was better than ICSA, which also supports our claim that the hybridization enhances its performance to manipulate high-dimensional optimization problems.

Table 5. The experimental results for 30-dimensional data.

Function	GA	DE	GSA	WOA	GWO	CSA	ICSA	SCCSA	ICSAGWO
UF-1	1.65E−02	8.2E−14	3.37E−16	1.11E−264	1.14E−103	1.00E+03	6.95E−28	9.22E−69	**0.00E+00**
UF-2	1.22E−02	1.5E−09	5.59E−08	6.21E−153	3.29E−60	1.69E+00	9.05E−17	8.25E−41	**0.00E+00**
UF-3	1.55E+02	6.8E−11	5.79E+02	1.37E+04	5.87E−26	2.49E+01	7.13E−27	4.31E−31	**0.00E+00**
UF-4	5.30E−01	**0.00E+00**	1.31E+00	3.20E+01	1.41E−25	3.52E+00	1.23E−14	2.15E−17	3.77E−318
UF-5	2.45E+01	**0.00E+00**	3.68E+01	2.68E+01	2.67E+01	6.45E+01	2.84E+01	5.91E+00	2.74E+01
UF-6	5.82E−03	**0.00E+00**	0.00E+00	6.56E−02	7.92E−01	8.71E−04	2.56E−01	4.14E−08	1.34E+00
UF-7	3.98E−03	4.63E−03	5.71E−01	1.00E−03	5.88E−04	2.86E−02	5.41E−05	1.34E−03	**4.10E-05**
MF-1	−2.57E+03	**−1.11E+04**	–	–	–	–	–	−3.08E+03	−2.94E+03
MF-2	2.68E−03	6.92E+01	2.39E+01	**0.00E+00**	3.32E−02	3.07E+01	**0.00E+00**	5.47E+00	**0.00E+00**
MF-3	1.25E−03	9.70E−08	1.31E−08	4.68E−15	9.65E−15	4.21E+00	7.16E−15	**8.88E−16**	**8.88E−16**
MF-4	6.56E−02	**0.00E+00**	1.48E+01	2.70E−03	9.86E−04	3.14E−02	**0.00E+00**	3.34E−02	**0.00E+00**
MF-5	2.56E−04	**7.90E−15**	1.36E−02	1.46E−02	4.92E−02	4.79E+00	1.67E−02	1.34E−02	8.18E−02
MF-6	1.92E−03	5.10E−14	6.99E−17	2.19E−01	6.34E−01	5.49E+00	2.79E−01	2.01E−02	1.50E+00
FMF-1	**9.98E−01**	**9.98E−01**	1.09E+00	1.95E+00	5.95E+00	9.98E−01	2.16E+00	9.98E−01	2.67E+00
FMF-2	6.33E−03	4.5E−14	2.27E−03	6.20E−03	2.30E−03	4.19E−04	3.71E−03	3.07E−04	3.19E−04
FMF-3	−1.03+00	**−1.03E+00**	**−1.03E+00**	**−1.03E+00**	**−1.03E+00**	**−1.03E+00**	**−1.03E+00**	**−1.03E+00**	**−1.03E+00**
FMF-4	–	3.98E−01	3.98E−01	3.98E−01	3.98E−01	3.98E−01	3.98E−01	–	3.99E−01
FMF-5	1.42E+01	3.00E+00	3.00E+00	3.00E+00	5.70E+00	3.00E+00	3.00E+00	3.00E+00	3.00E+00
FMF-6	−3.85E+00	–	−3.86E+00	−3.86E+00	−3.86E+00	−3.86E+00	−3.86E+00	−3.86E+00	3.86E+00
FMF-7	−3.21E+00	–	−3.32E+00	−3.15E+00	−3.26E+00	−3.29E+00	−3.18E+00	−3.26E+00	−3.25E+00
FMF-8	−6.09E+00	−1.02E+01	−6.92E+00	−9.39E+00	−5.98E+00	−8.31E+00	−5.22E+00	−8.22E+00	−5.05E+00
FMF-9	−6.10E+00	−1.04E+01	−1.02E+01	−9.02E+00	−1.04E+01	−1.01E+01	−5.09E+00	−8.94E+00	−5.09E+00
FMF-10	–	−1.05E+01	−1.03E+01	−8.69E+00	−1.05E+01	−1.03E+01	−5.13E+00	–	−5.13E+00

* **Note**: the optimal results in the table are highlighted in bold.

4.4 Experimental Series2: Comparison with Other CSA Algorithms in High Dimensional Data

In this sub-section, a comparative analysis of the performance of ICSAGWO was conducted, comparing it to algorithms such as CSA, ICSA on thirteen benchmark functions with 50, 100, 200, and 500-dimensional data.

Table 6 shows the experimental results of 50, 100, and 200-dimensional data, which ICSAGWO outperforms both the standard CSA and ICSA on nine of thirteen benchmark functions with 50 to 200-dimension sizes in UF-1 to UF-5, UF-7, and MF-2 to MF-4 functions. According to the performance on MF-6 function, ICSAGWO can explore the optimal results better for high-dimensional data as the dimensions are increasing to 100 and 200 dimensions. Therefore, ICSAGWO is relatively stable and consistent in manipulating high-dimensional data in all cases except UF-6, UF-7, and MF-5 functions.

Table 6. The experimental results for 50 to 200-dimensional data.

Function	50 Dim			100 Dim			200 Dim	
	CSA	ICSA	ICSAGWO	CSA	ICSA	ICSAGWO	ICSA	ICSAGWO
UF-1	1.55E$-$01	1.22E$-$27	**0.00E+00**	1.79E+01	2.39E$-$27	**0.00E+00**	6.71E$-$28	**0.00E+00**
UF-2	3.98E+00	7.38E$-$19	**0.00E+00**	1.10E+01	1.82E$-$25	**0.00E+00**	2.19E$-$43	**1.20E$-$316**
UF-3	2.51E+02	1.74E$-$26	**0.00E+00**	1.81E+03	5.40E$-$26	**0.00E+00**	5.58E$-$27	**0.00E+00**
UF-4	7.17E+00	9.99E$-$15	**1.03E$-$306**	1.05E+01	1.11E$-$14	**1.74E$-$296**	1.97E$-$14	**8.21E$-$286**
UF-5	1.71E+02	4.85E+01	**4.78E+01**	9.04E+02	**9.82E+01**	9.82E+01	1.98E+02	1.98E+02
UF-6	6.86E+01	**0.00E+00**	5.78E+00	1.86E+01	**0.00E+00**	1.76E+01	**0.00E+00**	4.31E+01
UF-7	6.12E$-$02	9.87E$-$05	**4.61E$-$05**	2.51E$-$01	1.35E$-$04	**2.75E$-$05**	6.41E$-$05	**3.16E$-$05**
MF-2	5.75E+01	**0.00E+00**	**0.00E+00**	1.20E+02	**0.00E+00**	**0.00E+00**	**0.00E+00**	**0.00E+00**
MF-3	4.96E+00	4.91E$-$15	**8.88E$-$16**	5.72E+00	1.84E$-$15	**8.88E$-$16**	**8.88E$-$16**	**8.88E$-$16**
MF-4	3.19E$-$01	**0.00E+00**	**0.00E+00**	1.15E+00	**0.00E+00**	**0.00E+00**	**0.00E+00**	**0.00E+00**
MF-5	5.76E+00	**3.30E$-$02**	2.84E$-$01	5.67E+00	**7.77E$-$02**	6.82E$-$01	**1.17E$-$02**	9.17E$-$01
MF-6	4.38E+01	**1.40E+00**	3.84E+00	1.17E+02	9.67E+00	**9.34E+00**	1.99E+01	**1.96E+01**

Table 7 presents the result of the evaluating performance as 500-dimensional data of ICSAGWO, which outperforms both CSA and ICSA on five benchmark functions as UF-1, UF-3, UF-5, and MF-2 to MF-4, respectively. Moreover, the consistent performance of increasing dimensional data of ICSAGOW as variation is small in mean value. Furthermore, the performance of ICSAGOW demonstrates a balance of exploration and exploitation abilities and behaving robust even in high dimensions.

Table 7. The experimental results for 500-dimensional data.

Function	CSA	ICSA	ICSAGWO	Function	CSA	ICSA	ICSAGWO
UF-1	2.16E+03	5.83E−29	**0.00E+00**	MF-2	–	–	0.00E+00
UF-2	–	–	1.53E−310	MF-3	5.72E+00	2.55E−15	**8.88E−16**
UF-3	9.28E+01	9.69E+00	**0.00E+00**	MF-4	2.04E+01	**0.00E+00**	**0.00E+00**
UF-4	–	–	7.12E−275	MF-5	–	–	1.09E+00
UF-5	3.10E+03	3.00E+03	**4.99E+02**	MF-6	3.74E+01	**1.46E−01**	4.97E+01
UF-6	3.10E+03	**7.60E−02**	1.18E+02				
UF-7	–	–	4.70E−05				

5   Conclusions and Future Work

Improving the weakness of CSA, this paper proposed a new hybrid algorithm modifying CSA with utilizing the strengths of GWO in the exploration ability. In addition, the concept of developing our proposed algorithm is to enhance the efficacy of CSA to manipulate high-dimensional data. To analyze the performance of the proposed algorithm, it is validated on three groups of benchmark functions. Interestingly, the proposed algorithm has shown to exhibit a robust search performance when handled with high dimensional data. Moreover, the proposed algorithm can achieve high performance in both exploration and exploitation and it can outperform the competitors with efficiency and robustness for supporting high dimensional data up to 500 with a comparison to CSA, GWO, etc. In future work, the authors will employ the proposed algorithm to validate its performance in real-world optimization problems, namely, traveling salesman problems, knapsack problems, feature selection problems, etc.

References

1. Liberti, L.: Introduction to global optimization. Ecole Polytechnique (2008)
2. Zhao, W., Wang, L., Zhang, Z.: Artificial ecosystem-based optimization: a novel nature-inspired meta-heuristic algorithm. Neural Comput. Appl. **32**(13), 9383–9425 (2019). https://doi.org/10.1007/s00521-019-04452-x
3. Ahmadianfar, I., Bozorg-Haddad, O., Chu, X.: Gradient-based optimizer: a new metaheuristic optimization algorithm. Inf. Sci. **540**, 131–159 (2020)
4. Halim, A.H., Ismail, I.: Combinatorial optimization: comparison of heuristic algorithms in travelling salesman problem. Arch. Computat. Methods Eng. **26**(2), 367–380 (2019)
5. Chouksey, M., Jha, R.K., Sharma, R.: A fast technique for image segmentation based on two Meta-heuristic algorithms. Multimedia Tools Appl. **79**(27–28), 19075–19127 (2020). https://doi.org/10.1007/s11042-019-08138-3
6. Hare, W., Nutini, J., Tesfamariam, S.: A survey of non-gradient optimization methods in structural engineering. Adv. Eng. Softw. **59**, 19–28 (2013)
7. Mühlenbein, H., Gorges-Schleuter, M., Krämer, O.: Evolution algorithms in combinatorial optimization. Parallel Comput. **7**(1), 65–85 (1988)

8. Krause, J., et al.: A survey of swarm algorithms applied to discrete optimization problems. In: Swarm Intelligence and Bio-Inspired Computation, pp. 169–191. Elsevier (2013)
9. Geem, Z.W., Kim, J.H., Loganathan, G.V.: A new heuristic optimization algorithm: harmony search. Simulation **76**(2), 60–68 (2001)
10. Dorigo, M., Di Caro, G.: Ant colony optimization: a new meta-heuristic. In: Proceedings of the 1999 Congress on Evolutionary Computation-CEC99 (Cat. No. 99TH8406), pp. 1470–1477. IEEE (1999)
11. Faramarzi, A., Heidarinejad, M., Mirjalili, S., Gandomi, A.H.: Marine Predators Algorithm: a nature-inspired metaheuristic. Expert Syst. Appl. **152**, 113377 (2020)
12. Mirjalili, S., Mirjalili, S.M., Lewis, A.: Grey wolf optimizer. Adv. Eng. Softw. **69**, 46–61 (2014)
13. Goldberg, D.E.: Genetic algorithms in search, Optimization, and Machine Learning (1989)
14. Askarzadeh, A.: A novel metaheuristic method for solving constrained engineering optimization problems: crow search algorithm. Comput. Struct. **169**, 1–12 (2016)
15. Rashedi, E., Nezamabadi-Pour, H., Saryazdi, S.: GSA: a gravitational search algorithm. Inf. Sci. **179**(13), 2232–2248 (2009)
16. Khan, A.T., Cao, X., Li, S., Hu, B., Katsikis, V.N.: Quantum beetle antennae search: a novel technique for the constrained portfolio optimization problem. Sci. China Inf. Sci. **64**(5), 1–14 (2021). https://doi.org/10.1007/s11432-020-2894-9
17. Mirjalili, S., Lewis, A.: The whale optimization algorithm. Adv. Eng. Softw. **95**, 51–67 (2016). https://doi.org/10.1016/j.advengsoft.2016.01.008
18. Islam, J., et al.: A modified crow search algorithm with niching technique for numerical optimization. In: 2019 IEEE Student Conference on Research and Development (SCOReD). 2019 IEEE Student Conference on Research and Development (SCOReD), Bandar Seri Iskandar, Malaysia, pp. 170–175. IEEE (2019). https://doi.org/10.1109/SCORED.2019.8896291
19. Mohammadi, F., Abdi, H.: A modified crow search algorithm (MCSA) for solving economic load dispatch problem. Appl. Soft Comput. **71**, 51–65 (2018)
20. Shi, Z., et al.: Improved crow search algorithm with inertia weight factor and roulette wheel selection scheme. In: 2017 10th International Symposium on Computational Intelligence and Design (ISCID), pp. 205–209. IEEE (2017)
21. Qu, C., Fu, Y.: Crow search algorithm based on neighborhood search of non-inferior solution set. IEEE Access **7**, 52871–52895 (2019)
22. Han, X., et al.: An improved crow search algorithm based on spiral search mechanism for solving numerical and engineering optimization problems. IEEE Access **8**, 92363–92382 (2020)
23. Yue, Z., Zhang, S., Xiao, W.: A novel hybrid algorithm based on grey wolf optimizer and fireworks algorithm. Sensors **20**(7), 2147 (2020)
24. Singh, N., Hachimi, H.: A new hybrid whale optimizer algorithm with mean strategy of grey wolf optimizer for global optimization. Math. Computat. Appl. **23**(1), 14 (2018)
25. Singh, N., Singh, S.B.: A novel hybrid GWO-SCA approach for optimization problems. Int. J. Eng. Sci. Technol. **20**(6), 1586–1601 (2017)
26. Farnad, B., Jafarian, A., Baleanu, D.: A new hybrid algorithm for continuous optimization problem. Appl. Math. Model. **55**, 652–673 (2018). https://doi.org/10.1016/j.apm.2017.10.001
27. Talbi, E.-G.: A taxonomy of hybrid metaheuristics. J. Heuristics **8**(5), 541–564 (2002)
28. Singh, N., Singh, S.B.: Hybrid algorithm of particle swarm optimization and grey wolf optimizer for improving convergence performance. J. Appl. Math. **2017**, 1–15 (2017)
29. Zhao, F., et al.: A hybrid algorithm based on self-adaptive gravitational search algorithm and differential evolution. Expert Syst. Appl. **113**, 515–530 (2018)
30. Arora, S., et al.: A new hybrid algorithm based on Grey wolf optimization and crow search algorithm for unconstrained function optimization and feature selection. IEEE Access **7**, 26343–26361 (2019)

31. Wolpert, D.H., Macready, W.G.: No free lunch theorems for optimization. IEEE Trans. Evol. Comput. **1**(1), 67–82 (1997)
32. Shi, Y., Eberhart, R.: A modified particle swarm optimizer. In: 1998 IEEE International Conference on Evolutionary Computation Proceedings. IEEE World Congress on Computational Intelligence (Cat. No. 98TH8360), pp. 69–73. IEEE (1998)
33. Holland, J.H.: Genetic algorithms. Sci. Am. **267**(1), 66–73 (1992)
34. Storn, R., Price, K.: Differential evolution–a simple and efficient heuristic for global optimization over continuous spaces. J. Global Optim. **11**(4), 341–359 (1997)
35. Jain, M., Rani, A., Singh, V.: An improved crow search algorithm for high-dimensional problems. J. Intell. Fuzzy Syst. **33**(6), 3597–3614 (2017). https://doi.org/10.3233/JIFS-17275
36. Khalilpourazari, S., Pasandideh, S.H.R.: Sine–cosine crow search algorithm: theory and applications. Neural Comput. Appl. **32**(12), 7725–7742 (2019). https://doi.org/10.1007/s00521-019-04530-0

A Fuzzy Rule Based Directional Approach for Salt and Pepper Noise Removal with Edge Preservation

Aritra Bandyopadhyay[1] and Devadatta Das[2(✉)]

[1] Supreme Knowledge Foundation Group of Institutions, Chandannagar, West Bengal, India
[2] Hooghly Engineering and Technology College, Hooghly, West Bengal, India
devadattadas2013@gmail.com

Abstract. Existing impulse noise reduction schemes generally deals with fat-tailed noise densities, using different non-linear filtering approaches while maintaining the texture information and edges along with handling the random impulses still is a challenging assignment in the field of image processing. Fuzzy filters are one of the approaches that can be appropriate for impulse noise elimination. This approach, not only can deal with the randomness of impulsive noisy pixels, as well as do handle the edge factors. This proposed system represents a restructured technique, in which thin-tailed and intermediate thin-tailed noises are removed by a fuzzy filter. At first, the scheme calculates a fuzzy derivative, so that it turns out to be less susceptible to local variations because of image structure; such as edges. After calculating fuzzy derivation, then fuzzy smoothing is performed and the membership functions are regulated based on the noise echelons. In first stage, fuzzy derivative is calculated for eight different directions and in second stage, fuzzy smoothing is performed by those fuzzy derivatives on the assistance of neighbouring pixel values. Those two phases are depended on fuzzy regulations which are created using membership functions. The proposed method is vastly effective on removing high density noises.

Keywords: Fuzzy filter · Salt and pepper noise · Directional approach

1 Introduction

In today's ecosphere, images are the most significant, and it is present everywhere in the modern digital arena. Images can be shared on any social networks like in Facebook, Twitter, Instagram etc. and also images can be saved, viewed, compressed, denoised and many more basic to complex operations can be performed on an image to deliver it the way a user wants it to be. When we perform some operations on any image, such as transmitting, compression, editing of images, noises can be present on the images having multi variant nature. Noise can be defined as the vague exception of shining or the information of color in images [1]. Due to unwanted movement of camera, faulty scanning process, circuit fault in scanners or digital cameras, these noises are produced [2]. That means a dark photo contains color noise in dark regions [3]. Out of the various

© Springer Nature Switzerland AG 2022
K. K. Patel et al. (Eds.): icSoftComp 2021, CCIS 1572, pp. 65–77, 2022.
https://doi.org/10.1007/978-3-031-05767-0_6

types of noise present in image, Gaussian noise is aroused by image capturing in poor lighten situation, whereas salt-and-pepper noise is a kind of impulsive noise which makes black and white spots seem in the images. Uniform noise is another type of noise which is produced by few distributing lighting pixels. In past several filters [1–20] where used to remove impulse noises. In Standard median filter [9] the window size was static but in AMF (Adaptive Median Filter) [18] the window size was flexible. In AMF, window size was increased so it led blurring effect. Another filter, PSMF (Progressive Switching Median Filter) [19] used a variant of switching median filter in recurrent manner. There were two parts of PSMF classified as, Noise detection and Noise filtering. PSMF reinstated the image with lessening the blurring consequence that seems in AMF at low noise density. When the fraction of noise level is higher, then the noisy image might not be improved reasonably. In DBA (Decision Based Algorithm), at first it was checked that a pixel was infected or not. After completing the detection process, if the pixel was found to be infected then it was substituted using median filtration [20]. It recovered the edges in better way, but if the neighbourhood pixels were replaced that caused flashing effect. In DBUTMF (Decision based un-symmetric Trimmed Median filter), if noise level is up to 70% then it got a healthier result. But it could not obtain the trimmed median value in case of higher noise densities. In MDBUTMF (Modified Decision based un-symmetric Trimmed Median filter), the infected image was investigated pixel wise and thereby the trimming procedure had been espoused [12]. It was the better version of DBUTMF [13]. In case if the entire pixels in the nominated window were found infected, then also the anticipated value can be attained by mean operation. When noise density is 80%–90% then it could not perform as it does in below noise densities. In BDND (Boundary Discriminative Noise Detection Technique), every pixel was replaced by its median value from its neighbourhood [5]. The BDND algorithm primarily classified the pixels of a confined window, centering on the present pixel, into three sets. Those are, minor intensity impulse noise, clean pixels and higher intensity impulse noise. Considering that the center pixel as uninfected, on condition that it was fitted to the "clean" pixel collection, or as "infected". The Boundary Discriminated Switching Bilateral Filter (BDSBF) [16] having intensively infected with universal noise, required two stages. Primarily, in boundary discrimination noise identification, a two-dimensional binary mapping was used. The designated binary map was recognized as binary vector, where 0's specifies equivalent pixel in the image as 'uninfected' and 1's indicates particular pixel as 'infected' pixel. Infected pixels are approximated by switching bilateral filter, by substituting each pixel in the image with a biased average of the intensities of adjacent pixels in the window. In another way, impulse noise was identified in medical images using information sets. The concept of information set came from the uncertainty in a fuzzy set using an entropy function. Uncertainty aspects can be effectively handled by fuzzy filters [10, 11]. Using the mentioned approach, NAISM filter [8] has been proposed, where the switching criterion sieved out the noisy pixels based on information set.

2 Proposed Methodology

This section involves detection of noisy pixels and correction of those detected noisy pixels.

2.1 Detection of Noisy Pixels

We have performed the noise detection procedure at the beginning. Firstly, we have taken the image matrix SMNIP of size 3 × 3, where X denotes the center of the matrix. We have chosen the said size of the window by trial-and-error method. Bigger odd numbered window had generated more detection errors while simulating that let us chose this window size. PIM_X denotes the pixel intensity value of X. Noise detection procedure is used for each of pixels in the image. SMNIPF denotes a binary flag image whose size is same as SMNIP, that contains '0' value. FLAG denotes the image matrix of the flag image. If (a, b) is the pixel of the matrix of the image SNIP, detected by the proposed noise detection algorithm as an un-infected, then '0' denotes the identical pixel position FLAG [a, b] of the flag image. If (a, b) pixel of the matrix are distinguished as infected, then '1' denotes the corresponding pixel position FLAG [a, b] of the flag image. Now in ensuing section, we have explained the proposed noise detection algorithm for 3 × 3 matrix in SMNIP:

Step 1: PIM_X denotes the intensity of the center pixel, if $PIM_X = 0$ then '1' is assigned at the corresponding pixel location of the flag image matrix FLAG [a][b].
Step 2: If $PIM_X = 255$ then also '1' is assigned at the corresponding pixel location of the flag image matrix FLAG [a][b].
Step 3: If $1 < PIM_X < 255$ then '0' is assigned at the corresponding pixel location of the flag image matrix FLAG [a][b]. '1' and '0' values are included by the output FLAG matrix that represents the corresponding infected and un-infected pixels intensities.

The flowchart for noise detection procedure is depicted in Fig. 1.

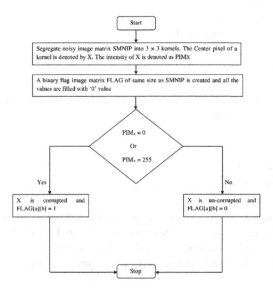

Fig. 1. Flowchart of noise detection procedure

2.2 Correction of Detected Noisy Pixels

See Fig. 2.

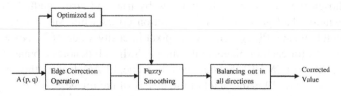

Fig. 2. Process of correction of detected noisy pixels

Here, the luminosity difference of a pixel and each of its neighbours is represented by A (p, q). By means of this luminosity difference value, we can detect the edges in actual direction. After detecting the edges, then we did fuzzy smoothing for calculating the average pixel value that is based on its neighbouring pixel value. We have executed the fuzzy smoothing process in eight different directions by calculating the average value. After calculating the average pixel value, we have taken those values for computing the correction circumstances. After calculating the correction value, we have added the correction value to the original value of the pixel. In next segment, we have applied edge correction for improved and smooth upshot. The directions of edges that is near to the center pixel (p, q) have to be detected first for further removal operation. Now considering 3×3 pixels block centering (p, q) pixel, the luminance difference value A (p, q) is found in eight different directions and the directions are I, J, K, L, IK, IL, JK, JL, as shown in Fig. 3.

$$AI (p, q) = (B (p, q - 1)) - (B (p, q))$$
$$AJ (p, q) = (B (p, q + 1)) - (B (p, q))$$
$$AIK (p, q) = (B (p + 1, q - 1)) - (B (p, q))$$

While, it would be the mean filter without edge conservation, by adding the average values of luminosity differences.

$$t1 = AIK (p, q),$$
$$t2 = AIK (p - 1, q - 1),$$
$$t3 = AIK (p + 1, q + 1).$$

On the above equations, if any two equations or all three equations are small then we can pretend that in the direction, there was no edge and the rectification value will be AIK (p, q). The final correction is added by the correction value AIK (p, q). Also, we would have added the fraction value to the final correction when we could show that there was an edge.

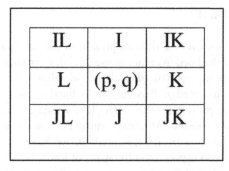

Fig. 3. Defining eight possible directions in a 3 × 3 kernel centering (p, q).

After completing the previous portion, now by using a fuzzy set, the difference is defined and the difference made to be small that is shown via Eq. 1 and Fig. 4.

$$\begin{cases} 1 - \frac{|a|}{sd}, \, if \, 0 < |a| \le sd \\ 0, \, if \, |a| > sd \end{cases} \tag{1}$$

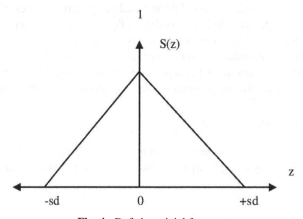

Fig. 4. Defining trivial fuzzy set

On the above Fig. 4, shows the small fuzzy set. In Fig. 4, the negative sd and positive sd is shown, where sd is defined by minimum standard deviation of all 8 × 8 pixel matrix of the image. Also, it can be said that it is the parameter of optimization decided in the stage of optimization. And now as we have taken those three differences t1, t2, t3 to be small, so now we can write the rule of fuzzy, "IF t1 is small and t2 is small and t3 is small then $A_M T$ (p, q) is small". T is vertical with the pixels, that pixels are present in AT (p, q), also t1, t2, t3 are vertical with the pixels that pixels are present in AD (p, q).

T ε {I, J, K, L, IK, IL, JK, JL}

Fuzzy Rule is used for every T that are present in the set, that means membership grade of $A_M T (p, q)$ is calculated in eight diverse directions in small set. In next section, we have conferred the optimization (finding sd), i.e., sd will be calculated. The circulation of the small MF threshold for edge discovery are defined by the calculated value of sd. By performing sampling process for the image, $K \times K$ image matrix is taken by us for computing the standard deviation. After captivating the matrix, we compute minimum standard deviation for all matrix. By this method, we have calculated the variation amount in minimum, which is present in the value of luma, for all $K \times K$ matrix in the image. Next, we have selected $K = 8$ to accomplish efficiency. For an edge, deviation is constantly greater than the calculated minimum deviation sd. Now for increasing the noise retrenchment, an amplification parameter is multiplied with standard deviation, i.e., sd = z * Standard Deviation. This was done for consecutive iterations. After completion of all of these, again we have to computed sd. Next in this section we have discussed the fuzzy smoothing. For the procedure, we have calculated the provision of rectification for the working value of the pixel. Now for every direction, we have used the span of the fuzzy rules. The main concept in this rule, if there is no edge in the definite direction, then on that direction, we use the crisp value for calculating the correction situations. Fuzzy derivative value is used in the first fragment for fabricating the edge conjecture. After finishing the first part, we have differentiated the positive and negative values for performing the fuzzy filtering in the second part. By using membership grade of $A_M T$ (p, q), we have found that how can we get the powerful value of the difference in the direction that will be calculated for final correction value. Now we have added the larger fraction difference value based on the larger membership grade. Whenever there was no edge then the variance value of the pixel is added on its ultimate correction.

$$C_T (p, q) = A_M T (p, q) * AT (p, q)$$

Here, $C_T (p, q)$ is the component of rectification in T direction and also it is a fraction of AT (p, q), in Fig. 5, by using fuzzy rules and membership functions, similar could be represented.

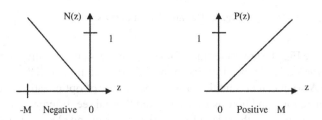

Fig. 5. Defining fuzzy smoothing process

In T direction, the rectification component is defined by C_T (p, q) and also it is the fraction of AT (p, q). The alike could be denoted by using the rule of fuzzy and functions of membership value. Here, we have defined the positive 'MF and negative MF, so in positive ranges, we can say that it is direct from [0 M] and in negative ranges, it is direct from [−M 0]. Where, extreme value of luma value is defined by M. For procuring the final rectification, we have specified the fuzzy rules in C_{Tp} = IF A_MT (p, q) direction, which is less. We can say C_{Tp} is positive when AT (p, q) is positive and also, we specified that C_{Tn} is negative when C_{Dn} = IF A_MT (p, q) is less and AT (p, q) is negative. Where, output term of positive rectification is demarcated by C_{Tp} and output term of negative rectification is defined by C_{Tn}. At a time, one of the two cannot be zero. In ultimate output we have taken small MF and positive or negative MF of the membership grade which is minimum. For example, we have C_{In} and C_{Ip} outputs for I direction. After calculating C_{Tp} and T_{Dn} in all directions, then the average value is calculated in all directions. For finding the ultimate rectification we can calculate the average value for all directions. AC (p, q) = $\sum$ T ε directions (C_{Tp}−C_{Tn})/8.

Here, {I, J, K, L, IK, IL, JK, JL} is defined by directions.

Now, we find the correction conditions with contribution, it will be done from those directions where edge is not existing on its vertical direction.

Under this section, noise correction algorithm is irradiated:

Step 1: At first, we have defined x = 0, that meant noiseless and then we have defined y = 1, that meant noisy.
Step 2: Now, we consider a 3 × 3 pixel block with pixel (p, q) that is present at the center of pixel block.
Step 3: After taking a 3 × 3 block of pixels, then we have to found A (p, q) for all eight directions, where A (p, q) is luma difference on the directions: I, J, K, L, IK, IL, JK, JL.
AI (p, q) = (B (p, q-1)) − (B (p, q))
AJ (p, q) = (B (p, q + 1)) − (B (p, q))
AIK (p, q) = (B (p + 1, q − 1)) − (B (p, q))
And likewise.
Step 4: Now, the average value of all pixels is calculated.
Step 5: We detect the edges for different neighbouring pixels by comparing with A (p, q) which is verticals with direction of rectification.
Step 6: After comparing, calculate,
t1 = AIK (p, q), t2 = AIK (p − 1, q − 1), t3 = AIK (p + 1, q + 1).
Step 7: Now, we consider a 5 × 5 pixel block with pixel (p, q) which is present at the center of pixel block.
Step 8: After taking a 5 × 5 block of pixels, then we have to check that grade of membership of A_MT (p, q) is small or not, so we used the 'if' declaration,
if t1 is small and t2 is small and t3 is small then grade of membership of A_MT (p, q) is small.

Step 9: Now, we calculate the standard deviation sd, sd $= z *$ Standard Deviation.
Step 10: After calculating the Standard Deviation, then we calculate the content of rectification in T direction and the content is fraction of AT (p, q), C_T (p, q) $= A_M T$ (p, q) *AT (p, q).
Step 11: For procurement of the ultimate rectification, we have stated the fuzzy rules for CTp and CTn direction.
So, we used the 'if' declaration,
if $A_M T$ (p, q) is small and AT (p, q) is positive then C_{Tp} is positive.
if $A_M T$ (p, q) is small and AT (p, q) is negative then C_{Tn} is negative.
Step 12: After implementation of the entire computation, at last we have to calculate, AC (p, q) $= \sum T \varepsilon$ directions $(C_{Tp} - C_{Tn})/8$.

The flowchart for noise correction procedure is depicted in Fig. 6.

The proposed architecture is mainly based on fuzzy set theory and fuzzy membership. We have used fuzzy rules in edge detection as well in the rest removal mechanism. The key advantage of the proposed method over the other image restoration method is in the application of fuzzy logic which provided us a tool with which we can deal with the uncertainties and vagueness in case of noise removal mechanism. The edge assumption is primarily comprehended by fuzzy derivative value. The main objective behindhand this method is to check that whether any edge is present in a particular direction or not. If not, then the derivative value is applied to calculate the rectification term. As we have to differentiate among the positive and negative outcome, two (positive and negative) linear fuzzy membership functions have been defined. For finding the ultimate rectification we have calculated the average value for all directions. By this way we can find more suitable and close predicted replacement for the impulse noise affected pixels.

3 Results and Discussion

After executing the proposed algorithm, output is shown in this section. The proposed method is implemented in MATLAB 2018a mounted in a Windows PC running with core i5 8th generation processor, and having 8 GB of RAM. Grayscale images such as Lena, Barbara, Goldhill are used as test images. The proposed method is compared up against BDND (Boundary Discriminative Noise Detection), DBA (Decision Based Algorithm), AMF (Adaptive Median Filter), SMF (Standard Median Filter). The metrics used for image quality assessment are Peak Signal to Noise Ratio (PSNR), Structural Similarity Index Measurement (SSIM) and Average Run Time (ART).

PSNR and MSE are defined below by respective equations.

$$PSNR = 10 \, log_{10} \frac{255^2}{MSE} \tag{2}$$

$$MSE = \left(\sum U, V \frac{(O(u, v) - d(u, v))^2}{UXV} \right) \tag{3}$$

Here, O $=$ Original Image, d $=$ De-noised image,
U $=$ Number of Rows and V $=$ Number of Columns.

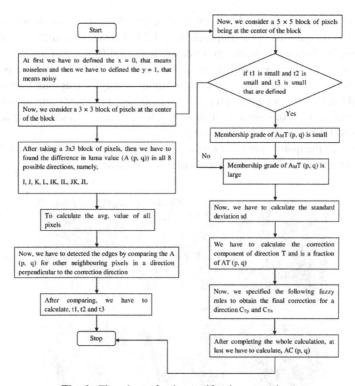

Fig. 6. Flowchart of noise rectification procedure

For computation of the structural similarity with the original and retrieved image, we used Structural Similarity Index Measurement (SSIM). The equation stated below defined the Structural Similarity Index Measurement (SSIM).

$$SSIM\,(e,f) = \frac{(2\mu_e\mu_f + G1)(2\sigma_{ef} + G2)}{\left(\mu_e^2 + \mu_f^2 + G1\right)\left(\sigma_e^2 + \sigma_f^2 + G2\right)} \tag{4}$$

Here, mean of the image e and f are μ_e and μ_f. σ_e and σ_f are defined as standard deviation of image e and f. G1, G2 are constants. The co-variance of e and f is defined as σ_{ef}.

Average run time (ART) of the total method was computed in MATLAB simulation. The time frame was represented in minutes.

Figure 7(a), (b) and (c) exhibited the approximated visual outputs of the noisy images at varied noise densities (ND). Figure 8 presented visual comparison of the prosed filter in comparison with the reputable state-of-the-art filters. As per the human-eye perception the visual outcome seemed to be respectable and superior compared to other said filters at even high noise densities. Table 1 and Table 2 exhibited excellent PSNR and SSIM results respectively up against the established filters where the noise density was high which portrayed capability of the proposed filter. Table 3 unveiled the average run time (in minutes) needed for executing the proposed method for different test images at various noise densities. As the key significance of the proposed filter has been to recuperate the quality of the restored image, the average run time may be lengthier than few other established filters in this area. The proposed method has functioned meticulously maybe by captivating some more amount of time which was authenticated by the admirable PSNR and SSIM results for different test images.

Through all the quantitative and qualitative assessments, it can be depicted that the proposed filter produced superior upshots at assorted noise densities.

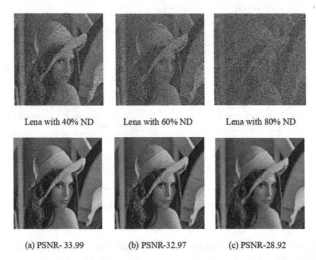

| Lena with 40% ND | Lena with 60% ND | Lena with 80% ND |

| (a) PSNR- 33.99 | (b) PSNR-32.97 | (c) PSNR-28.92 |

Fig. 7. Visual outcome of proposed filter for salt and pepper noise (a) 40% ND (b) 60% ND (c) 80% ND on Lena image

Table 1. Comparison of PSNR (db) values of some filters in contrast to proposed filter for Lena image at wide-ranging noise densities with respect to salt and pepper noise.

Filters	PSNR								
	10%	20%	30%	40%	50%	60%	70%	80%	90%
BDND [5]	36.27	35.23	34.86	34.25	33.21	31.90	29.56	27.90	25.05
DBA [20]	35.24	34.05	33.77	33.15	32.72	31.22	29.02	27.44	24.12
AMF [18]	33.14	32.98	32.54	32.05	31.28	30.19	28.42	26.69	21.26
SMF [9]	32.94	32.79	32.36	31.98	31.11	30.16	27.88	25.98	19.56
Proposed filter	37.97	36.43	35.51	34.86	33.67	32.97	29.72	28.92	25.93

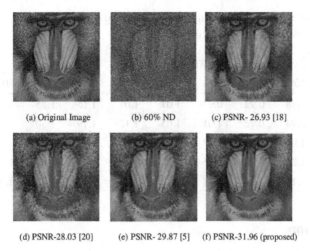

(a) Original Image   (b) 60% ND   (c) PSNR- 26.93 [18]

(d) PSNR-28.03 [20]   (e) PSNR- 29.87 [5]   (f) PSNR-31.96 (proposed)

Fig. 8. Visual outcome of proposed filter in comparison with various filters with respect to salt and pepper at 60% ND on Baboon image

Table 2. Comparison of SSIM values of some filters in contrast to proposed filter for Lena image at wide-ranging noise densities with respect to salt and pepper noise.

Noise %	SSIM				
	BDND [5]	DBA [20]	AMF [18]	SMF [9]	Proposed filter
10% ND	0.8354	0.8124	0.7747	0.3584	0.9102
20% ND	0.8214	0.7847	0.7454	0.3985	0.9001
30% ND	0.7984	0.7412	0.7004	0.4578	0.8921
40% ND	0.7549	0.6947	0.6654	0.5124	0.8754
50% ND	0.7245	0.6584	0.6231	0.5658	0.8524
60% ND	0.6249	0.5908	0.5832	0.4189	0.7783
70% ND	0.5865	0.5487	0.5124	0.3857	0.7325
80% ND	0.5138	0.5648	0.4589	0.3549	0.6984
90% ND	0.4785	0.5421	0.4154	0.3021	0.5887

Table 3. Average run time (ART) of the proposed method for various images at wide-ranging noise densities with respect to salt and pepper noise.

Images	Average rum time (ART) (in minutes)								
	10%	20%	30%	40%	50%	60%	70%	80%	90%
Lena	0.88	0.92	0.97	1.01	1.08	1.15	1.25	1.28	1.34
Baboon	0.94	0.96	1.01	1.05	1.12	1.19	1.30	1.36	1.41
Goldhill	1.06	1.09	1.13	1.19	1.24	1.30	1.35	1.39	1.48
Fingerprint	1.24	1.28	1.35	1.39	1.43	1.48	1.55	1.61	1.69
Barbara	1.16	1.21	1.27	1.35	1.41	1.45	1.52	1.57	1.62

4 Conclusion

We have observed that fuzzy filtering technique worked reasonably superior than the state-of-art filters. In the orthodox methods there are likelihoods of losing edge data. In the proposed method when the amplification factor is tuned up, then the flattening procedure is exaggerated thereby conserving the edges. For detecting the edges, we have used the difference of Fuzzy and successively repudiate the neighbours that leads to preservation of edges and effectually noise is abridged. The proposed architecture is limited to work on greyscale images and mainly designed to work in medium and high-density noises. As the key significance of the proposed filter has been to recover the quality of the restored image, the average execution time may be lengthier than few other reputed filters in this area.

References

1. Gonzalez, C.R., Woods, E.R.: Digital Image Processing, 2nd edn. Prentice Hall (2001)
2. Teddy, K.: Fingerprint enhancement by spectral analysis techniques. In: 31st Applied Imagery Pattern Recognition Workshop, pp. 16–18, WA, Washington DC (2002)
3. Armitage, W.D., Oakley, P.J.: Noise levels in color image enhancement. In: Visual Information Engineering, London, UK, pp. 105–108 (2003)
4. Fridman, P.: Radio astronomy image enhancement in the presence of phase errors using genetic algorithms. In: International Conference on Image Process, Thessaloniki, Greece, pp. 612–615 (2001)
5. Ng, P.E., Ma, K.K.: A switching median filter with boundary discriminative noise detection for extremely corrupted images. IEEE Trans. Image Process. 15(6), 1057–7149 (2006)
6. Tsai, Y.D., Lee, Y., Sekiya, M., Sakaguchi, S., Yamad, I.: A method of medical image enhancement using wavelet analysis. In: 6th International Conference on Signal Process. IJCA, Beijing, China, pp. 723–726 (2002)
7. Koli, A.M.: Review of impulse noise reduction techniques. Int. J. Comput. Sci. Eng. 4(02), 0975–3397 (2012)
8. Arora, S., Hanmandlu, M., Arora, S.: Filtering impulse noise in medical images using information sets. Int. J. Comput. Sci. Eng. 4(02), 0975–3397 (2012)
9. Huang, S.T.: Two-Dimensional Digital Signal Processing, 1st edn. Springer, Heidelberg (1981)

10. Tyan, C., Wang, P.P.: Image processing - enhancement, filtering and edge detection using the fuzzy logic approach. IEEE **2**(2) (1993). ISBN: 0-7803-0614-7
11. Plataniotis, K., Androutsos, D., Venetsanopoulos, A.: Adaptive fuzzy systems for multichannel signal processing. Proc. IEEE **87**(9), 1601–1622 (1999)
12. Esakkirajan, S., Veerakumar, T., Subramanyam, A.N., PremChand, H.C.: Removal of high density salt and pepper noise through modified decision based unsymmetric trimmed median filter. IEEE Sig. Process. Lett. **18**(5), 287–290 (2011)
13. Aiswarya, K., Jayaraj, V., Ebenezer, D.: A new and efficient algorithm for the removal of high density salt and pepper noise in images and videos. In: 2010 Second International Conference on Computer Modelling and Simulation, China. IEEE (2010). ISBN: 978-1-4244-5642-0
14. Paris, S., Kornprobst, P., Tumblin, J., Durand, F.: A gentle introduction to bilateral filtering and its applications. ACM SIGGRAPH **25**(3), 637–645 (2007)
15. Hore, A., Ziou, D.: Image quality metrices: PSNR vs. SSIM. In: International Conference on Pattern Recognition, Istanbul, pp. 2366–2369. OSA (2010)
16. Shobha, R.K., Satyanarayana, S.V.R.: Image denoising using boundary discriminated switching bilateral filter with highly corrupted universal noise. In: 2017 International Conference on Energy, Communication, Data Analytics and Soft Computing (ICECDS), India. IEEE (2017). ISBN: 978-1-5386-1888-2
17. Astola, J., Kuosamanen, P.: Fundamentals of Nonlinear Digital Filtering, 1st edn. CRC Press, Boca Raton (1997)
18. Hwang, H., Haddad, A.H.: Adaptive median filters: new algorithms and results. IEEE Trans. Image Process. **4**(4), 499–502 (1995)
19. Wang, Z., Zhang, D.: Progressive switching median filter for the removal of impulse noise from highly corrupted images. IEEE Trans. Circ. Syst. II Analog Digit. Sig. Process. **46**(1), 78–80 (1999)
20. Srinivasan, S.K., Ebenezer, D.: A new fast and efficient decision-based algorithm for removal of high density impulse noises. IEEE Sig. Process. Lett. **14**(3), 189–192 (2007)

Utility Distribution Based Measures of Probabilistic Single Valued Neutrosophic Information, Hybrid Ambiguity and Information Improvement

Mahima Poonia[(⊠)] and Rakesh Kumar Bajaj

Jaypee University of Information Technology, Solan 173234, India
mahimapoonia1508@gmail.com

Abstract. In the present communication, a novel concept of single valued neutrosophic information measure based on utility distribution and probabilistic randomness has been proposed. The proposed concept has been obtained by integrating the uncertainties caused by neutrosophic information, useful information (utility based) and probabilistic information. Further, in a similar integrating way, the divergence 'useful' information measure has also been proposed for the study of applicable mutual information. Consequently, the hybrid ambiguity and neutrosophic information improvement measures have been studied with the help of proposed 'useful' information measures. The applications in the field of decision models and constrained optimization have been outlined in the conclusions and future work.

Keywords: Neutrosophic entropy · Shannon entropy · Single valued neutrosophic information · Hybrid ambiguity · Information improvement · Utility distribution

1 Introduction

In the world of ambivalence, various problems arise due to insufficient data which decreases the limpidity of the concept and this makes it difficult to solve the problem using the crisp set theory. In literature the problem of the concept of fuzziness in the information was first given in 1965 by Zadeh's theory of Fuzzy set [1] and the degree of uncertainty created among the fuzzy set components was calculated in the year 1983 by Ebanks [2] and 1994 by Pal and Bezdek [3], using the theory named "Fuzzy entropy". The fuzzy entropy is a non-probabilistic measure based on the membership function. Later, in 1972 Luca and Termini [4] introduced a new measure of fuzzy entropy based on the Shannon function [10] and with the help of these two theories a new set of properties have been designed. These properties play a significant role in describing the fuzzy entropy. The fuzzy entropy by Luca et al. [4] is one of the simplest forms of entropy present

© Springer Nature Switzerland AG 2022
K. K. Patel et al. (Eds.): icSoftComp 2021, CCIS 1572, pp. 78–89, 2022.
https://doi.org/10.1007/978-3-031-05767-0_7

in the literature and is defined as,

$$M'(A) = -\sum_{i=1}^{n}[\Gamma_A(y_i)\log\Gamma_A(y_i) + (1 - \Gamma_A(y_i))\log(1 - \Gamma_A(y_i))]; \qquad (1)$$

where $\Gamma_A(y_i)$ denotes the degree of membership function and the properties of entropy measure are given below:

- $M'(A) = 0$ iff $\Gamma_A(y) = 0$ or 1.
- $M'(A)$ is maximum when $\Gamma_A(y) = 0.5$.
- $M'(A) = M'(A')$, where the sharpen version of A is A', i.e.
  $\Gamma_{A'}(x_i) \leq \Gamma_A(y_i)$ for $\Gamma_A(y_i) \leq 0.5$ & $\Gamma_A(y_i) \leq \Gamma_{A'}(y_i)$ for $\Gamma_A(y_i) \geq 0.5$
- $M'(A) = M'(\tilde{A})$ where $\tilde{A}$ is complement of A.

These properties are the necessary and sufficient conditions to form the fuzzy entropy measure. The Luca and Termini measure given by Eq. 1 also fulfills these conditions. In 1980, Kaufmann [5] proposed an entropy measure which played the basic for various new entropies in literature and is of form,

$$M'(A) = -\frac{1}{\log n}\sum_{i=1}^{n}\Gamma_A(y_i)\log\Gamma_A(y_i). \qquad (2)$$

In 1967, Havrda and Charvat [6] extended the concept of Kaufmann [5] and defined the following entropy measure;

$$M'(A) = \frac{1}{1-\alpha}\sum_{i=1}^{n}[(\Gamma_A^{\alpha}(y_i) + (1 - (\Gamma_A(y_i))^{\alpha} - 1]. \qquad (3)$$

Thus, with the help of fuzzy entropy, the quantity of information is obtained from the systems of fuzzy theory and the measure of information collected from this fuzziness is known as fuzzy information measure. This concept of information measure was further used by Joshi [7] in 2019 to generate a new measure based on Tsallis-Havrda-Charvat entropy. Later, in 2020, Li et al. [8] and Mahmood et al. [9] studied this concept using the structures of Gaussian kernels and Complex q-rang ortho pair fuzzy set respectively. The fuzzy information measure was first proposed by Shannon [10] in 1948, which is a probabilistic theory and contributes majorly to the communication sector. As per deliberation given by [11], the defined quantity of information conveyed is directly proportional to the probability of the probabilistic task. This implies that the information quantity defines the log of the event probability of A. i.e.;

$$M'(A) = -\log p(A);$$

where $p(A)$ denotes the probability and the average information over all the events is known as Shannon entropy.

Further, the concept of information entropy measure was extended to 'useful' information measure by Bhaker and Hooda [12] in 1993, who defined the generalized mean value characteristic measure for incomplete probability measure. This theory was later used by Hooda and Bajaj [13] in 2010 and they introduced a new 'useful' information measure for directed divergence of Zadeh's theory. Then, in 2016 the briefly description related to the overview of fuzzy information measure and generalized form of fuzzy entropy was presented by Ohlan [14] and in the same year Arora and Dhiman [15] contributed a new measure of fuzzy directed divergence and its applications in decision making problems in the literature. In 2018 & 2019, Sofi et al. [16,17] used the concept of parametric 'useful' fuzzy information measure for R - norm and obtained new properties with numerical examples respectively. These theories play a vital role in increasing the understanding of the concept but none of the above theories deals with the concept of neutrality. According to all of the above theories, the degree of membership function $\Gamma(y)$ was considered to define the measure and to deal with the difficulty. Therefore, to increase the range of solving the problems related to uncertainty, the degree of non-membership function $\Lambda(y)$ was calculated that is $\Lambda(y) = 1 - \Gamma(y)|y \in [0,1]$. This theory was explained in detail by Atanassov [18] in 1983 and is known as Intuitionistic fuzzy set. These two theories are efficient in solving various problems, but were found to be inefficient in solving the indeterminate situation or in other words the neutral situation, which was further extended by Smarandache.

Smarandache [19] introduced the theory of neutrality, widely known as the Neutrosophic set and this theory of neutrality plays a vital role in the case of Zadeh's theory of uncertainty. The concept of neutrosophy is defined by the degree of membership function of truth (Γ), neutrality (Λ) and falsity (θ). In other words, the triplet $(\Gamma, \Lambda, \theta)$ signifies the theory of neutrality and all the three components must lie in the interval $[0,1]$. The Smarandache concept of neutrosophy not only solved the problem of neutrality but also laid the foundation of various new concepts such as neutrosophic optimization, neutrosophic probability, neutrosophic entropy and neutrosophic logic. This increased its credibility in both the theoretical and different practical fields like data analysis [20], internet services [21] and many aspects of economic database [22]. Further, in 2005 Wang [23] proposed the concept of a Single valued neutrosophic set (SVNS) which have a major contribution in real life and engineering applications. The distance and entropy between the two single valued neutrosophic sets are calculated by Majumdar and Samanta [24]. In 2020, Singh presented two different applications of neutrosophic entropy using the thresholding segmentation [25] and clustering [26] algorithms to find the MR images related to the Parkinson's disease respectively. The results obtained using these two algorithms were further compared with the results already present in literature and are proved to be useful in the field of medicine. Later in literature, Patrascu [27] presented the extension of Shannon entropy for neutrosophic information set, in addition to the results obtained in para-consistent fuzzy information.

In this work, the concept of probabilistic occurrence and neutrosophic entropy have been put together along with utility distribution of the events to obtain a novel concept of 'useful' single valued neutrosophic probabilistic information measure. The proposed measure is a new kind of measure that will be helpful for the study of decision problems under utility distribution. In addition to this, some extended measures like hybrid ambiguity measure, analogous divergence measure and information improvement measure have also been discussed.

The current work is arranged in the following manner: Some basic definitions/preliminaries are given in Sect. 2. Section 3 contains the proposed entropy of the single valued neutrosophic information measure. The divergence of neutrosophic information is described in Sect. 4 of the current manuscript. The hybrid ambiguity is defined in Sect. 5. The information improvement measure of neutrosophic measure is explained in Sect. 6. Finally, the conclusions and future work are discussed in detail in Sect. 7.

2   Preliminaries

This segment of the current article contains the basic definitions for a better understanding of the concept. Consider M to be a universal set. Then,

Definition 1 [13] *(Ambiguity Measure).* "Suppose a subset A of M characterized as $\Gamma_A : M \to [0,1]$, where Γ_A denotes the grade of membership of $y \in M$ and described as,

$$\Gamma_A(x) = \begin{cases} 0 & \text{if } y \notin A \text{ and there is no ambiguity,} \\ 1 & \text{if } y \in A \text{ and there is no ambiguity,} \\ 0.5 & \text{if } \exists \text{ ambiguity whether } y \in A \& y \notin A. \end{cases} \quad "$$

Definition 2 [18] *(Intuitionistic Set).* "Consider a fixed set M. A in M signifies the IFS and is denoted by $M = \{< y, \Gamma_A(y), \Lambda_A(y) > | y \in M\}$ where $\Gamma_A \& \Lambda_A$ signifies the membership and non-membership functions respectively subject to $\Gamma_A \in [0,1], \Lambda_A \in [0,1] \& 0 \le \Gamma_A + \Lambda_A \le 1$."

Definition 3 [19] *(Neutrosophic Set).* "Consider a universal set M. Then, A denotes the neutrosophic set and is denoted by $M = \{< y, \Gamma_A(y), \Lambda_A(y), \theta_A(y) > | y \in M\}$ where $\Gamma_A, \Lambda_A \& \theta_A$ signifies the truth, neutrality and falsity functions respectively subject to $\Gamma_A, \Lambda_A, \theta_A \in]0^-, 1^+[$ satisfies condition $0^- \le \sup \Gamma_A(y) + \sup \Lambda_A(y) + \sup \theta_A(y) \le 3^+$."

Definition 4 [23] *(Single Valued Neutrosophic Set).* "Consider a universal set M, then the single valued set is denoted by $M = \{< y, \Gamma_A(y), \Lambda_A(y), \theta_A(y) > | y \in M\}$, where $\Gamma : M \to [0,1], \Lambda : M \to [0,1] \& \theta : M \to [0,1]$."

Definition 5 [10] *(Shannon Measure).* "Suppose $p_1, p_2, ..., p_n$ be the set of known probabilities of the events. Then, the measure $M(p_i)$ defined on this set is defined as,

$$M(p_i) = -k \sum_{i=1}^{n} p_i \log p_i;$$

where k is a positive constant.

3   Entropy of Single Valued Neutrosophic Information Measure

In this section of the study, the entropy of the single valued neutrosophic information measure is explained in detail with the help of theorem. The properties of single valued neutrosophic entropy is explained below.

Consider M' to be a set of all SVNSs and $A \in M'$. Then, the entropy of A, denoted by $E_{M'}(A)$, satisfies

1. $E_{M'}(A) = 0$ iff $\Gamma_A(y) = \Lambda_A(y) = \theta_A(y) = 0$ or 1.

2. $E_{M'}(A) = 1$, when $\Gamma_A(y) = \Lambda_A(y) = \theta_A(y) = 0.5$.

3. $E_{M'}(A_1) \geq E_{M'}(A_2)$ if $A_1 \subset A_2$ i.e.,
   $\Gamma_{A_1}(y_i) \leq \Gamma_{A_2}(y_i), \Lambda_{A_1}(y_i) \geq \Lambda_{A_2}$ & $\theta_{A_1}(y_i) \leq \theta_{A_2}(y_i)$.

4. $E_{M'}(A) = E_{M'}(\tilde{A})$ where $\tilde{A}$ is complement of A.

Next, the entropy measure given by Luca et al. [4] is extended and modified to find the entropy for the single value neutrosophic information measure. This measure takes the following form,

$$E_{M'_\Gamma}(A) = \frac{1}{2n \log(0.5)} \sum_{i=1}^{n} [\Gamma_A(y_i) \log(\Gamma_A(y_i)) + (1 - \Gamma_A(y_i)) \log((1 - \Gamma_A(y_i)))];$$

$$E_{M'_\Lambda}(A) = \frac{1}{2n \log(0.5)} \sum_{i=1}^{n} [\Lambda_A(y_i) \log(\Lambda_A(y_i)) + (1 - \Lambda_A(y_i)) \log((1 - \Lambda_A(y_i)))]; \quad (4)$$

$$E_{M'_\theta}(A) = \frac{1}{2n \log(0.5)} \sum_{i=1}^{n} [\theta_A(y_i) \log(\theta_A(y_i)) + (1 - \theta_A(y_i)) \log((1 - \theta_A(y_i)))].$$

Similarly, the 'useful' single valued neutrosophic information measure is obtained with some modification in 'useful' fuzzy information measure given by Hooda and Bajaj [13] earlier in literature and this measure takes the following form for truth, falsity and neutral membership functions i.e.,

$$E_{M_\Gamma'}(A; P; U) = \frac{1}{2n\log(0.5)} \frac{\sum_{i=1}^n u_i p_i [\Gamma_A(y_i)\log(\Gamma_A(y_i)) + (1 - \Gamma_A(y_i))\log((1 - \Gamma_A(y_i)))]}{\sum_{i=1}^n u_i p_i};$$

$$E_{M_\Lambda'}(A; P; U) = \frac{1}{2n\log(0.5)} \frac{\sum_{i=1}^n u_i p_i [\Lambda_A(y_i)\log(\Lambda_A(y_i)) + (1 - \Lambda_A(y_i))\log((1 - \Lambda_A(y_i)))]}{\sum_{i=1}^n u_i p_i}; \quad (5)$$

$$E_{M_\theta'}(A; P; U) = \frac{1}{2n\log(0.5)} \frac{\sum_{i=1}^n u_i p_i [\theta_A(y_i)\log(\theta_A(y_i)) + (1 - \theta_A(y_i))\log((1 - \theta_A(y_i)))]}{\sum_{i=1}^n u_i p_i};$$

where $u_i > 0$.

Theorem 1 *The measure (5) must satisfy the neutrosophic entropy measure properties given by (1–4).*

Proof **Axiom 1**: $E_{M_\Gamma'}(A; P; U) = 0$.
Then,

$$\frac{1}{2n\log(0.5)} \frac{\sum_{i=1}^n u_i p_i [\Gamma_A(y_i)\log(\Gamma_A(y_i)) + (1 - \Gamma_A(y_i))\log((1 - \Gamma_A(y_i)))]}{\sum_{i=1}^n u_i p_i} = 0.$$

$$\Gamma_A(y_i)\log(\Gamma_A(y_i)) + (1 - \Gamma_A(y_i))\log((1 - \Gamma_A(y_i)))] = 0.$$

Next, either $\Gamma_A(y_i) = 0$ or $1 \, \forall \, i = 1, 2, ..., n$.
This proves that it satisfies the crisp set property.

Axiom 2: $\Gamma_A(y) = \Lambda_A(y) = \theta_A(y) = 0.5$.
Putting this in Eq. (3), we get

$$\frac{1}{2n\log(0.5)} \frac{\sum_{i=1}^n u_i p_i [0.5\log(0.5) + (1 - 0.5)\log((1 - 0.5))]}{\sum_{i=1}^n u_i p_i} = 1.$$

Hence, $E_{M'}(A) = 1$, when $\Gamma_A(y) = \Lambda_A(y) = \theta_A(y) = 0.5$.

Axiom 3: If $A_1 \subset A_2$, then, $\Gamma_{A_1} \leq \Gamma_{A_2}$.
Also,

$$\Gamma_{A_1}(y_i)\log(\Gamma_{A_1}(y_i)) + (1 - \Gamma_{A_1}(y_i))\log((1 - \Gamma_{A_1}(y_i))) \leq \Gamma_{A_2}(y_i)\log(\Gamma_{A_2}(y_i))$$
$$+ (1 - \Gamma_{A_2}(y_i))\log((1 - \Gamma_{A_2}(y_i))).$$

This implies

$$M_\Gamma'(A_1; P; U) \leq M_\Gamma'(A_2; P; U).$$

Similarly, we can prove that

$$M_\Lambda'(A_1; P; U) \geq M_\Lambda'(A_2; P; U) \, \& \, M_\theta'(A_1; P; U) \leq M_\theta'(A_2; P; U).$$

Hence, this proves $E_{M'}(A_1) \geq E_{M'}(A_2)$ if $A_1 \subset A_2$.

Axiom 4: For the complement,
$M'_\Gamma(\tilde{A}; P; U)$

$$= \frac{1}{2n\log(0.5)} \frac{\sum_{i=1}^{n} u_i p_i [\Gamma_A^c(y_i) \log(\Gamma_A^c(y_i)) + (1 - \Gamma_A(y_i))^c \log(1 - \Gamma_A(y_i))^c]}{\sum_{i=1}^{n} u_i p_i}.$$

$$\implies M'_\Gamma(\tilde{A}; P; U)$$

$$= \frac{1}{2n\log(0.5)} \frac{\sum_{i=1}^{n} u_i p_i [(1 - \Gamma_A(y_i)) \log(1 - \Gamma_A(y_i)) + (\Gamma_A(y_i)) \log((\Gamma_A(y_i)))]}{\sum_{i=1}^{n} u_i p_i}.$$

$$\implies M'_\Gamma(\tilde{A}; P; U) = M'_\Gamma(A; P; U).$$

In similar manner, the condition can be obtained for the neutral and falsity membership functions.

Thus, we observe that all the axioms have been satisfied and this is a valid entropy measure.

It may be noted that these properties can also be satisfied with the help of some numerical examples. We consider Table 1 which shows the behavior of the proposed measure in case of the crisp set. In this case, the values of degree of truth membership function is 1 when membership is maximum whereas the values of indeterminacy and falsity are zero.

Secondly, in Table 2 when degree of membership is minimum then falsity component is 1 and the values of rest of the components are equal to zero i.e., $A = (1, 0, 0)$ or $(0, 0, 1)$. Consider the universe $U = (1, 2, 3, 4)$ with utilities $u_i = (u_1, u_2, u_3, u_4)$ and probabilities $p(A) = (0.1, 0.2, 0.3, 0.4)$.

Table 1. Behavior of proposed measure on crisp set in case of maximum membership

u_i	p_i	$(\Gamma_A(y_i), \Lambda_A(y_i), \theta_A(y_i))$	$(M'(\Gamma_A(y_i)), M'(\Lambda_A(y_i)), M'(\theta_A(y_i)))$
u_4	0.4	(1,0,0)	0
u_3	0.3	(1,0,0)	0
u_2	0.2	(1,0,0)	0
u_1	0.1	(1,0,0)	0

Table 2. Behavior of proposed measure on crisp set in case of minimum membership

u_i	p_i	$(\Gamma_A(y_i), \Lambda_A(y_i), \theta_A(y_i))$	$(M'(\Gamma_A(y_i)), M'(\Lambda_A(y_i)), M'(\theta_A(y_i)))$
u_4	0.4	(0,0,1)	0
u_3	0.3	(0,0,1)	0
u_2	0.2	(0,0,1)	0
u_1	0.1	(0,0,1)	0

Note: In similar manner, all the above properties of single valued neutrosophic information measure can be satisfied using the numerical example.

4 'Useful' Divergence Measure of Single Valued Neutrosophic Information Measure

In this segment of the study, we have described the properties of divergence and the divergence between the proposed neutrosophic information measures.
The divergence of two single valued neutrosophic sets is defined on the basis of the following parameters:

- Two positive and symmetric single valued neutrosophic sets are compared.
- Divergence is equal to zero when these two sets coincide.
- Divergence is inversely proportion to the similarity between the two sets. As the similarity increases the divergence decreases.

Consider two single valued neutrosophic sets A and B on the same similarity points y_i $(i = 1, 2, ..., n)$ and with neutrosophic vectors $(\Gamma_A(y_i), \Lambda_A(y_i), \theta_A(y_i))$ and $(\Gamma_B(y_i), \Lambda_B(y_i), \theta_B(y_i))$ where $(i = 1, 2, ..., n)$.

The simplest form of fuzzy divergence was introduced by Bhandari and Pal [28] as

$$I(A, B) = \sum_{i=1}^{n} [\Gamma_A(y_i) \log \frac{\Gamma_A(y_i)}{\Gamma_B(y_i)} + (1 - \Gamma_A(y_i)) \log \frac{(1 - \Gamma_A(y_i))}{(1 - \Gamma_B(y_i))}]. \quad (6)$$

Next, we consider two single valued neutrosophic fuzziness of A from B and the 'useful' measure for truth membership component of these two single valued neutrosophic directed divergence measure of A from B is given by:

$$I(A, B; P; U) = \frac{1}{2n \log(0.5)} \frac{\sum_{i=1}^{n} u_i p_i [\Gamma_A(y_i) \log(\Gamma_B(y_i)) + (1 - \Gamma_A(y_i)) \log((1 - \Gamma_B(y_i)))]}{\sum_{i=1}^{n} u_i p_i}. \quad (7)$$

Then, the 'useful' neutrosophic symmetric divergence measure is defined as,

$$\mathbf{J(A, B; P; U) = I(A, B; P; U) + I(B, A; P; U)}. \quad (8)$$

Theorem 2 *The proposed measure $I(A, B, P, U)$, i.e., $I(A, B; P; U) \geq 0$ if $\Gamma_A(y_i) = \Gamma_B(y_i)$ where $i = 1, 2, ..., n$ is a valid information measure.*

Proof Suppose

$$\sum_{i=1}^{n} \Gamma_A(y_i) = e, \quad \sum_{i=1}^{n} \Gamma_B(y_i) = f \quad \& \quad \sum_{i=1}^{n} u_i p_i = u.$$

Then,

$$\frac{1}{2n \log(0.5)} [\sum_{i=1}^{n} u_i p_i (\Gamma_A(y_i) \log \frac{\Gamma_A(y_i)}{\Gamma_B(y_i)}] \geq \frac{ue}{2n \log(0.5)} \log \frac{e}{f}. \quad (9)$$

In similar manner, we can prove that

$$\frac{1}{2n\log(0.5)}[\sum_{i=1}^{n} u_i p_i(1-\Gamma_A(y_i))\log\frac{(1-\Gamma_A(y_i))}{(1-\Gamma_B(y_i))}] \geq \frac{u(n-e)}{2n\log(0.5)}\log\frac{n-e}{n-f}. \quad (10)$$

Adding Eqs. (9) and (10),

$$\frac{1}{2n\log(0.5)}[\sum_{i=1}^{n} u_i p_i(\Gamma_A(y_i)\log\frac{\Gamma_A(y_i)}{\Gamma_B(y_i)} + \sum_{i=1}^{n} u_i p_i(1-\Gamma_A(y_i))\log\frac{(1-\Gamma_A(y_i))}{(1-\Gamma_B(y_i))}]$$

$$\geq \frac{u}{2n\log(0.5)}[e\log\frac{e}{f} + (n-e)\log\frac{n-e}{n-f}]. \quad (11)$$

Suppose

$$f(e) = \frac{1}{2n\log(0.5)}[e\log\frac{e}{f} + (n-e)\frac{n-e}{n-f}],$$

$$f'(e) = \frac{1}{2n\log(0.5)}[\log\frac{e}{f} + \frac{n-e}{n-f}],$$

$$f''(e) = \frac{1}{2n\log(0.5)}[\frac{1}{e} + \frac{1}{n-e}] > 0.$$

Thus $f''(e) > 0$, which proves that $f(e)$ is a complex function and have a minimum value when $e = f$. Secondly, $\sum_{i=1}^{n} u_i p_i > 0$.

Hence, $I(A, B; P; U) \geq 0$.

5   Notion of Hybrid Ambiguity 'Useful' Measure

In this segment of the current manuscript, hybrid ambiguity of the single valued neutrosophic information measure under utility distribution has been obtained.

Consider two single valued neutrosophic information sets A and B. The two conditions to be satisfied to find the hybrid ambiguity are given below:

– Then, entropy of set A defines ambiguity for the given set.

– Secondly, the difference between A and directed divergence of B is calculated by $I(A, B)$.

Hybrid ambiguity = Entropy of $A + I(A, B)$.

The hybrid ambiguity for the truth membership function is given below:

$$HA_\Gamma = \frac{1}{2n\log(0.5)}\frac{\sum_{i=1}^{n} u_i p_i[\Gamma_A(y_i)\log(\Gamma_A(y_i)) + (1-\Gamma_A(y_i))\log((1-\Gamma_A(y_i)))]}{\sum_{i=1}^{n} u_i p_i}$$

$$-\frac{1}{2n\log(0.5)}\frac{\sum_{i=1}^{n} u_i p_i[\Gamma_A(y_i)\log\frac{\Gamma_A(y_i)}{\Gamma_B(y_i)} + (1-\Gamma_A(y_i))\log\frac{(1-\Gamma_A(y_i))}{1-\Gamma_B(y_i)}]}{\sum_{i=1}^{n} u_i p_i}.$$

$$=\frac{1}{2n\log(0.5)}\frac{\sum_{i=1}^{n} u_i p_i[\Gamma_A(y_i)\log(\Gamma_B(y_i)) + (1-\Gamma_A(y_i))\log((1-\Gamma_B(y_i)))]}{\sum_{i=1}^{n} u_i p_i}.$$

$$\implies [I(A,B)]_\Gamma = \frac{1}{2n\log(0.5)} \frac{\sum_{i=1}^n u_i p_i [\Gamma_A(y_i)\log(\Gamma_B(y_i)) + (1 - \Gamma_A(y_i))\log((1 - \Gamma_B(y_i)))]}{\sum_{i=1}^n u_i p_i}.$$

Similarly, we can find for other two components for neutrosophic theory.

Remarks: It may be noted that we can establish a relation between entropy and directed divergence of two single valued neutrosophic sets as follows.
    If $\Gamma_B = 0.5$, then

$$[I(A,B)]_\Gamma = \frac{1}{2n\log(0.5)} \frac{\sum_{i=1}^n u_i p_i [\Gamma_A(y_i)\log\frac{\Gamma_A(y_i)}{0.5} + (1 - \Gamma_A(y_i))\log\frac{(1-\Gamma_A(y_i))}{0.5}]}{\sum_{i=1}^n u_i p_i}.$$

$$= \frac{1}{2} - \frac{1}{2n\log(0.5)} \frac{\sum_{i=1}^n u_i p_i [\Gamma_A(y_i)\log(\Gamma_A(y_i)) + (1 - \Gamma_A(y_i))\log((1 - \Gamma_A(y_i)))]}{\sum_{i=1}^n u_i p_i}.$$

Then, the relation obtained is given by

$$= \tfrac{1}{2} - (\text{Entropy of SVNS } A).$$

6 'Useful' Single Valued Neutrosophic Information Improvement Measure

The 'useful' information improvement measure for the three single valued neutrosophic sets under consideration can be explained as follows,
    Consider sets A and B, where set A is estimated from set B and was revised to set C. Then, the original and final ambiguity is given by $I'(A,B)$ and $I'(A,C)$. Then, the reduced ambiguity for truth membership function is given by

$$I'_\Gamma(A,B,C) = I'_\Gamma(A,B) - I'_\Gamma(A,C).$$

$$= \frac{1}{2n\log(0.5)} \left[\frac{\sum_{i=1}^n u_i p_i [\Gamma_A(y_i)\log(\Gamma_B(y_i)) + (1 - \Gamma_A(y_i))\log((1 - \Gamma_B(y_i)))]}{\sum_{i=1}^n u_i p_i} \right.$$
$$\left. - \frac{\sum_{i=1}^n u_i p_i [\Gamma_A(y_i)\log(\Gamma_C(y_i)) + (1 - \Gamma_A(y_i))\log((1 - \Gamma_C(y_i)))]}{\sum_{i=1}^n u_i p_i} \right].$$

$$= \frac{1}{2n\log(0.5)} \left[\frac{\sum_{i=1}^n u_i p_i [\Gamma_A(y_i)\log\frac{\Gamma_B(y_i)}{\Gamma_C(y_i)} + (1 - \Gamma_A(y_i))\log\frac{(1-\Gamma_B(y_i))}{(1-\Gamma_C(y_i))}]}{\sum_{i=1}^n u_i p_i} \right].$$

Similarly, we can obtain for the improved measure of the neutrality and falsity function of neutrosophic information measure.
    This is called the 'useful' single valued neutrosophic improved information measure.

7 Conclusion and Future Work

In the current work, we have successfully established the validity of the proposed measures named as the probabilistic single valued neutrosophic 'useful' information measure, 'useful' divergence measure, hybrid ambiguity and 'useful' information improvement measure of single-valued neutrosophic sets. All these measures have been explained and validated with the help of well established axioms and numerical example.

In future, the constrained optimization with a suitable applicability for the decision models can further be studied in view of the proposed measures. The concept of Lagrange's multipliers can accordingly be implemented for a formal derivations in future in reference with the work carried out by Hooda and Bajaj [13]. In the process moving forward in the direction of derivations, it is being proposed that we will need to take the multipliers in the following format for the different components of ambiguity,

$$u_i p_i \log \frac{\Gamma_A(y_i)}{1 - \Gamma_A(y_i)} = m.$$

References

1. Zadeh, L.A.: Fuzzy sets. Inf. Control **8**, 338–353 (1965)
2. Ebanks, B.R.: On measures of fuzziness and their representations. J. Math. Anal. Appl. **94**, 24–37 (1983)
3. Pal, N.R., Bezdek, J.C.: Measuring fuzzy uncertainty. IEEE Trans. Fuzzy Syst. **2**, 107–118 (1994)
4. De Luca, A., Termini, S.: A definition of non-probabilistic entropy in the setting of fuzzy set theory. Inf. Control **20**, 301–312 (1972)
5. Kaufmann, A.: Fuzzy Subsets. Fundamental Theoretical Elements 3. Academic Press, New York (1980)
6. Havrda, J.H., Charvat, F.: Quantification methods of classification processes concept of structural α entropy. Kybernetika **3**, 30–35 (1967)
7. Joshi, R.: A new picture fuzzy information measure based on Tsallis-Havrda-Charvat concept with applications in presaging poll outcome. Comput. Appl. Math. **39**(2), 1–24 (2020)
8. Li, Z., Liu, X., Dai, J., Chen, J., Fujita, H.: Measures of uncertainty based on Gaussian kernel for a fully fuzzy information system. Knowl.-Based Syst. **196**, 105791 (2020)
9. Mahmood, T., Ali, Z.: Entropy measure and TOPSIS method based on correlation coefficient using complex q-rung orthopair fuzzy information and its application to multi-attribute decision making. Soft. Comput. **25**(2), 1249–1275 (2021)
10. Shannon, C.: A mathematical theory of communication. Bellb Syst. Tech. J. **27**, 379–423 (1948)
11. Robert, A.: Information Theory. Dover Publications, New York (1990)
12. Bhaker, U.S., Hooda, D.S.: Mean value characteristic of useful information measures. Tamkang J. Math. **24**(4), 383–394 (1993)
13. Hooda, D.S., Bajaj, R.K.: 'Useful' fuzzy measures of information, integrated ambiguity and directed divergence. Int. J. Gen Syst **39**(6), 647–658 (2010)
14. Ohlan, A.: Overview on development of fuzzy information measures. IJARESM **412**, 17–22 (2016)
15. Arora, H.D., Dhiman, A.: On some generalised information measure of fuzzy directed divergence and decision making. Int. J. Comput. Sci. Math. **73**, 263–273 (2016)
16. Sofi, S.M., Peerzada, S., Baig, M.A.K.: Parametric generalizations of 'useful' R-norm fuzzy information measures. Int. J. Sci. Res. Math. Stat. Sci. **5**, 6 (2018)

17. Sofi, S.M., Peerzada, S., Baig, M.A.K.: A New two-parametric 'useful' fuzzy information measure and its properties. J. Mod. Appl. Stat. Methods **18**(2), 18 (2020)
18. Atanassov, K.: Intuitionistic fuzzy set. Fuzzy Sets Syst. **20**, 87–96 (1986)
19. Smarandache, F.: A Unifying Field in Logics. Neutrosophy: Neutrosophic Probability, Set & Logics. American Research Press, Rehoboth (1999)
20. Smarandache, F., Sunderraman, R., Wang, H., Zhang, Y.: Interval Neutrosophic Sets and Logic: Theory and Applications in Computing. HEXIS Neutrosophic Book, Ann Arbor (2005)
21. Khoshnevisan, M., Bhattacharya, S.: A note on financial data set detection using neutrosophic Probability. In: Smarandache, F. (ed.) Proceedings of the First International Conference on Neutrosophy, Neutrosophic Logic, Neutrosophic Set, Neutrosophic Probability and Statistics. University of New Mexico, pp. 75–80 (2002)
22. Khoshnevisan, M., Singh, S.: Neurofuzzy and neutrosophic approach to compute the rate of change in new economics. In: Smarandache, F. (ed.) Proceedings of the First International Conference on Neutrosophy, Neutrosophic Logic, Neutrosophic Set, Neutrosophic Probability and Statistics. University of New Mexico, pp. 56–62 (2002)
23. Wang, H., et al.: Single valued neutrosophic set. In: Proceeding in 10th on Fuzzy Theory and Technology, Salt lake city, Utah (2005)
24. Majumdar, P., Samanta, S.K.: On similarity and entropy of neutrosophic sets. J. Intell. Fuzzy Syst. **26**(3), 1245–1252 (2014)
25. Singh, P.: A neutrosophic-entropy based adaptive thresholding segmentation algorithm: a special application in MR images of Parkinson's disease. Artif. Intell. Med. **104**, 101838 (2020)
26. Singh, P.: A neutrosophic-entropy based clustering algorithm (NEBCA) with HSV color system: a special application in segmentation of Parkinson's disease (PD) MR images. Comput. Methods Programs Biomed. **189**, 105317 (2020)
27. Patrascu, V.: Shannon entropy for neutrosophic information. Infinite Study (2018)
28. Bhandari, D., Pal, N.R.: Some new information measures for fuzzy sets. Inf. Sci. **67**, 209–228 (1993)

Intelligent Friendship Graphs:
A Theoretical Framework

Indradeep Bhattacharya$^{(\boxtimes)}$ and Shibakali Gupta

University Institute of Technology, The University of Burdwan,
Burdwan 713104, West Bengal, India
indra485@gmail.com

Abstract. In terms of social networks, a friendship graph is nothing but an undirected graph where edges are bi-directional. One of the popular social networking sites is Facebook, where we could observe these bi-directional communications. For example, on Facebook, if a person will accept a friend request of another person, then an undirected edge will form between them so they can communicate with each other simultaneously. Consider a subset of a network or a sub-graph of an immense graph, which itself is very large to extract useful information from it. In this paper, we have proposed an intelligent edge contraction mechanism based on some well-defined logic. Firstly, we have discussed a novel partitioning algorithm, which generates information regarding unordered pairs (disjoint partitioning sets). The partitioning table will reduce the successive searching cost of using the adjacency matrix. After generating the partitioning table, we have demonstrated an intelligent edge contraction mechanism to diminish the dimension of a dense graph.

Keywords: Pairwise partitioning · Intelligent edge contraction ·
Optimized pairs generating function

1 Introduction

In this paper, we have tried to achieve the maximum mutual neighbors for a vertex. However, we need to select that vertex intelligently. There are lots of applications and advantages in adopting the maximum mutuals for a vertex in a graph. Publishing social network information for research objectives has raised serious concerns for individual privacy. There exist numerous privacy-preserving researches that can deal with different attack models. In this scenario, Sun et al. [1] proposed a novel privacy attack model and mentioned it as a mutual friend attack. In that model, the opposition could re-analyze a pair of friends by using their number of mutual friends (or mutual neighbors). In order to communicate this issue, they had introduced a unique anonymity concept, i.e., k-anonymity on the number of mutual friends (better known as k-NMF anonymity). According to the authors, it ensured that at least $(k-1)$

Supported by University Institute of Technology.

K. K. Patel et al. (Eds.): icSoftComp 2021, CCIS 1572, pp. 90–102, 2022.
https://doi.org/10.1007/978-3-031-05767-0_8

other friend pairs in the graph must share the same number of mutual friends. They had formed algorithms to achieve the k-NMF anonymity while conserving the actual vertex set in the sense that they allowed the occasional addition but no deletion of vertices. The authors had also given an algorithm to ensure the k-degree anonymity along with the k-NMF anonymity. Typically, various social networking systems, such as LinkedIn and Google Plus, also incorporate a feature that allows a user to query mutual friends between him/her and any other user he/she can reach using the available public search feature in the social networking systems. Although a mutual friend feature is really supportive in letting users find new friends and connect to them. However, Jin et al. [2] showed in their research that the mutual friend feature must increase significant privacy concerns as an attacker could use it to seek out some or all of the victim's friends. They had shown that by using mutual friend queries, an adversary could initiate privacy attacks (referred to as the mutual-friend-based attacks) to recognize friends and distant neighbors of targeted users or victims. In their research, they had analyzed these attacks and identified numerous attack structures that an attacker could use to build attacking procedures and recognize a victim's friends and his distant neighbors. In this context, we have developed an intelligent edge contraction mechanism along with a novel partitioning method (Pairwise Partitioning Algorithm). As we have mentioned earlier that our prime focus is to extract the maximum mutual neighbors during edge contraction. This contraction mechanism has several degrees of abstraction. The merged vertices of each degree of contraction have been replaced by respective dummy vertices and passed to the next degree of the contraction layer. The information regarding each contraction for a specific layer has been kept inside a table to realize the actual graph. It is very difficult for an advanced attacker to recognize the initial network without having the knowledge of the table of every layer (degree of contraction). Between each contraction phase, the partitioning mechanism has been applied. In mathematical computer science and graph theory, a graph partitioning problem defines the reduction of a graph to smaller graphs by partitioning its set of nodes into mutually exclusive groups. In the pairwise partitioning method, we have partitioned the unordered pairs of a graph into their respective partitioning sets based on the degree equivalency and the equality-inequality mechanism [5]. The outcome of this partitioning algorithm is a partitioning table, which reduces the continuous searching of the adjacency matrix. There are several other applications of graph partitioning method. Based on the basic theory of graph partitioning, Zhang et al. [4] proposed a streaming graph partitioning algorithm named, Akin. Their prime intention was to provide a preferable solution that was suitable for streaming graph partitioning in a distributed system. To diminish the edge-cut ratio, they made use of the similarity measure on the degree of vertices to gather structurally related vertices in the same partition as much as possible. In their research, they had displayed clearly that their proposed algorithm was able to achieve superior partitioning quality in terms of edge-cut ratio during maintaining a reasonable balance between all partitions. Their research-oriented improvements should be applicable to all real-life graphs.

Throughout the paper we have used the term edge-contraction. So, there is a need to understand about this concept. Let "G" be a graph and $e \in E(G)$ with $e = \{v_i, v_j\}$. Suppose, $W_{ij} \notin V(G)$. Contracting edge e in G results in the graph G/e obtained from G by removing edge e, replacing vertices v_i, v_j with new vertex w_{ij}, where w_{ij} is adjacent to the neighbors of v_i and v_j [6]. There are also numerous applications regarding edge contraction. One of them is the sparsification of motion-planning roadmaps. In this context, Shaharabani et al. [7] proposed a unique edge contraction algorithm to reduce the size of a motion-planning roadmap. Their algorithm disclosed the minimal effect on the quality of paths that could be extracted from the newly generated roadmap. The working phenomenon of the algorithm was to diminish a roadmap edge into a single vertex and the adjacency of the new vertex to the neighboring vertices of the contracted edge. They had compressed almost 99% of the relations and vertices at the cost of deterioration of average shortest path length by at most 2%.

2   A Basic Overview on Pairwise Partitioning

The pairwise partitioning mechanism differs slightly from the traditional graph partitioning algorithms [8]. In this partitioning method, we have dealt with the unordered pairs of a graph depending on their degree equivalency. At this point, we need to know about the meaning of degree equivalency. Let, λ represents the degree of each vertex and its mathematical notation is provided below –

$$\lambda(v) = k \tag{1}$$

Here "v" represents any node, and "k" implies any positive integer that holds the degree of a vertex. Suppose four vertices, v_1, v_2, v_3 and v_4 construct a graph G such that, $\lambda(v_1) = \lambda(v_3) = a$, and $\lambda(v_2) = \lambda(v_4) = b$, where a and b represent the degree of the mentioned vertices [9] respectively. As v_1 and v_3 having the same value of λ, so they should make one set, and another set should be formed using v_2 and v_4. After getting these sets, we need to apply the optimized pairs generating function (Eq. (2)). In our previous manuscript [5] based on the applications of pairwise partitioning, we had successfully demonstrated the overall functioning of the algorithm. However, we had made a slight mistake while adjusting optimized pairs generating function. In this paper, we have rectified that issue (i.e., Eq. (2)) and demonstrated the functioning of the algorithm.

$$P(S^n, i, j) = \bigcup_{i=1}^{n} \{(S_i \times S_i) \cup \bigcup_{j=1}^{n-1} (S_i \times S_{j+1})\} - \bigcup_{j=1}^{n-1} \{\bigcup_{i=1}^{n} (S_{j+i} \times S_j)\} \tag{2}$$

In Eq. (2), S_i or S_j denotes any set, and n denotes the number of sets. It eliminates unnecessary symmetric pairs (unordered pairs) to reduce the number of computations. Equation (2) forms an intermediate R-set, where for any pair $\{u_i, v_j\}$, if i equals j then we need to remove that pair from the intermediate set (R-set) to construct the $R_{Optimized}$ set. Before moving towards the interpretation of the pairwise partitioning algorithm, we need to realize the meaning of unordered pairs. As we know that undirected graphs are composed of unordered pairs, which means pairs {x, y} and {y, x} are identical. Now, we are all set to apply the equality-inequality detection mechanism. This mechanism can be applied with the help of an adjacency matrix. In this case, a flag; named "equality" has been employed for detection. For each pair of vertices, if there exists an edge between two vertices, the equality flag will be set to 1. The degree of each vertex from that pair is then converted to its equivalent binary, and an X-NOR operation should be performed to get its partitioning set, else if there exists no edge between two vertices equality flag will be set to 0, and an X-OR operation should be performed (inequality detection) to get its equivalent partitioning set. After getting all such partitions, we can easily determine, out of them, which ones are essential. In digital electronics [3], we use X-NOR logic as equality detector and X-OR as inequality detector [10]. To evaluate the respective partitioning sets for each unordered pair, these logical detectors play an important role.

2.1 Explanation of Pairwise Partitioning

In this section, with the help of a random undirected graph (Fig. 1), we have demonstrated the functioning of the partitioning algorithm briefly. According to the algorithm (Algorithm 1), $\lambda(v_1) = 2$, $\lambda(v_2) = 2$, $\lambda(v_3) = 4$, $\lambda(v_4) = 2$, and $\lambda(v_5) = 2$. Therefore, two sets are sufficient to allocate all the vertices in the graph and they are, $s_1 = \{v_1, v_2, v_4, v_5\}$, and $s_2 = \{v_3\}$. By Eq. (2) and other mathematical logics (conditions), we must get the following optimized R set. Here, $R_{Optimized} = \{\{v_1, v_2\}, \{v_1, v_4\}, \{v_1, v_5\}, \{v_2, v_4\}, \{v_2, v_5\}, \{v_4, v_5\}, \{v_1, v_3\}, \{v_2, v_3\}, \{v_4, v_3\}, \{v_5, v_3\}\}$. In this scenario, we can apply equality-inequality mechanism to get exact partitioning set for each pair in $R_{Optimized}$. For example, in the case of $\{v_1, v_2\}$, we see the vertices are not connected by an edge i.e., $adjacency(\{v_1, v_2\}) = 0$ (inequality occurs). Therefore, the degree of each vertex in this pair should be converted to its equivalent binary and then X-OR operation should be performed between them so, $binary(\lambda(v_1)) = 0010$, and $binary(\lambda(v_2)) = 0010$. Therefore, $d([binary(\lambda(v_1)) \oplus binary(\lambda(v_2))]) = 0$

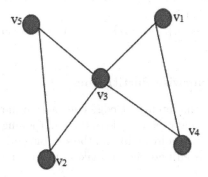

Fig. 1. A random undirected graph with five vertices and six edges.

(decimal equivalent X-OR result), which implies this pair should be inserted into P_0. Like this way, pair $\{v_2, v_5\}$ should be inserted into P_{15}. As they are related to each other therefore the equality flag becomes 1, and X-NOR operation should be performed. The entire process will be continuing until all the pairs will get their respective partitioning sets. The table (Table 1) referenced in this section, portrays the partitioning results. The proposed algorithm (Algorithm 1) generates all possible pairs, which have been categorized into essential and non-essential pairs (as mentioned in the given table). The given partitioning table (Note: A partitioning table can be thought of as a data structure, where all the necessary information related to a graph is stored in the form of simplified pairs within specific partitioning sets. Status information is associated with each partitioning set in the partitioning table) (Table 1) provides the details of essential and non-essential partitioning sets containing vertex pairs. The essential and non-essential criterion is based on the equality flag and should be determined easily with the assistance of the adjacency matrix. The pairs within the essential and non-essential sets having adjacency values 1 and 0 respectively. From Table 1, we get, $P_0 \cup P_9 \cup P_{15} = R_{Optimized}$, and $P_0 \cap P_9 \cap P_{15} = \phi$. Therefore, the partitioning of the above graph has been achieved successfully. Let E_P, and NE_P denote the set of essential and non-essential partitioning sets respectively. Consider the above example where,

$$E_P = P_9 \cup P_{15} \tag{3}$$

and,

$$NE_P = P_0 \cup \phi \tag{4}$$

Table 1. Experimental results about pairwise partitioning supported to Fig. 1

Partitioning set	Selected pairs	Status
P_0	$\{\{v_1, v_2\}, \{v_1, v_5\}, \{v_2, v_4\}, \{v_4, v_5\}\}$	Non-essential partitioning set
P_9	$\{\{v_1, v_3\}, \{v_2, v_3\}, \{v_4, v_3\}, \{v_5, v_3\}\}$	Essential partitioning set
P_{15}	$\{\{v_1, v_4\}, \{v_2, v_5\}\}$	Essential partitioning set

2.2   Algorithm of Pairwise Partitioning

Before moving towards the intelligent edge contraction method, in this section, we have provided the actual pseudo code of the partitioning algorithm. From the above discussion, we may acquire an idea of the stopping criteria of the algorithm. The pseudo-code referenced underneath makes our understanding even clearer.

Algorithm 1. Pairwise Partitioning

Input An undirected graph G (V, E)

1: **procedure** PAIRWISE PARTITIONING $(S, P, \lambda, v, N, R_{Optimized})$
2:     $N, n \leftarrow$ *number of vertices, number of sets*
3:     $v \leftarrow$ *any vertex in the graph*
4:     $\lambda \leftarrow$ *acts as an operator to the degree of any vertex*
5:     $S \leftarrow$ *set of similar degree vertices*
6:     $R_{Opt}, P \leftarrow$ *optimized pairs' set, set of all partitioning sets*
7:     **for** $i \leftarrow 1$ to N **do**
8:         $k_i \leftarrow \lambda(v_i)$
9:     **end for**
10:    *make-set (v_i) according to the same degree of vertices, i.e.*
11:    **if** $(\lambda(v_a) = \lambda(v_b))$ **then**
12:        $S \leftarrow \{v_a, v_b\}$
13:    **end if**
14:    *apply generalized Cartesian product or Eq. (2), i.e.*
15:    $R_{temp} = \bigcup_{i=1}^{n}\{(S_i \times S_i) \cup \bigcup_{j=1}^{n-1}(S_i \times S_{j+1})\} - \bigcup_{j=1}^{n-1}\{\bigcup_{i=1}^{n}(S_{j+i} \times S_j)\}$
16:    *for any pair $\{u_i, v_j\} \in R_{temp}$*
17:    **if** $((i = j)$ or $\{u_j, v_i\} \in R_{temp})$ **then**
18:        *remove pair $\{u_i, v_j\}$*
19:    **end if**
20:    *for each pair $\{x, y\} \in R_{Opt}$*
21:    **if** $(adjacency\{x, y\} = 1)$ **then**
22:        $equality = 1$
23:        $p_{X-NORresult} \leftarrow \{binary(\lambda(v_x)) \odot binary(\lambda(v_y))\}$
24:    **else**
25:        $p_{X-ORresult} \leftarrow \{binary(\lambda(v_x)) \oplus binary(\lambda(v_y))\}$
26:    **end if**
27:    **if** $((p_1 \cup p_2 \cup p_3 \cup .... \cup p_n) = R_{Opt}$ and $(p_1 \cap p_2 \cap p_3 \cap .... \cap p_n) = \phi)$ **then**
28:        *success*
29:    **end if**
30: **end procedure**

---

3   Intelligent Contraction

In this section, we have discussed the novel contraction algorithm and defined some intelligent logic to achieve the maximum mutual neighbors for a vertex. Now, the question is, why should we choose maximum contraction for a vertex? We can give a satisfactory answer to this question with the help of an example (Fig. 2). In Fig. 2, we could observe that while contracting edge e_1, we get G/e_1, which has only four edges after the initial contraction. On the other hand, by contracting edge e_2, G/e_2 generates six edges in the initial contraction. The reason behind it is that, in the case of G/e_1, the merged vertices (i.e., v_2 and v_5) have two neighbors in common whereas in G/e_2, there exists only one neighbor i.e., v_5 for vertices v_2 and v_3. In this case, G/e_1 is comparatively more sparse than that of G/e_2 due to the above reason. In the long run, it may help to rationalize any dense graphs in a better way (for example, transportation and road networks). In this paper, we have mentioned the security-related application [11] of our intelligent contraction method.

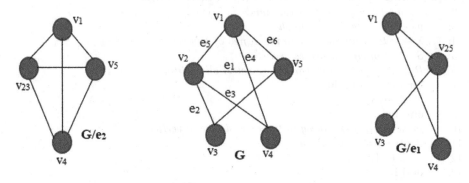

Fig. 2. A comparison between G/e_1 and G/e_2 with respect to the actual graph G.

3.1 Prerequisites for Intelligent Contraction

Before proceeding towards the contraction algorithm, we need to realize some prerequisites. So, in this context, we have defined some logic which must be implemented to achieve maximum common neighbors for a vertex and to reduce the dimension of any graph structure (friendship network).

i In each degree of contraction, initially we need to apply the pairwise partitioning algorithm to generate the set of the essential partitioning sets (i.e., E_P) and select the highest degree vertex or vertices "v_i". Search all possible unordered pairs in E_P, where v_i should be present. That is:

$$\alpha_{v_i} = \{\{v_i, u_1\}, \{v_i, u_2\}, ..., \{v_i, u_k\}\} \; or, \; \alpha_{v_i} = \bigcup_k \{\{v_i, u_k\}\}$$

here, α_{v_i} denotes a set of unordered pairs, where v_i must be present. Similarly, we need to seek out the possible unordered pairs for each u_k in E_P.

ii Let us create a set δ_{v_i} with the assistance of α_{v_i}, which includes all the neighbors of v_i. Similarly, we need to form δ_{u_k} for each u_k neighbors of v_i.

iii After step two, we must apply a check to find out maximum common neighbor between v_i and each u_k neighbors of v_i. The mathematical support has been referenced as follows:

$$\max(|(\delta_{v_i} \cap \delta_{u_1})|, |(\delta_{v_i} \cap \delta_{u_2})|, ..., |(\delta_{v_i} \cap \delta_{u_k})|)$$

iv At this point, we need to select that pair $\{v_i, v_j\}$, which has the highest common elements (neighbors) and apply the intelligent contraction algorithm. If they fail during contraction operation, we must remove that pair from the respective α_{v_i} set and also update the respective δ_{v_i} set as well. Then, repeat step three.

v If the maximum degree vertex or vertices (i.e., v_i) fails/fail to contract with any of its neighbors, then we need to determine the successive maximum degree vertex or vertices, and repeat step one for each maximum (2nd maximum, 3rd maximum, etc.) degree.

vi The contracted vertex or vertices of the form, w_{ij} cannot be contracted further within the same degree of contraction. It should be replaced by dummy vertex before transferring towards the next degree of contraction.

vii Suppose, during contraction mechanism, w_{ij} may occur in a relation, i.e., $\{v_i, w_{ij}\}$. In this context, the α_{v_i} set should be updated:

$$\alpha'_{v_i} = \alpha_{v_i} \setminus \{v_i, w_{ij}\}$$

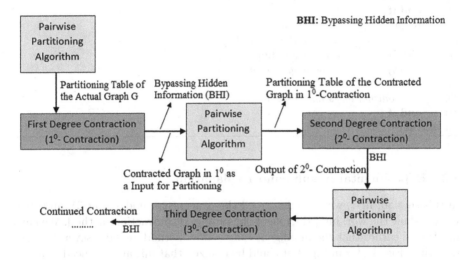

Fig. 3. A conceptual model for intelligent contraction, where we have shown different degree of contraction layer.

In the experimental section, we have demonstrated a detailed understanding of the conceptual model (Fig. 3) regarding the intelligent contraction. If there doesn't exist at least a vertex, which can take part in the contraction mechanism, the algorithm will stop.

3.2   The Contraction Algorithm

In the earlier section, we have discussed the necessary prerequisites for this intelligent edge contraction algorithm. Based on those prerequisites, we have defined the intelligent contraction mechanism underneath.

Algorithm 2. Intelligent Edge Contraction

1: **for** any unordered pair of the form $\{v_i, v_j\}$ **do**
2:      $\alpha_{v_i} = search\_pair(\{v_i, u_k\})$ such that; $\{v_i, u_k\} \in E_P$
3:      $\alpha_{v_j} = search\_pair(\{v_j, u_m\})$ such that; $\{v_j, u_m\} \in E_P$
4:      $\alpha'_{v_i} = \alpha_{v_i} \backslash \{v_i, v_j\}$
5:      $\alpha'_{v_j} = \alpha_{v_j} \backslash \{v_j, v_i\}$
6:      *Create the δ-sets for v_i and v_j such that*
7:      $\delta_{v_i} = \bigcup_k u_k$ and $\delta_{v_j} = \bigcup_m u_m$
8:      **if** $(\delta_{v_i} \cap \delta_{v_j}) \neq \phi$ **then**
9:          **if** $deg(v_i) \geq deg(v_j)$ **then**
10:              $\delta'_{v_i} = \delta_{v_i} \backslash (\delta_{v_i} \cap \delta_{v_j})$
11:          **else**
12:              $\delta'_{v_j} = \delta_{v_j} \backslash (\delta_{v_i} \cap \delta_{v_j})$
13:          **end if**
14:      **end if**
15:      **if** $|(\delta_{v_i} \cap \delta_{v_j})| \geq |\delta'_{v_i}|$ **then**
16:          $w_{ij} = v_i \cup v_j$; *merging of v_i and v_j*
17:      **else if** $|(\delta_{v_i} \cap \delta_{v_j})| \geq |\delta'_{v_j}|$ **then**
18:          $w_{ij} = v_i \cup v_j$; *merging of v_i and v_j*
19:      **else**
20:          *Contraction is not possible*
21:      **end if**
22: **end for**

3.3   Experimental Results and Explanation

In this section, first, we have described the conceptual model (Fig. 3), and then we will show the working principle of our proposed contraction method. We have mentioned earlier that the intelligent contraction method includes several phases (shown in Fig. 3). From Fig. 3, it could be realized that initially, we need to apply the pairwise partitioning algorithm to achieve the partitioning table of the actual graph. The concept regarding the partitioning table has been provided earlier. After generating the initial partitioning table, we must pass the information into the first-degree contraction layer or phase. The merged vertex/vertices i.e., $w_{ij} \notin V(G)$, should be replaced by a corresponding dummy vertex and proceeded towards the succeeding partitioning table generating module. It produces the information regarding the contracted graph in $1°$-contraction phase and passes to the immediate contraction phase. Like this way, it would be continuing until the contraction algorithm will stop. The intermediate outcomes for every contraction phase are some hidden information, which should be kept inside a table securely. Depending upon the size of the graph [12], we could adjust the number of phases of contraction, which must provide a severely sparse graph, as compared to the initial dense one.

In this experimental module, we have shown up to $3°$-contraction phase of a dense graph. In Fig. 4, we could observe there are four distinct phases, denoted by I, II, III, and IV. Here, "I" indicates the actual dense graph, "II" and "III" imply the contracted input graphs for the second and third-degree contraction phases.

In the case of Fig. 4(I), we could see that vertex v_1 has the highest degree in the graph G. According to the prerequisites of our proposed algorithm (Algorithm 2), we need to select v_1, create the α-set for v_1. The description of the α-set for vertex v_1 has been referenced underneath.

$$\alpha_{v_1} = \{\{v_1, v_8\}, \{v_1, v_9\}, \{v_1, v_7\}, \{v_1, v_6\}, \{v_1, v_5\}, \{v_1, v_2\}\}$$

It could be verified that there are 6 possible neighbors of v_1, and out of them $\delta_{v_1} \cap \delta_{v_7}$, and $\delta_{v_1} \cap \delta_{v_5}$ provide maximum common elements. In this case, we have considered $\{v_1, v_7\}$ for the intelligent edge contraction method. Fortunately, it succeeds during the intelligent edge contraction mechanism. According to the contraction algorithm, we get w_{17} such that, $w_{17} = v_1 \cup v_7$ and $w_{17} \notin V(G)$. The new vertex w_{17} cannot take part in the current contraction phase ($1°$-contraction phase). In that case, the intelligence of the algorithm will choose v_5 as the next vertex for contraction. However, it fails to merge with its neighbors. Sequentially, the rest of the vertices i.e., v_2, v_3, v_4, v_6, v_9, and v_{10} also fail to contract with their neighbors. According to our conceptual model, this is the right time to replace the merged vertex w_{17} with a dummy vertex (for example, v_{11} in Fig. 4(II)). The new graph (Fig. 4(II)) should be used in the immediate contraction phase (i.e., the $2°$-contraction phase). The intermediate outcome (Table 2) of the first phase of contraction has been referenced underneath. By using the δ'_{v_i}-set of v_i and δ'_{v_j}-set of v_j we can detect the infrequent neighbors of v_i and v_j. Similarly, we can define the $2°$, and $3°$ contraction phase. In the case of the $2°$-contraction, $v_{12} = w_{11\,3} = v_{11} \cup v_3$ (Fig. 4(III)). The hidden information (Table 3) regarding the second phase of contraction has been referenced as follows:

Table 2. Experimental Results Regarding $1°$ contraction (hidden information: one)

Merged vertex (w_{ij})	Original vertex (v_i)	Original vertex (v_j)
w_{17}	v_1	v_7

Table 3. Experimental Results Regarding $2°$ contraction (hidden information: two)

Merged vertex (w_{ij})	Original vertex (v_i)	Original vertex (v_j)
$w_{11\,3}$	v_{11}	v_3

Figure 4(IV) implies our final contraction phase, where $w_{12\,2} = v_{12} \cup v_2$. It depicts a severe sparsification as compared to the initial graph G.

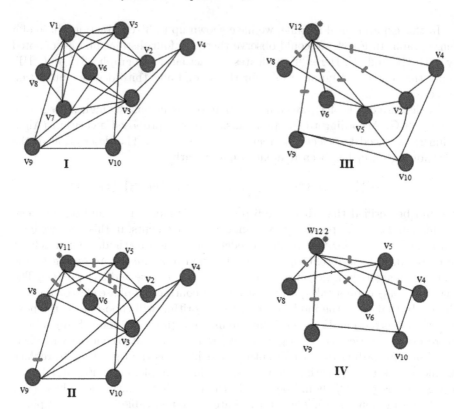

Fig. 4. Different phases of the intelligent contraction based on the conceptual model, where the red mark represents the dummy vertex, green and blue stripes denote the contracted common edges and infrequent edges respectively. (Color figure online)

In this experimental analysis, we have taken the hidden information of the contracted graph up to the third-degree contraction phase. Our final observed results regarding the third phase of the contraction have been provided below (Table 4).

Table 4. Experimental Results Regarding $3°$ contraction (hidden information: three)

Merged vertex (w_{ij})	Original vertex (v_i)	Original vertex (v_j)
$w_{12\,2}$	v_{12}	v_2

4 Possible Application Regarding the Intelligent Contraction

In this section, we have discussed the possible application based on our proposed approach. To prevent the mutual friend-based attack [2], we could take the help of the intelligent contraction method. As we have mentioned earlier, that our proposed conceptual model has been classified into several contraction phases, where each contraction phase returns some hidden information in the form of a table. A network administrator only knows this hidden information, but the attacker does not. As the contracted network representation contains several dummy vertices, therefore, it is very difficult for an attacker to realize the actual network without having the knowledge of the intermediate phase-information. In our example, we have evaluated up to $3°$ contraction, which is severely sparse and it contains several hidden information.

5 Conclusions and Future Scope

In this paper, initially, we have discussed the concept and working principles of the pairwise partitioning algorithm, where we have defined the optimized pairs generating function that removes the unnecessary symmetric pairs during computation. The pairwise partitioning algorithm is the prime ingredient to interpret the conceptual model of the intelligent contraction method, as it reduces the continual search of the adjacency matrix. The unordered pairs within the set of essential partitioning sets (i.e., E_P) always having adjacency value one. After describing the pairwise partitioning algorithm, we have discussed the intelligent contraction method. We have demonstrated the conceptual model, which is required to understand the working fundamentals regarding the intelligent edge contraction method. We have elaborately discussed the working of the conceptual model and shown the possible application of our proposed theory. The intelligent edge contraction is nothing but a theoretical idea, which identifies the maximum mutual neighbors for a vertex to diminish the dimension of the actual dense graph in a fewer number of contraction phases.

In our future research, we will try to prove our given statement that is, the intelligent contraction algorithm takes a fewer number of contraction phases than any other contraction algorithm. Also, we will try to implement our algorithm in the field of network security.

References

1. Sun, C., Yu, P.S., Kong, X., Fu, Y.: Privacy preserving social network publication against mutual friend attacks. In: 13th International Conference on Data Mining Workshops, Dallas, TX, USA, pp. 883–890. IEEE (2013). https://doi.org/10.1109/ICDMW.2013.71
2. Jin, L., Joshi, J.B.D., Anwar, M.: Mutual-friend based attacks in social network systems. Comput. Secur. **37**, 15–30 (2013). https://doi.org/10.1016/j.cose.2013.04.003

3. Ghosh, B., et al.: A novel approach to realize of all optical frequency encoded dibit based XOR and XNOR logic gates using optical switches with simulated verification. Opt. Spectrosc. **124**(3), 337–342 (2018). https://doi.org/10.21883/OS.2018.03.45655.82-17

4. Zhang, W., Chen, Y., Dai, D.: AKIN: a streaming graph prtitioning algorithm for distributed graph storage systems. In: 18th IEEE/ACM International Symposium on Cluster, Cloud and Grid Computing (CCGRID), Washington, DC, USA, pp. 183–192. IEEE (2018). https://doi.org/10.1109/CCGRID.2018.00033

5. Bhattacharya, I., Gupta, S.: A novel partitioning algorithm to process large-scale data. In: Pan, I., Mukherjee, A., Piuri, V. (eds.) Proceedings of Research and Applications in Artificial Intelligence. AISC, vol. 1355, pp. 163–171. Springer, Singapore (2021). https://doi.org/10.1007/978-981-16-1543-6_15

6. Courcelle, B.: Clique-width and edge contraction. Inf. Process. Lett. **114**(1–2), 42–44 (2014). https://doi.org/10.1016/j.ipl.2013.09.012

7. Shaharabani, D., Salzman, O., Agarwal, P.K., Halperin, D.: Sparsification of motion-planning roadmaps by edge contraction. In: IEEE International Conference on Robotics and Automation, Karlsruhe, Germany, pp. 4098–4105. IEEE (2013). https://doi.org/10.1109/ICRA.2013.6631155

8. Sanders, P., Schulz, C.: Engineering multilevel graph partitioning algorithms. In: Demetrescu, C., Halldórsson, M.M. (eds.) ESA 2011. LNCS, vol. 6942, pp. 469–480. Springer, Heidelberg (2011). https://doi.org/10.1007/978-3-642-23719-5_40

9. Ali, P., Mukwembi, S., Munyira, S.: Degree distance and vertex-connectivity. Discrete Appl. Math. **161**(18), 2802–2811 (2013). https://doi.org/10.1016/j.dam.2013.06.033

10. Uthayakumar, T., Raja, R.V.J., Porsezian, K.: Realization of all-optical logic gates through three core photonic crystal fiber. Opt. Commun. **296**, 124–131 (2013). https://doi.org/10.1016/j.optcom.2012.12.061

11. Stylianopoulos, C., Almgren, M., Landsiedel, O., Papatriantafilou, M.: Multiple pattern matching for network security applications: acceleration through vectorization. J. Parallel Distrib. Comput. **137**, 34–52 (2020). https://doi.org/10.1016/j.jpdc.2019.10.011

12. Zhao, Y.: More on the minimum size of graphs with given rainbow index. Discussiones Mathematicae Graph Theory **40**(1), 227–241 (2020)

Modeling of the Koch-Type Fractal Wire Dipole Antenna with the Random Forest Algorithm

Ilya Pershin and Dmitrii Tumakov[✉]

Federal University, Kazan, Russia
dtumakov@kpfu.ru

Abstract. A wire dipole is considered, the arms of which have the geometry of a Koch-type curve of the first, second, and third iterations. The aim of the work is to analyze the applicability of an algorithm based on the composition of decision trees for solving the direct problem (the problem of determining the electrodynamic characteristics of the antenna) and the inverse problem (the problem of reconstructing the antenna geometry). A comparison with the linear regression model for both problems is made. More than a twofold improvement in accuracy is shown for determining all electrodynamic characteristics. Estimates are obtained for the average absolute error when reconstructing the geometric parameters of the antenna from its electrodynamic characteristics. It is concluded that the error for the random forest is less than for a linear model from 2 to 5 times.

Keywords: Antenna simulation · Fractal antenna · Wire dipole · Machine learning · Random forest

1 Introduction

Wire antennas are widely used in various communication systems [1, 2]. For many years, wire antennas have remained quite promising devices [3], since changing the geometry makes it possible to make the size of the antennas smaller and improve the electrodynamic characteristics of the antenna. One of the base approaches to the miniaturization of antennas is to complicate the geometry of the radiator, thus increasing the path of the current along the conducting surface [4]. For example, for this, they use the addition of multiple cuts of the correct geometry on the radiator [5–7] or form fractal radiators [8–11]. The approach based on the fractalization of the radiator is one of the effective ways to complicate the antenna geometry [12–14]. A well-studied class of fractal wire antennas is the Koch-type wire dipole [15], which has a geometry similar to the geometry of a conventional fractal Koch wire dipole [16–18].

Designing an antenna with complex geometry is computationally time-consuming. To achieve the best geometric parameters of the antenna design and the desired antenna characteristics, various methods are used based on the sequential complication of geometry [19, 20], genetic algorithms [21, 22], differential evolution algorithms [23, 24], and other methods [25]. To speed up the process of antenna design an approach based on regression models is used [26–28]. These models usually represent nonlinear functions of the geometric parameters of the dipole arm and describe the relationship between

© Springer Nature Switzerland AG 2022
K. K. Patel et al. (Eds.): icSoftComp 2021, CCIS 1572, pp. 103–115, 2022.
https://doi.org/10.1007/978-3-031-05767-0_9

the electrodynamic characteristics of the antenna and the geometric parameters of its arm. For a wire dipole of the Koch type, a relationship was established between various electrodynamic and geometric characteristics, and regression models were constructed [15, 29, 30]. The resulting models can be used to quickly design the antenna geometry according to the specified electrodynamic characteristics [31].

Today, the use of machine learning methods is the most promising direction for the rapid design of various antenna devices [32–35]. In this paper, we consider the application of the composition of decision trees [36] for solving both the problem of calculating the electrodynamic characteristics from given geometric characteristics, and the inverse problem. The accuracy of application of the proposed method for various iterations of Koch-type fractal wire antennas is estimated. The main goal of the work is to apply an algorithm based on the composition of decision trees to solve the direct and inverse problems of antenna modeling. For this, the first three prefractals of the Koch-type dipole are considered.

The work consists of the introduction, the part for describing the algorithm for constructing prefractals and the decision tree algorithm, the part in which the direct and inverse problems are solved, and the conclusion. The work presents tables showing the absolute errors of the algorithm in the case of various iterations of the Koch prefractal.

2 Koch-Type Prefractals

A Koch-type fractal is not a classic fractal. Its geometry is similar to that of the Koch fractal. The Koch fractal curve can be constructed using the following iterative scheme [37, 38]:

$$K_0 = [0, 1], \quad K_n = \bigcup_{i=1}^{4} A_i(K_{n-1}), \quad K = \lim_{n \to \infty} K_n,$$

where $A_i, i = 1..4$ – affine transformations of the plane:

$$A_1\begin{pmatrix} x \\ y \end{pmatrix} = \frac{1}{3}\begin{pmatrix} x \\ y \end{pmatrix}, \quad A_2\begin{pmatrix} x \\ y \end{pmatrix} = \frac{1}{3}\begin{pmatrix} 1/2 & -\sqrt{3}/2 \\ \sqrt{3}/2 & 1/2 \end{pmatrix}\begin{pmatrix} x \\ y \end{pmatrix} + \begin{pmatrix} 1/3 \\ 0 \end{pmatrix},$$

$$A_3\begin{pmatrix} x \\ y \end{pmatrix} = \frac{1}{3}\begin{pmatrix} 1/2 & \sqrt{3}/2 \\ -\sqrt{3}/2 & 1/2 \end{pmatrix}\begin{pmatrix} x \\ y \end{pmatrix} + \begin{pmatrix} 1/2 \\ 1/2\sqrt{3} \end{pmatrix}, \quad A_4\begin{pmatrix} x \\ y \end{pmatrix} = \frac{1}{3}\begin{pmatrix} x \\ y \end{pmatrix} + \begin{pmatrix} 2/3 \\ 0 \end{pmatrix}.$$

The first K_1 and second K_3 iterations of the Koch curve are shown in Fig. 1. The K_1 set is commonly referred to as the Koch fractal generator. The Koch generator can be specified as a line connecting five points (vertices) in series

$$\{x_0 = (0, 0), x_1 = (0.333, 0), x_2 = (0.5, 0.288), x_3 = (0.666, 0), x_4 = (1, 0)\}.$$

In the general case, affine transformations $A_i : R^3 \to R^3$ are of the following view:

$$A_i\begin{pmatrix} t_i \\ x_i \\ y_i \end{pmatrix} = \begin{pmatrix} a_i & 0 & 0 \\ c_{i1} & D_{i1} & D_{i2} \\ c_{i2} & D_{i3} & D_{i4} \end{pmatrix}\begin{pmatrix} t_i \\ x_i \\ y_i \end{pmatrix} + \begin{pmatrix} e_i \\ d_{i1} \\ d_{i2} \end{pmatrix}.$$

For Koch-type curve, the matrices D_{ij} have the form [31]

$$D_{i1} = D_{i4} = \begin{pmatrix} 0.333 & 0 \\ 0 & 0.333 \end{pmatrix}, \quad D_{i2} = \begin{pmatrix} 0.167 & -0.289 \\ 0.289 & 0.167 \end{pmatrix},$$

$$D_{i3} = \begin{pmatrix} 0.167 & 0.289 \\ -0.289 & 0.167 \end{pmatrix}.$$

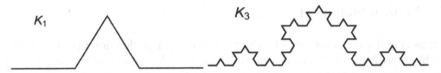

Fig. 1. First and third iterations of a Koch curve

In what follows, we will assume that all i satisfy the following conditions:

$$A_i \begin{pmatrix} t_0 \\ x_0 \\ y_0 \end{pmatrix} = \begin{pmatrix} t_{i-1} \\ x_{i-1} \\ y_{i-1} \end{pmatrix}, \quad A_i \begin{pmatrix} t_4 \\ x_4 \\ y_4 \end{pmatrix} = \begin{pmatrix} t_4 \\ x_4 \\ y_4 \end{pmatrix}.$$

Then, the following is obtained:

$$a_i = t_i - t_{i-1}, \quad e_i = t_{i-1},$$

$$\begin{pmatrix} c_{i1} \\ c_{i2} \end{pmatrix} = \begin{pmatrix} x_i - x_{i-1} \\ y_i - y_{i-1} \end{pmatrix} - \begin{pmatrix} D_{i1} & D_{i2} \\ D_{i3} & D_{i4} \end{pmatrix} \begin{pmatrix} x_4 - x_0 \\ y_4 - y_0 \end{pmatrix},$$

$$\begin{pmatrix} f_{i1} \\ f_{i2} \end{pmatrix} = \begin{pmatrix} x_{i-1} \\ y_{i-1} \end{pmatrix} - \begin{pmatrix} D_{i1} & D_{i2} \\ D_{i3} & D_{i4} \end{pmatrix} \begin{pmatrix} x_0 \\ y_0 \end{pmatrix}.$$

We will consider a wire dipole symmetric with respect to the feeding point, located in the center of coordinates. The antenna arms have a geometry that coincides with the first-order Koch prefractal (Fig. 2). Initiating Koch-type prefractals differ from Koch prefractals only in the position of the central vertex A. Let us change the coordinates of this vertex at the right half of the dipole $A = (A_x, A_y)$ in a certain interval. Note that the vertex $A = (3.75$ cm, 2.475 cm$)$ corresponds to the classical Koch curve. The vertex $B = (-A_x, A_y)$ will change symmetrically with respect to vertex A.

Thus, we get a set of antennas which are different only in the position of the vertices A and B. First, we need to define the parameters that characterize the geometry of the antenna. Let the parameters be half the length of the dipole, $L = l_0 + l_1 + l_2 + l_3$ (cm), the ratio of the length of the lateral arm at vertex A: l_2/l_1 and the angle at the specified vertex, θ (radians). The selected parameters will be referred to as input parameters. Note that the linear antenna length varies from -7.5 cm to 7.5 cm, and $l_0 = l_3 = 2.5$.

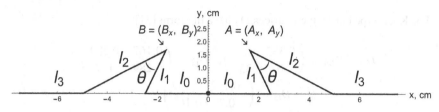

Fig. 2. Balanced wire Koch-type dipole

3 Problem Statement

Let us consider the possibility of using the composition of decision trees to solve the problems arising in the design of a wire dipole antenna having a complex arm geometry. In particular, we consider the problem of calculating the electrodynamic characteristics from the geometric characteristics of the dipole (direct problem) and the inverse problem, which consists of calculating the geometric characteristics from the given electrodynamic characteristics of the dipole.

Let us form three sets of the considered dipoles. The geometry of the arms of each plurality of antennas represents the first, second, and third iterations of the Koch-type prefractals, respectively. For the first set, the vertex coordinates A changes from 7.25 mm to 7.1775 cm with a step of 2.5375 mm. The coordinates of the vertex A for the second set change along the x-axis from 7.5 mm to 6.45 cm with a step of 3 mm, and along the y-axis from 3 cm to 6.5625 cm with a step of 1.875 mm. For the third set, x-coordinates be changed from 2.25 cm to 5.1 cm with a step of 1.5 mm, and y-coordinates from 1.5 cm to 5.625 cm with a step of 2.0625 mm. Calculations of the electrodynamic characteristics of the obtained antennas were performed in the frequency range from 100 MHz to 3 GHz with a step of 14.5 MHz in the FEKO program. For the first and second sets of dipoles, the wire thickness is d = 2 mm, for the third set d = 1 mm. From each set of antennas, antennas were selected that have no self-intersection in geometry and have a bandwidth greater than zero. Thus, 640 antennas were selected for the first iteration, 235 for the second, 307 for the third.

Each element of the dataset has the following set of geometric characteristics: coordinates of the central vertex (A_x, A_y), arm length L, slope angle θ, ratio l_2/l_1. As the electrodynamic characteristics, we use the base frequency FR, the bandwidth BW, the reflection coefficient S11, the resistance R and the Gain. As a metric of the quality of the model, we will use the mean absolute error (MAE)

$$MAE = \frac{\sum_{i=1}^{n} |y_i - x_i|}{n}. \tag{1}$$

4 Choosing the Optimal Prefractal Iteration Using the Decision Tree

When developing a fractal antenna, the prefractal order, which the geometry of the antenna arm should have to achieve optimal values of the electrodynamic characteristics, has a great influence. The prefractal order determines the complexity of the antenna fabrication; therefore, the choice of the optimal value of the order of the prefractal is an important task. In this section, we consider the problem of classifying the prefractal order of the antenna geometry by its electrodynamic characteristics and a given antenna height A_y. As the electrodynamic characteristics, we choose the base frequency FR, the bandwidth BW and the reflection coefficient S11. The output is the iteration number of the antenna (first to third). To solve this problem, we use the decision tree algorithm.

The decision tree is a supervised machine learning algorithm that allows recovering nonlinear dependencies of arbitrary complexity [39]. The idea of constructing a decision tree is to optimally partition the training sample X from the point of view of a predetermined quality functional Q, which is usually based on some criterion of information, such as Gini impurity, entropy, etc. In our implementation, we use Gini impurity as a criterion of information

$$I_G(p) = 1 - \sum_{i=1}^{j} p_i^2, \tag{2}$$

where p_i is a fraction of elements marked as class i, and j is a number of classes in the dataset. After completing the construction of the tree, each leaf is assigned an answer. In the case of solving the classification problem, this may be the most common class in the worksheet, and in the case of regression, it may be the median, mean, or some predetermined function. In our decision tree implementation, we use the most common class for the classification problem.

We evaluate the quality of the constructed classification model using 5-fold cross-validation and calculating the average weighted F-score by each class as a quality metric:

$$F_1 = 2\frac{precision * recall}{precision + recall} = \frac{TP}{TP + \frac{1}{2}(FP + FN)}. \tag{3}$$

Figure 3 shows a decision tree with a bounded depth of 5. With a tree depth of 13, the optimal mean value of the weighted F-measure of 0.981 is achieved on 5-fold cross-validation (Figs. 4, 5 and 6).

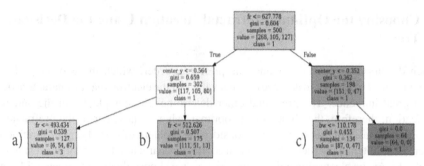

Fig. 3. An example of the constructed decision tree for the problem of classifying the iteration of the prefractal according to the given electrodynamic characteristics and height (the first three levels of the decision tree)

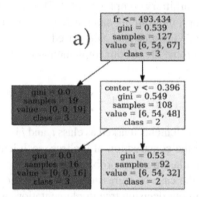

Fig. 4. An example of the constructed decision tree for the problem of classifying the iteration of the prefractal according to the given electrodynamic characteristics and height (depth 3–5). This is a continuation of the left side of the decision tree from Fig. 3

5   Solving a Direct Problem Using the Random Forest

The random forest method is an implementation of bootstrap aggregation for a number N of basic algorithms (in our case, a decision tree) $b_i(x)$, where x is a set of input data. We construct a method $\tilde{\mu}$ that generates a random subsample $\tilde{X} \subset X$. Thus, the final composition is the average value of the basic algorithms

$$a_N(x) = \frac{1}{N} \sum_{i=1}^{N} b_i(\tilde{\mu}(X)). \qquad (4)$$

For each iteration, using the random forest algorithm, we construct a set of models, each of which accepts the coordinates of the central vertex (A_x, A_y) and the arm length L as input, and outputs one of the following electrodynamic characteristics at the output: base frequency FR, bandwidth transmission BW, reflection coefficient S11, resistance R, gain Gain. Let's build linear regression models as the baseline. Let us calculate the MAE improvement factor K, which is equal to the ratio of the MAE for the linear model to

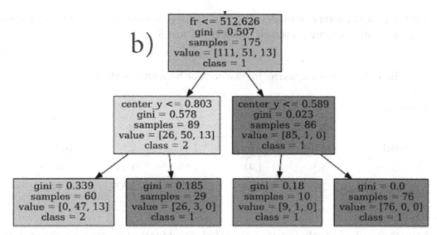

Fig. 5. An example of the constructed decision tree for the problem of classifying the iteration of the prefractal according to the given electrodynamic characteristics and height (depth 3–5). This is a continuation of the center of the decision tree from Fig. 3

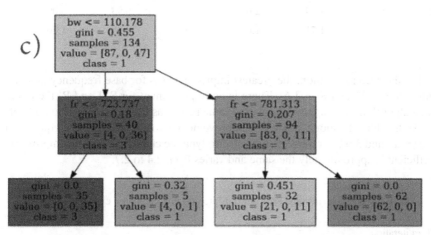

Fig. 6. An example of the constructed decision tree for the problem of classifying the iteration of the prefractal according to the given electrodynamic characteristics and height (depth 3–5). This is a continuation of the right side of the decision tree from Fig. 3

the MAE for the decision tree model. The 5-fold cross-validation results for the random forest for each iteration are presented in Table 1, 2 and 3.

Table 1. Mean average error for first iteration by geometric characteristics.

First iteration					
	FR	BW	S11	R	Gain
Linear model	14.64	7.2	2.41	2.62	0.0045
Random forest	5.36	1.41	1.17	1.04	0.0013
K	2.73	5.09	2.06	2.53	3.44

From the results of Table 1, it can be concluded that for the first iteration, the greatest improvement by the random forest algorithm occurs for the bandwidth (K = 2.73). At the time, the worst improvement – two-fold occurs for the S11.

Table 2. Mean average error for second iteration by geometric characteristics.

Second iteration					
	FR	BW	S11	R	Gain
Linear model	7.33	3.26	0.4	0.89	0.0015
Random forest	4.29	2.03	0.37	0.93	0.0012
K	1.71	1.6	1.07	0.96	1.26

For the second iteration, the greatest improvement is for base frequency and bandwidth (K = 1.71 and K = 1.6). There is no improvement for S11 and R, the random forest algorithm gives approximately the same results as the linear regression model.

For the third iteration, the greatest improvement is found for the base frequency and gain (K around 3.6). For the rest of the electrodynamic characteristics, the improvement coefficient is approximately the same and varies from 2.4 to 2.7.

Table 3. Mean average error for third iteration by geometric characteristics.

Third iteration					
	FR	BW	S11	R	Gain
Linear model	12.46	4.13	1.66	1.81	0.0033
Random forest	3.45	1.54	0.65	0.75	0.0009
K	3.6	2.69	2.55	2.41	3.66

In general, according to the analysis of Tables 1, 2 and 3, it can be concluded that the random forest algorithm gives an improvement of more than two times for all considered electrodynamic ones in the case of the first and third iterations. For the second iteration, the least improvement is observed. This is due, among other things, to the fact that the dataset contains 640 and 307 examples for the first and third iteration, respectively, and only 235 examples for the second iteration.

Now let's build a multi-iteration model using the random forest algorithm for all data at once. As input data, we use the coordinates of the central vertex (A_x, A_y), the length of the dipole arm L. Let's test the resulting model both on all data and on data for each iteration. The results of cross-validation are displayed in Table 4.

Table 4. Mean average error for all iterations by geometric characteristics.

	FR	BW	S11	R	Gain
Linear model	18.88	7.51	2.07	2.66	0.005
Random forest	4.63	1.54	0.87	0.95	0.00122
K	4.07	4.87	2.37	2.8	4.1

From Table 4, it can be concluded that more than a fourfold improvement in prediction quality is achieved for bandwidth, gain and base frequency. For S11 and R, more than a twofold improvement was obtained.

Let us estimate the importance of characteristics using permutation importance [40]. The idea of permutation importance for the x_i characteristic is the difference in the model quality estimate between the initial values of the x_i characteristic and the shuffle ones (Fig. 7).

From the analysis of Fig. 7, we can conclude that the largest contribution is made by the length of the dipole arm and the dipole height (y-coordinate of the central vertex of the initial iteration).

6 Reconstruction of the Dipole Geometry (Inverse Problem)

Let us now consider the reconstruction of the antenna geometry from the electrodynamic characteristics discussed above: the base frequency FR, the bandwidth BW, the reflection coefficient S11, the resistance R and the Gain. Let's build a model using the random forest algorithm to predict the coordinates of the central vertex (A_x, A_y) and the length of the dipole arm. In Table 5, we display the MAE for each predicted characteristic.

Previously, using permutation importance, it was shown that the height A_y of the dipole arm and the length of the dipole arm are of the greatest importance. From Table 5, we can conclude that for the second and third iterations, using the random forest algorithm, better quality models were obtained than for the first iteration.

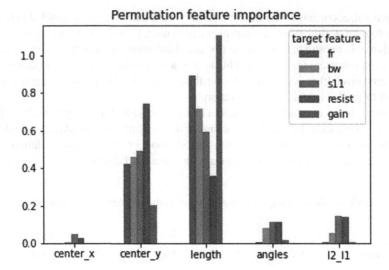

Fig. 7. Permutation importance for direct problem for all iteration

Table 5. Mean average error for each iteration by electrodynamic characteristics.

Iteration number	A_x	A_y	Length
First	0.0474	0.012	0.001
Second	0.0347	0.0097	0.0017
Third	0.0468	0.0054	0.0017

7 Analysis of the Results

In the previous sections, the application of the random forest algorithm was shown to solve the direct and inverse problems of antenna modeling. The results obtained are shown for the solution of the direct problem are given in Tables 1, 2, 3 and 4. The results indicate that the accuracy of the proposed algorithm is superior to the accuracy of the linear model algorithm. The result is especially visible for the antenna arm, which is the first or third iteration of the Koch-type curve. The solution to the antenna design problem (inverse problem) using the random forest algorithm also has high accuracy. This is evidenced by the results shown in Table 5.

The worst result is obtained for the reflection coefficient and the antenna impedance. However, the average improvement in the accuracy of the algorithm is more than 2.3 times better than the results of the linear model.

Thus, the algorithm proposed in the work can be used both for modeling the operation of a wire fractal antenna and for designing the antenna according to the given electrodynamic characteristics. It should also be noted that the calculations are carried out with the tree depth equal to 13. However, by increasing the tree depth, even more accurate results can be achieved.

8 Conclusion

The application of the random forest method for solving the problem of obtaining electrodynamic characteristics from given geometric characteristics of a wire dipole of the Koch type with a complex geometry of the arms is considered. It is shown that the random forest algorithm more than twice calculates the characteristics of the linear model. Moreover, for some cases, the refinement reaches more than 5 times. It is concluded that the electric length of the dipole and the height of the central vertex of the zero iteration have the greatest influence on the value of the electrodynamic characteristics.

The inverse problem, the problem of reconstructing the antenna geometry from the given electrodynamic characteristics, has been solved. It is shown that, for all iterations, the mean average error for the dipole length is less than 0.002, for the y-coordinate of the central vertex, on the order of one hundredth, and for the x-coordinate of the central vertex, on the order of several hundredths.

As further research, it may be interesting to consider other approaches of machine learning methods to design and simulate fractal antennas, for example, they can be neural networks of various architectures. First of all, it is advisable in order to identify the optimal design algorithms for this class of antennas.

Acknowledgement. This paper has been supported by the Kazan Federal University Strategic Academic Leadership Program.

References

1. Balanis, C.A.: Antenna Theory: Analysis and Design. Wiley, Hoboken (1997)

2. Tran, D.-D., Ha, D.-B., Nayyar, A.: Wireless power transfer under secure communication with multiple antennas and eavesdroppers. In: Duong, T.Q., Vo, N.-S. (eds.) INISCOM 2018. LNICSSITE, vol. 257, pp. 208–220. Springer, Cham (2019). https://doi.org/10.1007/978-3-030-05873-9_17

3. Singh, K., et al.: Fractal antennas: a novel miniaturization technique for wireless communications. Recent Trends Eng. 2(5), 172–176 (2009)

4. Fujimoto, K., Morishita, H.: Modern Small Antennas. Cambridge University Press, Cambridge (2013)

5. Markina, A., Tumakov, D.: Designing the four-tooth-shaped microstrip antenna for Wi-Fi applications. In: Proceedings - 2019 1st International Conference on Control Systems, Mathematical Modelling, Automation and Energy Efficiency, pp. 25–30 (2019)

6. Markina, A., Tumakov, D.: Designing a single-band monopole six-tooth-shaped antenna with preset matching. In: Proceedings - 2020 IEEE East-West Design and Test Symposium, pp. 1–6 (2020)

7. Markina, A., et al.: Designing the symmetrical eight-tooth-shaped microstrip antenna for Wi-Fi applications. In: Proceedings - 2018 IEEE East-West Design and Test Symposium, pp. 491–495 (2018)

8. Kubacki, R., et al.: Minkowski island and crossbar fractal microstrip antennas for broadband applications. Appl. Sci. 8(3), 334 (2016)

9. Trinh-Van, S., et al.: High-gain waveguide-fed circularly polarized Spidron fractal aperture antenna. Appl. Sci. 9(4), 691 (2019)

10. Kumar, A., Pharwaha, A.P.S.: An optimal multiband compact modified crinkle fractal antenna for wireless applications. In: Proceedings - 6th International Conference on Signal Processing and Integrated Networks, Noida, India, pp. 927–931. IEEE (2019)

11. Wang, Y., et al.: A miniaturised LS Peano fractal antenna for partial discharge detection in gas insulated switchgear. Sens. Transducers 4(2), 19–25 (2020)

12. Karpukov, L.M., et al.: The properties of the fractal wire antennas. In: Proceedings of MMET International Conference, pp. 310–312 (2009)

13. Wagh, K.H.: A review on fractal antennas for wireless communication. Rev. Electron. Commun. Eng. 32, 37–41 (2015)

14. Anguera, J., et al.: Fractal antennas: an historical perspective. Fractal Fractional 4(1), 3 (2020)

15. Tumakov, D.N., et al.: Modeling of the Koch-type wire dipole. Appl. Math. Model. 51, 341–360 (2017)

16. Li, Y., et al.: The analysis and comparison of the electromagnetic radiation characteristic of the Koch fractal dipole. In: Proceedings of International Symposium on Antennas, Xi'an, China, pp. 15–18. IEEE (2012)

17. Rani, M., et al.: Variants of Koch curve: a review. In: IJCA Proceedings on Development of Reliable Information Systems, Techniques and Related Issues, pp. 20–24, 2012

18. Karim, M.N.A., et al.: Log periodic fractal Koch antenna for UHF band applications. Prog. Electromagn. Res. 100, 201–218 (2010)

19. Azarm, B., et al.: Novel design of dual band-notched rectangular monopole antenna with bandwidth enhancement for UWB applications. In: Iranian Conference on Electrical Engineering 2018, Mashhad, Iran, pp. 567–571. IEEE (2018)

20. Yang, J., et al.: Design of miniaturized dual-band microstrip antenna for WLAN application. Sensors 16(7), 983 (2016)

21. Orankitanun, T., Yaowiwat, S.: Application of genetic algorithm in tri-band U-slot microstrip antenna design. In: 17th International Conference on Electrical Engineering/Electronics, Computer, Telecommunications and Information Technology 2020, pp. 127–130. IEEE (2020)

22. Mishra, R.G., et al.: Optimization and analysis of high gain wideband microstrip patch antenna using genetic algorithm. Int. J. Eng. Technol. 7(1.5), 176–179 (2018)

23. Kaur, J., et al.: Design and optimization of a dual-band slotted microstrip patch antenna using Differential Evolution Algorithm with improved cross polarization characteristics for wireless applications. J. Electromagn. Waves Appl. **33**(11), 1427–1442 (2019)
24. Khanna, R., Kaur, J.: Optimization of modified T-shape microstrip patch antenna using differential algorithm for X and ku band applications. Microw. Opt. Technol. Lett. **60**(1), 219–229 (2018)
25. Ankita, E., Nayyar, A.: Review of various PTS (Partial Transmit Sequence) techniques of PAPR (Peak to Average Power Ratio) reduction in MIMO-OFDM. Int. J. Innov. Technol. Explor. Eng. **2**(4), 199–202 (2013)
26. Abgaryan, G.V., et al.: Application of correlation and regression analysis to designing antennas. Rev. Publ. **4**, 1–13 (2017)
27. Tumakov, D.N., et al.: Fast method for designing a well-matched symmetrical four-tooth-shaped microstrip antenna for Wi-Fi applications. J. Phys. Conf. Ser. **1158**, 042029 (2019)
28. Markina, A.G., Tumakov, D.N.: Designing a dual-band printed monopole symmetric tooth-shaped antenna. Lobachevskii J. Math. **41**, 1354–1362 (2020). https://doi.org/10.1134/S19 95080220070264
29. Pershin, I., Tumakov, D.: Relationship between base frequency of the Koch-type wire dipole and various dimensions. In: 2020 IEEE East-West Design & Test Symposium, Varna, Bulgaria, pp. 1–6. IEEE (2020)
30. Abgaryan, G., Tumakov, D.: Relationship between lacunarity and bandwidth of a Koch-type wire antenna. Amazonia Investiga **7**(15), 88–98 (2018)
31. Tumakov, D., et al.: Miniaturization of a Koch-type fractal antenna for Wi-Fi applications. Fractal Fractional **4**(2), 25 (2020)
32. Misilmani, H.M., et al.: A review on the design and optimization of antennas using machine learning algorithms and techniques. Int. J. RF Microw. Comput. Aided Eng. **30**(10), e22356 (2020)
33. Kayumov, Z., Tumakov, D., Markina, A.: Application of neural networks to simulate a monopole microstrip four-tooth-shaped antenna. In: Garg, D., Wong, K., Sarangapani, J., Gupta, S.K. (eds.) IACC 2020. CCIS, vol. 1368, pp. 106–119. Springer, Singapore (2020). https://doi.org/10.1007/978-981-16-0404-1_9
34. Kaur, M., Singh, J.: Giuseppe Peano and Cantor set fractals based miniaturized hybrid fractal antenna for biomedical applications using artificial neural network and firefly algorithm. Int. J. RF Microw. Comput. Aided Eng. **30**(3), e22000 (2019)
35. Kumar, A., et al.: Optimization of antenna parameters using neural network technique for Ku-band applications. In: 7th International Conference on Signal Processing and Integrated Networks, Noida, India, pp. 706–709. IEEE (2020)
36. Bishop, C.M. (ed.): Pattern Recognition and Machine Learning. Springer, New York (2006)
37. Barnsley, M.F., Harrington, A.N.: The calculus of fractal interpolation functions. J. Approx. Theory **57**, 14–34 (1989)
38. Igudesman, K., et al.: New approach to fractal approximation of vector-functions. Abstr. Appl. Anal. **2015**, 278313 (2015)
39. Breiman, L., et al.: Classification and Regression Trees. Wadsworth, Belmont (1984)
40. Altmann, A., et al.: Permutation importance: a corrected feature importance measure. Bioinformatics **26**, 1340–1347 (2010)

Comparative Analysis of Deep Learning Techniques for Facemask Detection

Ghazala Furqan[✉], Najme Zehra Naqvi[✉], and Arunima Jaiswal[✉]

Department of Computer Science and Engineering, Indira Gandhi Delhi Technical University for Women, Madarsa Road Opposite St. James Church, Kashmere Gate, Delhi, India
{ghazala024mtcse19,najmezehra,arunimajaiswal}@igdtuw.ac.in

Abstract. Pandemic caused owing to widespread of corona-virus has changed our lives upside down. Covering the face area particularly nose and mouth is the prime need of the hour. Any negligence of not wearing the mask or incorrectly wearing the mask can be hazardous. This necessitates the need of understanding the real importance of wearing the mask appropriately in order to avoid the spread of Covid 19. Knowing the present population of the country, manual monitoring of the individuals is quite difficult. So, this research puts forward the use of deep learning techniques for automatic facemask detection using techniques such as capsule network, ResNet50, Mobile-Net architecture, and Convolution Neural Network. The techniques are validated on the merged dataset taken from MaskedFace-Net dataset and Kaggle (publicly available) based on the performance measures namely accuracy, precision, recall and F1-score. Amongst all, the results showed that capsule neural network achieved superlative performance with the accuracy of around 99% in comparison to other aforesaid deep learning techniques.

Keywords: Deep learning · Convolution network · Capsule network · Facemask

1 Introduction

As per WHO [1] there have been 187,296,646 confirmed Covid-19 positive cases, amongst whom total 4,046,470 deaths have been declared. The source of the spread of Covid 19 virus are the infected droplets which can be transmitted to a healthy human through his nose and mouth. In order to reduce the spread of the virus, WHO provided the guidelines to cover the face and mouth region along with other precautions. One infected person can infect a group of people, and in result those people can spread the virus further. If the infected person wears a mask, the chances of infecting the others in the surrounding is very less [2]. Hence, it shows the importance of face masks in preventing the spread of Covid-19. The health authorities of various nations are implementing the rule of wearing face mask to prevent the spread of Covid 19 [3]. But there are some people who are having casual approach and negligence towards protection from the virus. This leads threat to the people who are following appropriate behavior specially wearing a proper mask along with other precautions, because of which the need of the mask detection arises. The technique to detect whether a person is wearing a mask or not

© Springer Nature Switzerland AG 2022
K. K. Patel et al. (Eds.): icSoftComp 2021, CCIS 1572, pp. 116–126, 2022.
https://doi.org/10.1007/978-3-031-05767-0_10

is called face mask detection. It is like object recognition from an image. Object recognition is divided into image classification and object localization and object detection [4]. Image classification refers to classifying the class of the object present in an image. In object localization the location of object/objects in the given image is located and a boundary box is drawn around it. In computer vision object detection refers to identifying an object in a digital image. Multiple object detection techniques are introduced over a period of time. Face Detection refers to finding the faces in an image. Using various computer technologies, the facial region of a human can be detected. Facial Detection can be done on photographs, videos or in real time.

Face mask detection is a technique, which helps in identifying whether a person is wearing a mask or not. Face Mask Detection can also be done by using the various machine learning and deep learning techniques.

This paper presents the implementation of face mask detector using Deep Learning Techniques, and showing the comparative study of these models. First the datasets are merged and the distorted images were removed manually. Next, Image augmentation technique is used to increase the training dataset, so that the performance of the model can be enhanced. The rest of the paper is as follows. Section 2 is dedicated to the review of related works that has been done in the past. Section 3 gives a description of the datasets used. Section 4 contains the details of proposed model. In Sect. 5 result of the research are mentioned. Lastly, we concluded our work.

2 Related Work

Generally, the classification of the face mask is done in two categories (face mask worn or face mask not worn), using traditional machine learning techniques, but we are classifying into three categories - face mask correctly worn, incorrectly worn face mask or face mask not worn using deep learning techniques. The recent work that has been done for face mask detection has been discussed in this section.

Ting D.S.W. et al. [5] discussed the possible digital solution that can be used for controlling the spread of COVID 19. The IoT platforms can be used to access the data in order to monitor the Covid-19 pandemic. Big data can be used to model the studies and through which the countries can be guided to make healthcare policies. Another technology is using internet and social media to make the general public aware about the situation of the world. Lastly, AI and deep learning can be used to improve the detection of Covid 19.

[6] designed a face mask detection model using hybrid deep transfer learning along with machine learning methods. The Resnet 50 is used to extract the features, and Support Vector Machine (SVM), decision trees, and ensemble algorithm were used for face mask detection. Three datasets were used for the research. The SVM classifier showed the best result among the mentioned techniques, achieved 100% in LFW, 99.49% in SMFD, and 99.64% accuracy in RMFD.

[7] proposed a reliable method for masked face recognition, first the masked region is discarded and pre-trained CNN is used to extract the features from the obtained region, instead of fully connected layer, Bag-of-feature paradigm is applied on the feature maps of the last convolution layer. Multi-layer perceptron is used for classification process. By using third feature map with 60 codewords in the last convolution layer the best recognition rate of 91.3% was achieved.

The authors in [8] proposed a masked face detection approach using LLE-CNN. They addressed two issues of masked face detection, first is limited available region for the processing due to the hidden facial region and second is unavailability of a proper dataset. They introduced a proper dataset of masked face. The proposed model outperforms six other techniques by at least 15.6%.

3   Dataset

Two pre-existing datasets-MaskedFace-Net [9] and Face Mask Lite Dataset (source: Kaggle) are combined for the research. MaskedFace-Net contains 133,783 images of correctly and incorrectly masked images.

Face Mask Lite Dataset contains 20000 images of masked and no masked faces from which we took no masked face images. From both the dataset, 3000 images per class are used for training and 1000 images per class are used for testing purpose. Figure 1 shows sample of the images of the dataset (faces with correctly worn mask, incorrectly worn mask which has been taken from MaskedFace-Net, the faces without mask taken from Face Mask lite Dataset). Figure 2 shows the distribution of the images among three defined classes.

Fig. 1. Sample images of the dataset

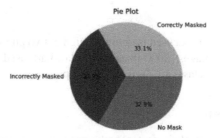

Fig. 2. Distribution of data among 3 classes.

The merged dataset contains three classes that is correctly worn mask, incorrectly worn mask and no mask worn.

4 Proposed Work

A lot of creative ideas has been proposed for face mask detection using various techniques, but they were simply categorized in two categories (Mask is worn/ Mask not worn), whereas, it can be classified as mask correctly worn, mask not worn and mask incorrectly worn, so that the people can be guided accordingly to wear the mask. Figure 3 shows the proposed model. In order to recognize whether a person is wearing a mask correctly or incorrectly or not wearing a mask, an image dataset is provided as input to the system. Next step is to pre-process the image, and then the data augmentation is done, and finally the model is trained.

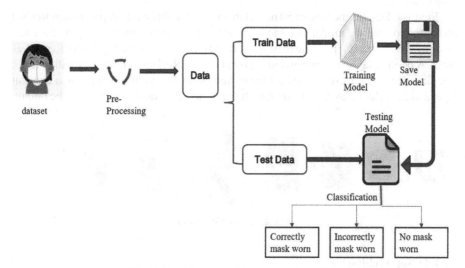

Fig. 3. Architecture of the proposed model

4.1 Data Collection

As mentioned above the dataset used is a combination of two publicly available datasets, from which total 9000 images (3000 images per class) are used for training and 3000 images (1000 images per class) are used for testing purpose.

4.2 Data Pre-processing

Data preprocessing is extremely important as the model is trained on the provided dataset. First corrupted images are removed from the dataset. Next step is to generate various images of the same picture with various angles by rotating, changing the color to grey scale and rescaling, so that the model can be trained on various poses of the same person, for which data augmentation is used.

4.3 Application of Deep Learning Techniques

The learning process that is used in this project is supervised learning, where a dataset is provided to the model to learn, and infer the output itself. For the best experience, four different models are implemented (Resnet 50, Mobile-Net Architecture, Convolution Neural Network and Capsule Neural Network) for comparative performances.

Residual Network 50
Due to the vanishing gradient problem training extremely, deep networks were difficult. But with the help of ResNet very deep neural network with more than 150 layers can be trained efficiently, hence the training of extremely deep neural network becomes an easy task [11].

Figure 4, Depicts the Resnet 50 model that is used in this work. A pre trained ResNet 50 is used to train the model and transfer learning is used to learn the weight of the last layer of the network. The images of the size $224 \times 224 \times 3$ are fed to the ResNet as input. A pooling layer is connected to the output of the ResNet, followed by flatten layer. The first dense layer use Rectified Linear Activation Function followed by a dropout layer and then Dense layer with soft max function which is used to normalize the output.

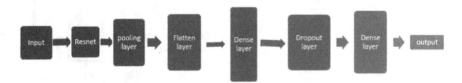

Fig. 4. ResNet50 model

Mobile Net Architecture
It is Convolution Neural Network that can be used for mobile and embedded vision applications. To make light weight and low latency neural networks, they have a stream-lined architecture, which is based on depthwise separate convolutions instead of usual

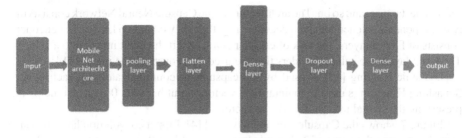

Fig. 5. Mobile-Net architecture model

convolutions. The layers of mobile net are generally followed by batch normalization and rectified linear activation function, then fully connected layer, then sends the data to softmax function to classify it [12]. Figure 5 show the mobile net architecture used.

Convolution Neural Network
It is a Deep Learning algorithm which inputs images, assign them the learnable baises and weights to the different objects available in the picture. The pre- processing needed in a CNN is very less in comparison with other classification algorithms [13]. Figure 6 depicts a CNN. Before feeding the images to the CNN they are converted into grayscale. The first layer is a convolution layer with a kernel size of 3 × 3 and 32 filters using rectified linear activation function with an input size of 24 × 24 × 1. Max pooling is then used to reduce the spatial dimensions of the output. Last convolution layer has 64 filter and a kernel of 3 × 3, again max pooling is used to choose the best features. To improve convergence a dropout layer is used with a rate of 0.25, followed by flatten layer. Dense layer with Rectified Linear Activation Function is added followed by another dropout layer with a rate of 0.5 to converge the output. Lastly a dense layer is added with Softmax activation function to squash the matrix into output probabilities.

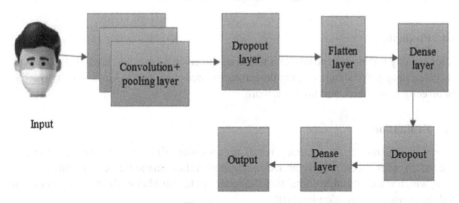

Fig. 6. CNN model

Capsule Neural Network
Capsules are vector specifying the functions of the item and its likelihood. these functions can be any of the instantiation parameters like position, length, orientation, deformation,

speed, hue, texture, and so on. The architecture of the Capsule Neural Network consists of two components an encoder and a decoder. Together, they contain 6 layers. The encoder consists of First 3 layers, the task of encoder is to convert the input image into a vector. The last 3 layers are called decoders that are used to reconstruct the images [14].

Dynamic Routing [15] is used by the capsules to communicate with each other. Squashing Function is used to maintain the vector output between 0–1, it also helps in preserving the spatial information of the vector.

Figure 7, shows the Capsule Neural Network [14]. First a convolution layer is connected to receive the input of 3 channels kernel size of 9×9 and stride of 1×1. Next layer is of primary capsule consisting 8 layers with input channel 256 kernel size 9×9 and stride of 2×2 followed by conventional Digitcaps layer, lastly, the decoder layer is added consisting linear function with input samples of size 48 and output samples of size 512 with a bias value true, followed by Rectified Linear Activation Function having an inplace value of true. Again, a linear function is added with 512 input sample size and 1024 output samples size followed by Rectified Linear Activation Function. Again, a linear function with input sample size 1024 and output sample size 2352 with true bias is added followed by sigmoid function.

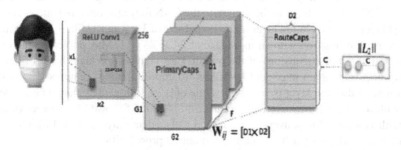

Fig. 7. Capsule neural network

4.4 Prediction

After training the model, the model is tested on the given dataset and predicts the class of the given image. We are classifying the data into three class, that is Mask worn correctly, incorrectly worn mask and no mask worn.

4.5 Evaluation

To evaluate the model, the accuracy, recall, precision and F1-Score of the models are calculated. In order to compute the above-mentioned evaluation metrics, confusion matrix is used, which is a cross table containing the record of the number of the true classification and the predicted classification [10].

5 Experiment Result

The proposed system first pre-process the input data set by manually removing the noisy images and using image augmentation technique to get all the possible inputs of an image. After the preprocessing training models are built and trained on the input set. For testing the trained model, testing data set is fed into the system, which then classify whether a person is wearing mask correctly, incorrectly or not wearing a mask. To evaluate the models, various criteria such as accuracy, precision, recall, F1 score, and testing time are considered. For training and testing google colaboratory tool is used along with GPU. After testing all the models on approximately 3000 images, it is found that the performance of the capsule neural network is marginally better than the other models (Table 1).

Table 1. Results obtained by application of deep learning techniques

S. no	Models	Accuracy	Precision	Recall	F1-score
1	ResNet50	98.88%	98.89%	98.89%	98.89
2	Mobile-Net architecture	91.19%	91.32%	91.11%	91.22%
3	CNN	98.39%	98.49%	98.49%	98.49%
4	Capsule Neural Network	99.0%	99.51%	99.41%	99.51%

ResNet 50 architecture takes around 41 min to train the model, and the testing parameters approximately equal to 98.9% and the time for execution is 1 min 35 s. Figure 8, shows the above-mentioned parameter plotted against the epoch in a ResNet 50 model. For Mobile-Net architecture the time taken to train the model is around 41 min, and the testing accuracy is 91.2%, precision is 91.3%, recall is 91.1%, F1 Score is 90.12% and the time for execution is 1 min 44 s. Figure 9, shows the above-mentioned parameter plotted against the epoch in a Mobile Net architecture model. For CNN, the time taken to train the model is around 43 min 46 s, and the testing accuracy is 98.39%, precision, recall, F1 Score are 98.49% and the time for execution is 1 min 35 s. Figure 10, shows the above-mentioned parameter plotted against the epoch in a CNN model. For Capsule Neural Network the time taken to train the model is around 34 min, and the testing accuracy is 99%, precision, recall, F1 Score are 99.51% and the time for execution is 1 min 35 s. Figure 11, shows the above-mentioned parameter plotted against the epoch in a Capsule Neural Network model.

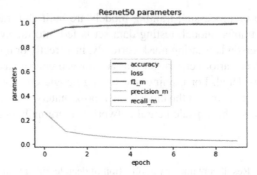

Fig. 8. Plot of the ResNet 50 parameters

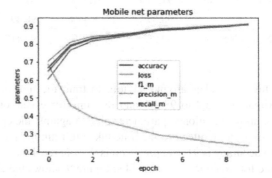

Fig. 9. Plot of Mobile Net architecture parameters

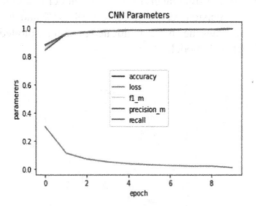

Fig. 10. Plot of CNN parameters

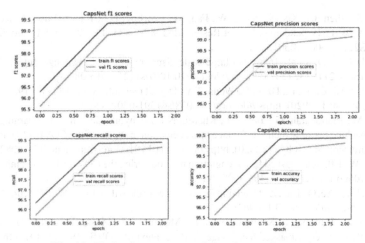

Fig. 11. Evaluation measures of capsule neural network.

6 Conclusion

In this paper, we briefly discussed the motivation of the research, literature review, then we gave a description of the data set and we explained the learning methods for face mask detection. Using deep learning tools, the Capsule Neural Network has achieved high results. Also, a comparative study for different face mask detection methods is also discussed. Upon implementation of the abovesaid techniques and comparative results of implementation, the Capsule Neural Network model is the most effective technique for detection of face mask with 99% accuracy.

7 Future Scope

In future the Capsule Neural Network can be implemented for the real time. Once the data in respect of the same is generated, then with the help of the Dataset, segregation of unmasked people will be done accordingly from the public at large. Once the model identifies who is not wearing Face Mask or wearing the Face Mask but the same is not wearing properly, in that case, the person can easily penalize or counselled for the necessity of face mask. And by all the efforts, to some extent spread of Virus can be stopped and more lives could be saved from the deadly Covid 19 virus.

The model can further be trained to classify how wrong the mask is worn (uncovered nose and mouth, uncovered nose and uncovered chin) as the used dataset have this information as well.

References

1. WHO Coronavirus Disease (COVID-19) Dashboard. https://covid19.who.int/
2. Liu, X., Zhang, S.: COVID-19: face masks and human-to-human transmission. Influenza Respir. Viruses https://doi.org/10.1111/irv.12740

3. Feng, S., Shen, C., Xia, N., Song, W., Fan, M., Cowling, B.J.: Rational use of face masks in the COVID-19 pandemic. Lancet Respir. Med. **8**(5), 434–436 (2020). https://doi.org/10.1016/S2213-2600(20)30134-X

4. Russakovsky, O., et al.: ImageNet large scale visual recognition challenge. Int. J. Comput. Vis. **115**(3), 211–252 (2015). https://doi.org/10.1007/s11263-015-0816-y

5. Ting, D.S.W., Carin, L., Dzau, V., Wong, T.Y.: Digital technology and COVID-19. Nat. Med. **26**(4), 459–461 (2020). https://doi.org/10.1038/s41591-020-0824-5

6. Loey, M., Manogaran, G., Taha, M., Khalifa, N.E.: A hybrid deep transfer learning model with machine learning methods for face mask detection in the era of the COVID-19 pandemic. Measurement **167**, 108288 (2020). https://doi.org/10.1016/j.measurement.2020.108288

7. Hariri, W.: Efficient masked face recognition method during the COVID-19 pandemic (2020). https://doi.org/10.21203/rs.3.rs-39289/v1

8. Ge, S., Li, J., Ye, Q., Luo, Z.: Detecting masked faces in the wild with LLE-CNNs, pp. 426–434 (2017). https://doi.org/10.1109/CVPR.2017.53

9. Cabani, A., Hammoudi, K., Benhabiles, H., Melkemi, M.: MaskedFace-Net - a dataset of correctly/incorrectly masked face images in the context of COVID-19. Smart Health (2020). ISSN: 2352-6483. https://doi.org/10.1016/j.smhl.2020.100144

10. Grandini, M., Bagli, E., Visani, G.: Metrics for multi-class classification: an overview (2020)

11. He, K., Zhang, X., Ren, S., Sun, J.: Deep residual learning for image recognition, pp. 770–778 (2016). https://doi.org/10.1109/CVPR.2016.90

12. Howard, A.G., et al.: MobileNets: efficient convolutional neural networks for mobile vision applications (2017). arXiv:1704.04861

13. O'Shea, K., Nash, R.: An introduction to convolutional neural networks. ArXiv e-prints (2015)

14. Hinton, G.E., Krizhevsky, A., Wang, S.D.: Transforming auto-encoders. In: Honkela, T., Duch, W., Girolami, M., Kaski, S. (eds.) ICANN 2011. LNCS, vol. 6791, pp. 44–51. Springer, Heidelberg (2011). https://doi.org/10.1007/978-3-642-21735-7_6

15. Sabour, S., Frosst, N., Hinton, G.: Dynamic routing between capsules (2017)

Relating Machine Learning to the Real-World: Analogies to Enhance Learning Comprehension

Vishnu S. Pendyala$^{(\boxtimes)}$ ⓘ

San Jose State University, San Jose, CA 95192, USA
vishnu.pendyala@sjsu.edu

Abstract. Machine learning is an exciting field for many, but the rigor, math, and its rapid evolution are often found to be formidable, keeping them away from studying and pursuing a career in this area. Similarity has been substantially explored in machine learning algorithms such as in the K-nearest neighbors, Kernel methods, Support Vector Machines, but not so much in human learning, particularly when it comes to teaching machine learning. In the course of teaching the subject to undergraduate, graduate, and general pool of students, the author found that relating the concepts to real-world examples greatly enhances student comprehension and makes the topics much more approachable despite the math and the methods involved. This paper relates some of the concepts, artifacts, and algorithms in machine learning such as overfitting, regularization, and Generative Adversarial Networks to the real world using illustrative examples. Most of the analogies included in the paper were well appreciated by the students in the course of the author's teaching and acknowledged as enhancing comprehension. It is hoped that the material presented in this paper will benefit larger audiences, drawing more learners to the field, resulting in enhanced contributions to the area. The paper concludes by suggesting deep learning for automatically generating similarities and analogies as a future direction.

Keywords: Machine learning · Nearest neighbors · Learning by analogy

1 Introduction

Machine learning is rapidly evolving as the mortar of modernization. The diverse applications of machine learning are making it ubiquitous in the areas of modernization. Detailed studies such as by the World Economic Forum [1] confirm the increased need for manpower skilled in machine learning and related areas to drive the fourth industrial revolution [1]. The study brings out the need to retrain the existing workforce in addition to training the budding engineers in the new skills. However, the study of machine learning continues to be a daunting prospect for young students and experienced professionals alike. There is a serious need to simplify the discourse on the subject and make it interesting to wider communities. Human beings and machines learn by analogy. Human beings relate new knowledge to what they already know to help in the assimilation of the new knowledge. It is not possible to easily comprehend an abstract concept, completely

© Springer Nature Switzerland AG 2022
K. K. Patel et al. (Eds.): icSoftComp 2021, CCIS 1572, pp. 127–139, 2022.
https://doi.org/10.1007/978-3-031-05767-0_11

new from thin air if it does not resemble or relate to any known metaphor. This is similar to how machines learn to classify the test data, unseen till then, by relating it to the training data that is used to build a model. Analogies, therefore, play a critical role in comprehending complex topics.

The fact that machine learning can be applied to fundamental aspects of our civilization and social fabric, such as the truthfulness of information [2–5] clearly shows the impact it can have on transforming our lives and how intertwined it is with the world we live in. It should therefore not be too difficult to relate the machine learning topics to real-world phenomena in ways that enquiring minds can easily understand. As an experiment, the author took upon himself to teach children and adults ages 8 and up, two areas that have a profound impact on life: the ancient wisdom of the Bhagavad Gita and the cutting-edge technology of machine learning, drawing parallels between the two [6, 7]. Both of them comprise philosophically deep and intellectually challenging concepts. It turns out that there are quite a few underlying principles that are common to both and apply to the real world as well. The talks on the topics were well understood and appreciated by the participants, including two 8-year-old children. It proves that machine learning can be made interesting to many and not intimidating.

The concepts of similarity, data proximity, and nearest neighbors have found tremendous applications in machine learning. Given that the dot product of vectors inherently has the semantics of similarity and matrix multiplications are essentially a series of dot products, it can be concluded that a significant part of machine learning is inherently, learning by analogy. While classification algorithms such as K-Nearest Neighbors and Non-linear Support Vector Machines using Kernel Methods use the concept of data proximity directly, any algorithm that uses the dot product or matrix multiplication is directly or indirectly leveraging the notion of similarity. In that sense, dot product can be interpreted, using a real-world analogy, as a "doting" product that measures how much a vector is like (or fond of) another vector. Matrix multiplication can be explained as a series of "doting" products.

A doctorate in philosophy is often a required or preferred qualification for many teaching positions for a valid reason. Teachers are expected to analyze and explain the underlying philosophy of the subject they teach. This is best done in small increments or deltas, building upon what is already known. Analogies aid in this delta learning. This paper presents some of the analogies that can simplify the concepts and enhance the comprehension of the topics in machine learning. From a literature survey based on searches using relevant keywords, there is no evidence of enough work done in this direction. The next section summarizes the current relevant literature. The subsequent sections discuss some key concepts and artifacts of machine learning, relating them to real-world phenomena and philosophies. The paper ends with a conclusion and future directions.

2 Literature Survey

The literature survey was carried out by using search queries such as '*teaching "machine learning" "real world"*,' '*understanding "machine learning"*' and '*"machine learning" analogy "real world"*'. The search results include textbooks such as [8], which explain how machine learning can be applied to the real world, but not how the machine learning concepts are similar to the real-world notions, indicating that the work in this paper is unique and novel. In an extensive 124-page write-up, Mehta et al. [9] draw some parallels between Physics and machine learning to explain various concepts. Using what they call a "physics-inspired pedagogical approach," they point out that similar to Physics, machine learning emphasizes empirical results and intuition. They compare the cost function to "energy," some of the steps in Stochastic Gradient Descent to the momentum-based methods, the pooling step in Convolutional Neural Networks to the decimation step of Renormalization Group (RG).

The human brain and machine learning algorithms both take high-dimensional data as input and perform classification tasks. In [10], the author touches upon multiple areas of intersection of neuroscience and machine learning when giving insights into his laboratory's research program to decipher the algorithms in biological computations that go on in human and animal brains. Analogies accelerate the pace of innovation. Demonstrating this important hypothesis, the authors of [11] use recurrent neural networks to mine idea repositories, specifically, an online crowdsourced product innovation website, Quirky, to generate analogies. The inspiration from these analogies caused the participants in their experiments to generate better ideas. Drawing an analogy to human learning in a teacher-student setup, authors of [12] propose a machine learning framework for various DNN models that uses far lesser training data and executes faster, in fewer iterations, but still achieves similar performance. Acknowledging the difficulty in teaching machine learning to design students, the authors of [13] describe their attempts at using the Lego Mindstorm NXT platform that the students are more at home with, to explain the concepts.

Machine learning models, to a substantial extent, are opaque, lacking explainability. Addressing this issue, the author of [14] describes how this is a problem for socially significant applications of machine learning. The opacity not only makes it difficult to interpret the results but makes it harder for the students to get deeper insights. Relating the algorithms to real-world experiences, as we describe in this paper alleviates the comprehension difficulties. There is hardly any published research that delves into the problem of teaching machine learning effectively to any population [15]. Detailing the learning objectives and strategies for helping students achieve those objectives, the author of [15] presents insights into educating creative practitioners such as musicians on machine learning topics. The functioning of the human brain has inspired the development of neural networks and deep learning frameworks. The authors of [16] argue that relating infant and toddler psychology to algorithms used in computer vision such as Convolutional Neural Networks may result in newer principles of learning.

Teaching machine learning in K-12 schools is increasingly gaining attention for a good reason. Software skills were introduced in K-12 schools a few years ago to prepare the students for the social environment that then was increasingly becoming computer savvy. A similar need is felt today to prepare the young students for the revolution that machine learning and Artificial Intelligence are bringing in. It is therefore imperative that we invent better ways to teach machine learning to all ages. In their detailed work mapping visual tools to ten years of education in K-12 schools, von Wangenheim et al. [17] present an extensive survey of the tools that can be used to teach machine learning to K-12 school students. Google's Teaching Machine (TM2) is one of them, which is already popular with students. Authors of [18] compare the deep learning that happens in human learning to the deep learning that is part of machine learning to explain the corresponding similarities and differences between the two.

2.1 Contribution

This paper presents a number of real-world analogies to simplify and explain machine learning concepts and paradigms. From a literature survey and to the best of the author's knowledge, this work is the first of its kind. A future direction the author plans for this work is to explore similarity further and using deep learning, automatically generate analogies for difficult topics in any area of study from a corpus of real-world information such as Wikipedia.

3 Machine Learning Parallels to Real World

Learning, whether in humans or machines follows similar principles. A newborn learns to recognize his father and mother using features and labels. When the newborn is pointed to a person labeled "dad," the baby collects the features of this person through her sensory organs. The features may be that the person labeled "dad" is taller, has facial hair, wears shirts, with a deeper, low-pitch voice, has short hair on head, and so on, whereas the person labeled "mom" may be a bit shorter, has a high-pitch voice, does not have any facial hair, but has long hair on head and so on. As the baby sees more persons, her brain realizes that these features are not constant. They vary. The variability of features, such as the length of the hair or the pitch of the voice is not predictable. A new aunt walks in with a randomly varying voice pitch. The feature varies randomly. In mathematical terms, the baby starts to model each feature as a random variable as she examines more and more people. The label also varies and is the target of the learning exercise, so is called the target variable.

3.1 Logistic Regression

In the above analogy, people are rows of data, their features are columns in the table, and the feature is modeled as a random variable. As more people walk in, the baby realizes that not all features are equal. One of the uncles had long hair too, implying the weight for the feature, 'length of the hair' may not be as much, say as the weight for the feature, 'has facial hair'. Once the analogy is laid out, math automatically follows. A feature, modeled as a random variable is associated with a Probability Density Function (PDF). The 'label' depends on the features, each of which is associated with a weight. The simplest dependency is when the label varies linearly with each feature. The label can therefore be expressed as a weighted sum of features, as we studied in high school. It must be noted that a weighted sum is a dot product of the transpose of the vector of weights and the vector of features. Thus similarity, which the dot product is often a measure for, is fundamental to machine learning. If the label can take only two classes such as 'man' and 'woman,' the label, also called the target variable, can be modeled as a probability of a person being a man or a woman. The weighted sum, therefore, needs to be converted into a value between 0 and 1 to represent a probability and an established way to do it is by passing it through a logistic function. The machine learning model in this case is appropriately called Logistic Regression. The label in this case is a logistic function of the weighted sum of the features.

3.2 Loss Function

The above analogies map the material world in the problem domain to mathematical artifacts. Therefore, all the features involved in the domain need to be converted to numbers so that we can do math with them. When the conversion is done, we need an apples-to-apples comparison. For instance, if our problem is to classify a person as a man or a woman based on height and weight, a difference of 5 inches. in height and difference of 30 lb in weight are numerically very different. If Euclidean distance is considered as a measure of similarity, both 5 and 30 are squared, giving both the differences the same importance, and the square root of their sum is used to compare the respective persons. This is erroneous because the weight difference is unduly influencing similarity. The numbers therefore need to be normalized.

A reputed engineering school, which accepts students from various states of the country based on their scores given in their respective states normalizes the scores. This is because some states may be liberal whereas some are more specific. The numbers representing the various features of the data similarly need to be scaled appropriately using normalization techniques. Once the features are all normalized numbers, the next problem is to compute the weights. From the discussion so far, it is clear that we cannot always predict the label accurately, given the nature of random variables and probability. Therefore, a machine learning algorithm such as Logistic Regression comes with a cost. In real world, from a business perspective, the loss is computed as the difference between the sale price and the cost price. In the case of machine learning algorithms, a simple way to think of loss is to view it as the difference between the actual label and the label predicted by the algorithm. It is more a measure of how erroneous the model functioned.

Loss can be seen as a feedback to the system. A business views loss as an important lesson to learn from and takes measures to avoid or minimize the loss. In the case of machine learning algorithms, loss avoidance is not possible because 100% accuracy is never possible. Loss can only be minimized. Loss in machine learning is also a feedback or a lesson back to the system. Just like a mild rebuke may not always work on a student, a simple loss does not have enough effect. The loss, therefore, needs to be a bit more involved than a simple difference between the actual label and the predicted label. A slightly more sophisticated loss is when the difference is squared. There are more sophisticated loss "functions," that can be used to make the algorithm learn faster, but the fundamental factor in all these loss functions is the basic difference between the actual label and the predicted label. It must be noted that the loss function cannot be made complex beyond a level, just like excessively harsh feedback beyond a certain point can demoralize students and have a negative effect. Well-balanced feedback, on the other hand, can motivate the student to learn better. Same applies to the loss functions in machine learning. We need to use a loss function of appropriate complexity and semantics.

The goal of a machine learning algorithm is to minimize the loss, similar to that of a business. We know that the predicted label is a function of the weighted sum of the features. By the time we compute the value of the loss function, we know the values of the features and the actual labels. The only unknown in the loss function is therefore the weights. Loss function becomes an equation with weights as the unknown variables. Finding the weights can therefore be treated as a multivariate maxima-minima problem in partial differential calculus, with the goal being to minimize the loss. This can be accomplished by equating the partial derivatives of the loss function with respect to the weights for each feature, to 0, just like in any other minima problem in calculus.

3.3 Universal Approximation Theorem

We often encounter self-starters and serial entrepreneurs, who have a goal in mind and achieve that goal or at least come close to it over some time against all odds. Such go-getters rise quickly in career and are widely accepted. Once they master the art of succeeding, it is just a matter of applying the learned model to different scenarios. Determined students prepare for a certification or competitive exam over a period and ace the exams. In such cases, a goal can be seen as one or more outputs. The student aiming at the goal takes inputs such as textbooks, courses, blogs, and videos, works with them repeatedly over some time, possibly going back and forth reading books, watching videos, and eventually masters the concepts.

Artificial Neural Networks (ANN) can be thought of similarly. They accept inputs, work with them for some time and produce the desired results, approximately. Computer Science is mostly applied math. The artifact that takes inputs and produces outputs is called a function in math. Since ANNs can approximately do that for any combination of inputs and outputs, they are considered universal function approximators. Universal because this applies to any combination of the inputs and outputs. Function because they behaved like a math function that takes inputs and produces outputs. We know that the results are approximate. The self-starters master the art of succeeding. ANNs master the art of learning to produce any desired output from the given inputs over many back and forth iterations over the entire dataset called epochs.

3.4 Bias

Logistic Regression helps determine the label or class of the given data. This is just one way to come up with the label and as described above, is not free of errors or 100% accurate. Sometimes, we tend to judge a person's behavior and 'label' the person without giving enough thought and consideration. Such simplistic assumptions are called biases. For instance, we often find a bias against a particular age group, gender, or ethnicity arising out of simplistic assumptions. The fact is that the behavior of human beings is far beyond the simplistic assumptions made based on age group, gender, or ethnicity. Therefore, our preconceived hypothesis based on simplistic assumptions is prone to error. Similarly, when a machine learning algorithm makes a simplistic assumption, it suffers from bias, which is one type of error. For instance, Logistic Regression makes a simplistic assumption that the label is linearly dependent on the features and classes can be determined by a predefined function. This results in an error attributable to the bias.

Table 1 below lists a few common assumptions in machine learning and the real world, which result in bias.

Table 1. Simplifying assumptions in ML models and the real world which result in bias (error)

Machine learning models	Real world
One line or hyperplane fits all data points to use linear regression for prediction	People of a particular gender or ethnicity or religion are all behaviorally the same
Features of the data domain are all independent of each other	Judging a person, place, or thing by the first or the best or the worst impression
Each data point is independent of all other data points or observations	Overconfidence in oneself and the decisions one makes

3.5 Overfitting, Variance, and Occam's Razor

There is another type of error in machine learning that is attributable to overfitting. This occurs when the machine learning algorithm closely models the data. It is similar to when people give undue importance to their personal beliefs, values, and perceptions without seeing the big picture. In such cases, two people starting their lives together do not get along after a few days if their value systems are different. In machine learning parlance, the pair has overfitted to the limited value system and behavior they are used to. Algorithms do a similar mistake when they attach undue importance to the limited data they are given. This could manifest as large weights or complex mathematical functions to model the data. Occam's razor, which is widely applicable to many phenomena applies to machine learning as well, because of which, we need to consider model complexity when evaluating a machine learning model and prefer simpler models, which is one way to avoid overfitting.

We often find persons who are highly admired or liked in their respective circles in which they grew up but find it difficult to adjust and get acceptance when they join a newer circle such as by way of a marriage. Similarly, machine learning models, which perform well with the training data based on which they are modeled, but do not perform as accurately with new data, called test data, are said to suffer from "variance." Variance is the other type of error that impacts machine learning models. Variance is typically a result of overfitting the machine learning model to the training data. When the variance component of the error is absent or minimum, the model is said to generalize well, similar to the case of a progressive upbringing of a person.

3.6 Regularization

Spirituality suggests that one should not be too attached to anything in this world. That is, the weights we attach to physical happenings and the impact we perceive should not be high. The Sanskrit word, "Hari" that is often used for meditation and spirituality means "reduce". Spirituality reduces the weights we attach to the physical happenings around us. In machine learning, this functionality is achieved by what is called "regularization." Regularization reduces weights by adding a smaller function, called the regularizer, to the loss function at the time of finding its minima. When this new loss function with an added regularizer function is minimized using partial derivatives, the resulting weights are smaller compared to the weights obtained without adding the regularizer. This is similar to how a person depressed because of intense worry is suggested to develop a minor avocation to distract herself from deep, depressing thoughts, which are weighing heavy on the mind. Depression is a result of attaching huge weights to one or more happenings or features of life. The weights can be reduced by adding an activity that lightens up the day.

4   Machine Learning Types and Algorithms

Machine learning algorithms are associated with considerable math that can intimidate new learners. Drawing parallels with the real world is important for improving comprehension in this area. Table 2 lists a few analogies that help simplify the intuition behind machine learning concepts, types, and algorithms.

4.1   Supervised and Unsupervised Learning

Learning is said to be supervised, when the data is already labeled, much akin to a child learning to recognize the world using the labels that the parents attach to the entities that they want their children to know about. A child is supervised to learn about the world around her through the labels such as "good" and "bad" for a behavior or entity. Using these labels, the child develops a habit of attaching relative importance or weights to the features of these entities or behaviors she observes. These weights remain in the child's mind as impressions or models. A person can be modeled and predicted based on the weights or impressions in his mind. When the weights or impressions in a person's mind are fully known, the person is mostly predictable. The advertisement world on the Internet runs on this premise. Internet companies such as Google and Facebook predict what advertisements a visitor to their websites may like based on the information they can gather about the visitor. This information is a result of the impressions in the web surfer's mind and the weights she attaches to the various entities she encounters surfing the web.

Similarly, a Supervised machine learning model typically comprises the weights it learns from the labeled data, also called the training data. A machine learning algorithm has learned the model from the training data just like a person learns the relative importance or weights from the labeled entities and behaviors he has been trained on since childhood. When a child grows up to college age and starts to stay in a dorm by herself, parents are no longer available to label the entities such as her classmates or their behavior as good or bad. That is when the child's learning becomes unsupervised. A simple unsupervised task for the child may be to form groups of friends. In machine learning, this unsupervised process of forming groups of data is called clustering. Unlabeled data is grouped into clusters based on similarity.

Most of the learning in a human being's life happens unsupervised. It took many years for the researchers to realize that unsupervised learning holds the true promise of human-level machine intelligence. Recent Turing Award winner, Yann LeCun [19] in 2020, called a form of unsupervised learning, termed self-supervised learning as the future of machine learning. A simple analogy with human learning may have revealed this much earlier. This is one simple example to indicate the power of similarity and analogies in learning and discovery.

Table 2. A few parallels between machine learning and the real-world

Machine learning concept	Real-world/Simpler analogy	Similarity
Supervised learning	Parents labeling what is good and bad for children	The world is already classified
Unsupervised learning	Students forming groups without any supervision	Lack of labels
Matrix multiplication	Series of "doting" products	Similarity of two vectors/entities
Maximum margin classifier	Arbiter	Equidistant from either class
Artificial neural network	Self-starters, Achievers	Mastering the art of achieving targets
Hidden layers in artificial neural networks	Departments in an organization	Division of labor
GAN	Akinator game	Adversarial nature
Lazy learning	Last-minute exam preparation	No preparation in advance
Boosting	Student improving exam over exam	Focus on past failures
Overfitting	Narrow-minded	Undue importance to limited artifacts
Regularization	Spirituality	Reduce weights attached to features
PCA	Caricature	Capturing the variance

4.2 Support Vector Machines

An arbiter or a judge of a court of law is required to be equally unattached to the disputing parties. This concept is imbibed in the philosophy of Linear Support Vector Machines. A decision boundary in binary classification is essentially acting as an arbiter between two classes. The decision boundary in Support Vector Machines (SVM) is the Maximum Margin Hyperplane (MMH), which is equidistant from the data points at the boundaries of both classes.

4.3 Lazy and Eager Learners

Students tending to procrastinate and put off preparations until the last minute before the exam are lazy learners. In machine learning, algorithms such as the k-nearest neighbors (KNN) do not process the data until the test data arrives, qualifying them as lazy learners. On the other hand, most machine learning models such as logistic regression, SVM, and ANNs do most of the work before the test data arrives, much like the proactive students who prepare for the exams well in advance. In the latter scenario, the models in the mind in the case of the student or some serialized fashion in the case of the machine learning model are ready to be used when it is time to test. Also, when the training data is noisy, if using the KNN algorithm, the value of k needs to be high. This is analogous to when rumors and inaccurate information is floating around such as on the Web, one needs to depend on more information sources than when the information is all true and accurate.

4.4 Ensemble Methods

Ensemble methods combine the efforts of multiple models to cover their individual lacunae, much like in the civilized world where multiple people come together to overcome their individual weaknesses and leverage group synergies. It is an established finding in management that diversity improves team dynamics and the overall quality of the output. Much in the same way, the trees in a random forest are designed to be as different as possible. In sequential ensemble methods like boosting, there is increased focus on the past mistakes, much akin to the way students focus on the mistakes they committed in their previous exams to score better in the forthcoming exams.

4.5 Principal Component Analysis

When drawing a caricature of a celebrity, a cartoonist exaggerates the unique features of the celebrity, while downplaying the common aspects. For instance, if the celebrity has a relatively long nose, the cartoonist exaggerates the nose so that the celebrity can be easily identified from the caricature even if it is not a true depiction of the celebrity. It is much easier and faster to draw a caricature than a portrait. It still serves the purpose of identifying the celebrity in the context. What the cartoonist did in her mind is a Principal Component Analysis (PCA) of the celebrity's face. She identified where the celebrity varies the most and predominantly captured those principal components of the celebrity's face. In machine learning, principal components are determined by computing the eigenvectors of the covariance matrix of the attributes.

Many cultures and people have favorite directions. For instance, some pray facing east, while some others pray facing in the direction of their holiest shrine. Similarly, square matrices have favorite directions. This is the direction of their eigenvectors. When a square matrix is multiplied by a vector, the result is usually a vector with a changed direction and magnitude. The square matrix spends its transformational energy in both rotating and scaling the original vector to produce a new vector. However, when the same square matrix is multiplied by the vectors in its favored directions, all of the square matrix's transformational energy is spent in scaling the vector, increasing its magnitude, while doing nothing to rotate it. The resultant vector is stretched in the favorite direction but its direction is the same as the original vector. The first principal component of a given dataset is in the direction of the most variance in the data. That direction is naturally the most favored direction of the covariance matrix of the dataset, which is the direction of its first eigenvector.

4.6 Artificial Neural Networks and Deep Learning

Artificial Neural Networks (ANN) are biologically inspired and offer many analogies from the real world. For instance, each hidden layer in an ANN accomplishes a minor task in the overall solution to a problem, reminiscent of the division of labor in the industry. The hidden layers can be thought of as divisions in an organization, each responsible for a chunk of the overall mission of the organization. Each neuron in an ANN can be compared to an ant. A single neuron, like a single ant, may not be able to achieve much, but a collection of neurons can do wonders much like a colony of ants can kill the strongest serpent, when they work in tandem.

Interspersing the ReLU activation function in ANNs provides the much-needed non-linearity and boosts the power of ANNs significantly, much similar to how short breaks and context switches can help refresh minds and energize thoughts. Quite a few deep learning frameworks can be compared to parts of our brain that have a similar function. For instance, Convolutional Neural Networks (CNN) used for computer vision are comparable to the occipital lobe of the brain and Recurrent Neural Networks (RNN) have a similar function as the frontal lobe of the brain.

More advanced deep learning frameworks such as Generative Adversarial Networks (GAN) can be explained using analogies too. The discriminator acts like a parent teaching a toddler to write the letters of the alphabet. The generator, like the toddler, generates noise or gibberish at first. The discriminator gives feedback for the generator to better itself much like the parent does to the toddler. After many iterations, both the generator and the toddler learn to produce almost accurate output. Another analogy is the game of Akinator, sometimes called Bulls-eye, where one player or the system in the case of Akinator, thinks of a person and the other player keeps guessing who the person in the first player's mind by asking questions that can only be answered in either a yes or a no. After many iterations of questions and yes/no answers, the second player can guess the character correctly.

5   Conclusion and Future Directions

Time and again, software concepts continue to draw inspiration from the real world. Deliberately or unconsciously, many computer science artifacts bear a striking resemblance to the happenings in the real world. Similarity is probably the single most important underlying principle of machine learning. From a linear predictor to advanced deep learning frameworks, all use dot products. Dot product that is ubiquitously used in machine learning is a measure for similarity. It can therefore be concluded that machines predominantly learn by way of similarity. However, this fundamental way of learning remains unexplored to a significant extent in human learning of difficult topics like machine learning itself. This paper attempted to start filling that gap. When the subject is challenging as is the case with machine learning, it helps to draw parallels to the concepts that the students are already familiar with to help explain the underlying philosophy.

Accordingly, a number of real-world analogies for machine learning concepts have been discussed in this paper. All the analogies are based on human intuition and ingenuity. A future direction for this work is to evaluate the feasibility of automatic generation of analogies, not just for machine learning topics, but for any advanced subject with hard-to-understand concepts. The similarity is a fundamental notion in machine learning. Using the right type of topic, language modeling, and NLP techniques, it may be possible to discover similarities automatically between topics using deep learning, particularly given the universal approximation theorem. It is hoped that this first of its kind work will open up exploring the similarity angle of human learning using both automation and human ingenuity.

References

1. World Economic Forum: Data Science in the New Economy. http://www3.weforum.org/docs/WEF_Data_Science_In_the_New_Economy.pdf. Accessed 20 Sept 2020
2. Pendyala, V.: Veracity of Big Data: Machine Learning and Other Approaches to Verifying Truthfulness, 1st edn. Apress, San Jose (2018)
3. Pendyala, V.S., Figueira, S.: Towards a truthful world wide web from a humanitarian perspective. In: 2015 IEEE Global Humanitarian Technology Conference (GHTC), 8 October, pp. 137–143. IEEE (2015)
4. Pendyala, V.S.: Securing trust in online social networks. In: Sahay, S.K., Goel, N., Patil, V., Jadliwala, M. (eds.) SKM 2019. CCIS, vol. 1186, pp. 194–201. Springer, Singapore (2020). https://doi.org/10.1007/978-981-15-3817-9_12
5. Pendyala, V.S.: Evolving a truthful humanitarian world wide web. https://scholarcommons. scu.edu/eng_phd_theses/18. Accessed 20 Sept 2021
6. India West: Free classes on Bhagavad Gita and other ancient scriptures with technology interludes. http://www.indiawest.com/calendar/free-classes-on-bhagavad-gita-and-other-anc ient-scriptures-with-technology-interludes/event_a549b5b8-bef7-11ea-a4bf-308d99b28daf. html. Accessed 20 Sept 2020
7. Pendyala, V.S.: [Video Playlist]. Bhagavad Gita 101 with technology Interludes for ages 7 and up. https://www.youtube.com/playlist?list=PLrbG5zg_L7VKj79UBsBAXbpT1hn4R kG9a. Accessed 20 Sept 2020
8. Brink, H., Richards, J.W., Fetherolf, M., Cronin, B.: Real-World Machine Learning. Shelter Island, Manning (2017)
9. Mehta, P., et al.: A high-bias, low-variance introduction to machine learning for physicists, p. 810:1–24. Physics reports (2019)
10. Helmstaedter, M. The Mutual Inspirations of Machine Learning and Neuroscience. Neuron **86**, 25–28 (2015)
11. Hope, T., Chan, J., Kittur, A., Shahaf, D.: Accelerating innovation through analogy mining. In: Proceedings of the 23rd ACM SIGKDD International Conference on Knowledge Discovery and Data Mining, pp. 235–243 (2017)
12. Fan, Y., Tian, F., Qin, T., Li, X.Y., Liu, T.Y.: Learning to teach. arXiv preprint arXiv:1805. 03643 (2018)

13. Van Der Vlist, B., et al.: Teaching machine learning to design students. In: International Conference on Technologies for E-Learning and Digital Entertainment, pp. 206–217. Springer, Berlin (2008). https://doi.org/10.1007/978-3-540-69736-7

14. Burrell, J.: How the machine 'thinks': understanding opacity in machine learning algorithms. Big Data Soc. 1–12 (2016)

15. Fiebrink, R.: Machine learning education for artists, musicians, and other creative practitioners. ACM Trans. Comput. Educ. (TOCE) **19**(4), 1–32 (2019)

16. Smith, L.B., Slone, L.K.: A developmental approach to machine learning? Front. Psychol. **8**, 2124 (2017)

17. Gresse von Wangenheim, C., Hauck, J.C.R., Pacheco, F.S., Bertonceli Bueno, M.F.: Visual tools for teaching machine learning in K-12: a ten-year systematic mapping. Educ. Inf. Technol. **26**(5), 5733–5778 (2021). https://doi.org/10.1007/s10639-021-10570-8

18. Webb, M.E., et al.: Machine learning for human learners: opportunities, issues, tensions and threats. Educ. Tech. Res. Dev. **69**(4), 2109–2130 (2020). https://doi.org/10.1007/s11423-020-09858-2

19. Hinton, G., LeCunn, Y., Bengio, Y.: AAAI'2020 keynotes Turing award winners event. https://www.youtube.com/watch?v=UX8OubxsY8w. Accessed 10 Aug 2021

Sentiment Analysis of Twitter Data Using Machine Learning Approaches

Vishal Gaba$^{(\boxtimes)}$ (ID) and Vijay Verma (ID)

National Institute of Technology Kurukshetra, Kurukshetra, Haryana, India
gabavishal77@gmail.com

Abstract. Twitter is one of the most widely used social media microblogging platforms. It generates a large amount of opinioned data every day in the form of tweets. This makes it a suitable choice for sentiment analyses and many other studies. This work investigates different machine learning approaches for analyzing sentiments of tweets aiming at enhanced accuracy. In particular, we have suggested a framework that is based on a hybrid neural network model by combining Convolutional Neural Network (CNN) and Bi-Directional Long Short Term Memory (BiLSTM). Twitter API is used to collect tweets with specific keywords related to sports, politics, news, healthcare, and technology. While fetching data, hashtags and emojis are used, which are later converted into text form for assessing their effect on overall performance. The textual data is encoded using modified TF-IDF and Glove Vector embedding in Neural Network (NN) models with vectors pre-trained on Twitter data. A comparative analysis is also presented for the various machine learning algorithms (SVM, Naïve-based, Decision tree, KNN) with different datasets (IMDB, CMRD, 1.6M Kaggle, and custom-built datasets) based on accuracy and other metrics such as f-score, precision and recall. A considerable improvement is observed when hashtags and emojis are utilised with the hybrid CNN-BiLSTM model.

Keywords: Sentiment analyses · Machine learning · CNN · LSTM · Feature selection · Hashtag · Emoji · Hybrid classification

1 Introduction

Twitter is a famous microblogging platform that allows its users to send and receive short posts called tweets. It has been used in various studies due to its ever-increasing rate of data generation and a huge pool of users. In June 2019, Twitter reached more than 348M monthly active users, which created about 500M tweets a day [1]. This makes it ideal for any study that involves behaviour analysis or one that needs to analyse social or industrial impact. Sentimental analysis is one such study that aims to determine public's opinion. Patterns identified in such an analysis could provide meaningful analyses. Such studies have applications in various domains such as marketing, political domain, product feedback, behavioural analyses etc. [7, 11]. Recent findings [6, 9] also show that trend changes are also linked with the sentiment of a topic. Thus, it could also improve

K. K. Patel et al. (Eds.): icSoftComp 2021, CCIS 1572, pp. 140–152, 2022.
https://doi.org/10.1007/978-3-031-05767-0_12

the efficiency of trend detection algorithms. For this, sentiment prediction must give us results with decent accuracy. Although there are existing techniques for sentiment analyses, analyses of best techniques and identifying scope for improvements in currently adopted methods could prove to be of great significance. A general approach for collecting and processing data is shown in Fig. 1. After getting text data, we must ensure it is relevant and in a proper format. For this, firstly, subjectivity analyses need to be done to keep opinions and remove factual information, after which the text is classified for sentiment polarity, i.e. positive or negative. This is popular as a two-step classification approach. Recent approaches use text classifiers such as Support Vector Machine (SVM), Naïve Bayes (NB), Logistic Classifier and Random Forest (RF) for sentimental analyses purpose [7, 24]. Studies analyzed show good accuracy for SVM and NB for text analyses, with a preference for NB, especially for large data and for the Twitter platform.

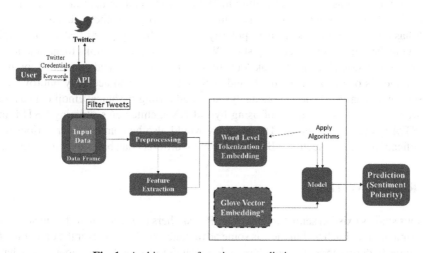

Fig. 1. Architecture of sentiment prediction system

While training various models for sentiment prediction, the text is first converted in numerical form. Word embedding provides meaningful information about words and their correlation. Table1 shows a representation of word embedding. This embedding represents words in vector form, and words with similar vectors are more likely to be related to each other. For instance, 'cat' and 'dog' will have similar vectors (both are animals), and 'good' and 'bad' will have opposite vectors. This would introduce meaningful information about words and their correlation rather than just a count of their occurrences. Various approaches for sentimental analysis have also come up, such as RNN (Recurrent Neural Network), CNN, and their hybrid architectures. Combining information from these two architectures, i.e., LSTM and CNN, is a promising approach as it could utilize features from both models [2, 6]. Such methods in text classification would help increase our classification accuracy of the test data. These architectures, related methods and their effects on a general and specific dataset are to be analyzed.

Table 1. Word Embedding using Glove [20] 25d with a 6-word sentence. This would later be used as weights while training NN models.

	1	2	3	4	5	...	...	24	25
The weather is good today!	−0.01	0.02	0.21	0.17	−0.44	...	...	−0.84	−0.31
	−0.86	−0.70	−0.70	−1.26	−0.20	...	...	0.03	0.09
	−0.13	−0.20	−0.13	−0.57	−0.30	...	...	0.25	−0.06
	−0.54	0.60	−0.15	−0.02	−0.14	...	...	0.93	0.02
	−0.83	0.02	−0.06	−0.92	−0.22	...	...	0.45	0.17
	0.40	−0.88	−0.23	−0.35	−0.10	...	...	1.20	1.16

Utilizing hashtags, a special feature in Twitter is yet another approach to get popularity [7]. Emoji - symbolic depiction of emotions, objects or symbols can be used along with hashtags to improve sentiment polarity accuracy like in [7, 8] instead of simply removing them in the preprocessing stage. We also make use of emojis to automatically annotate our dataset. We plan to make our study more exhaustive and propose to compare certain factors over various methods and analyse their use and accuracy improvement on sentiment analysis. The factors analysed include using hashtags, emoji data as text in tweets, as well as the effect of using hybrid NN architectures utilizing LSTM and BiLSTM to increase accuracy. The accuracy would also be compared for various text classification methods such as NB, SVM, CNN for different databases over these factors.

2   Related Work

Various studies exist on sentiment analyses. Researchers use varied in different stages of experimentation. Each technique has some advantage or drawback over the other one [6, 7]. Major tradeoffs include computational cost and accuracy. Recent approaches have shown good results from Machine Learning (ML) models such as CNN and LSTM for text classification as well [6].

2.1   Data Collection and Classification Models

Twitter API is used to collect data. This can also be done over a period of days or even months using online methods such as Amazon Web Services Elastic Cloud Compute (AWS EC2) [6, 8] or in one go for a smaller dataset [11]. There are a variety of ways and algorithms adopted for sentiment analyses by different authors. The author in [7] uses unigram, hashtag polarity, and a lexicon build for online reviews as features to get polarity. Accuracy of individual features and all features combined over NB, SVM, LR and RF algorithm was analysed. All features, when combined, give highest accuracy with RF and LR. NB, combined with the hashtag feature, was the model of choice due to its high accuracy and low computational cost. This was followed by SVM.

2.2 Sentiment Analysis on Twitter

Several approaches exist for performing sentiment analysis, specifically on Twitter. There are many variations in the techniques applied over various stages, from fetching the data and extracting features to finally building a classifier model for sentiment prediction. Preprocessing is also a crucial stage; features such as hashtags and emoticons must be preserved to be used as features in later stages, as indicated in various studies. For instance, in [7, 8], hashtag terms and emoticons are used for enriching the text instead of removing them while preprocessing. User mentions and retweets were not considered as they did not add any semantic value to the data. Consider following (Table 2):

Table 2. Example of terms and corresponding conversions

Emoticons to corresponding emotion:	:), :D, :-) → Happy ‖ :(, :-(, >:(→ Sad
Hashtag keywords to separated words:	#Congratulations for your achievement! #GodBless → "Congratulations for your achievement! God Bless"

The domain of these analyses can include election campaigns, stock prediction or marketing, according to which. Depending on domain and use, factors such as location or time may be of importance. For instance, in [4] author uses increasing or decreasing positive sentiment for a particular political figure to detect the result of election campaigns in specific areas. A unique approach of using hashtags as features was seen in [7], where the sentiment of a tweet was decided based on the sentiment polarity of the hashtag. Specific tags were appended to tweets based on the count of positive and negative hashtags. The count of these tags helps in annotating the dataset. Another approach is populating a dataset's sentiment values based on emoticons as in [7]. This results in a large dataset within a short period of time. Including this approach in [7] and the use of emoji makes our study more comprehensive. In [11], username was used to download data, with dataset size reduced further after preprocessing phase. Here, to detect a trend, TF-IDF technique is used, after which machine learning algorithms are used to predict mood. In [6], where a CNN-LSTM hybrid was used for sentiment classification.

3 Challenges and Problem Statement

There are certain challenges associated with online data for further processing and accurate analysis. These may be specific to each online platform. This section addresses some of these challenges. Tweet data obtained from API could be from any location, of any language and from any user. So, we need to specify API parameters suitable to our requirement as in [8]. We search for data on a topic by using specific and manually selected keywords related to the topic. Data thus obtained is in real-time but cannot be used directly for further analyses due to noise. Unused data from Twitter additionally includes user mentions, likes, retweets, count of followers, shares etc. [3] Most twitter data consist of such elements instead of data in word form [13]. Other elements, such as hashtags and emoticons, are also treated as noise in most analyses but are argued to cause a significant increase in accuracy when tweaked to be included in an acceptable form with the collected data. Moreover, such data is not the only challenge for noisy data, but objective or more factual tweets would also act as noise since they would not have any significant sentiment value. Some features and their related challenges were dealt with while preparing data and for further analysis and results. These include "Retweet, Likes and Shares", "Hashtags", "Emoticons", "Negation Words", and "Abbreviations". Retweets are a copy of other users' opinions and cannot be concluded as the opinion of the user who has posted the tweet. Hashtags, words preceded by the # symbol, although removed in certain cases while processing data, helps in sentiment classification and trend detection. Emoticons, though, are not text, but the meaning of symbols can be added to enrich the text. Negation words such as not may change the sentiment value but may not be detected. E.g. 'Not good' may still be treated as good in sentiment analyses. Abbreviations can be converted in full form for analyses, e.g. 'lol' expresses laughter. It may not be difficult for humans to understand a sentence and its context to interpret its meaning and sentiment. It is still challenging for a computer machine to do the same. So, any techniques which detect occurrences of words and rank them (e.g., TF-IDF) may still not be able to get the meaning or context of the sentence. Twitter imposes further challenges with its limit of 250 words. Thus, an accurate interpretation of sentiment remains a challenging task.

4 Proposed Method

This study aims to utilize techniques for sentiment polarity and trend detection. Certain unique techniques were identified while going through the existing tools and methods. A few notable contributions are explained below, with their implementation and usage.

4.1 Data Preprocessing

Data preprocessing deals with preparing the data in a form suitable for further analysis. We have combined the best techniques from existing studies, including hashtags and emojis as text to enhance outcomes. Details of data preprocessing and its stages are explained in Sect. 6.1.

4.2 Annotation of the Dataset

To train our classification models, the dataset needs to be annotated with a target class. Manual annotation for each tweet as positive, neutral or negative becomes a tedious and time-consuming task, especially for large datasets. For our study, we have adopted a unique approach to automatically annotate the dataset. We automatically annotate data based on whether a particular emoji is present in the tweet. Tweets with emojis related to happiness are marked as one, and tweets with emojis related to anger or sadness are marked as 0. Our approach is novel in terms of data collection as well as data annotation. This approach saves time and effort and thus, allows us to increase the size of our dataset.

4.3 Modifications with TF-IDF

TF-IDF deals with the frequency of words in a given document and occurrences over all documents giving higher value to those words with higher occurrences within a document and fewer occurrences across multiple documents. This technique also removes conjunctions and parts of speech that are not so significant for our analyses, like 'is', 'and', 'the' etc.

Tf-IDF may be regarded as a good input for training models in many NLP tasks, such as detecting important topics in a review or even training sentiment analysis models [24]. In our approach, a novel technique is developed to improve its use for sentiment classification models. According to our approach, words with larger sentiment intensity must be given larger weightage than words having no sentiment intensity. E.g. "I won't say the location, but instead, the arrangement was terrible". Here the words 'terrible' must be given more weightage even if it has a low TF-IDF score. The following formula is used to add value to tf-idf values:

$$Tf' = Tf + (S/5) * SD \tag{1}$$

where S is sentiment intensity from a scale of -5 to $+5$, SD is the Standard Deviation of the TF-IDF matrix. This increases the Tf-IDF value linearly, according to the sentiment intensity value of the word. Afinn library gives sentiment intensity on a 5-point scale. Vader and Text Blob are also be used to increase words vocabulary and analysis of Afinn.

5 Tools and Techniques

For our analysis, we use Python version 3.6.9 as our programming language. Modules and packages in python promote modularity and code reuse. We use Colab, a free cloud service that allows us to run code on a browser and is well suited for ML applications. Colab also has a free Graphics Processing Unit (GPU) feature for faster execution. Additionally, Natural Language Toolkit 2.0 (NLTK) is used to help in building the classifier. It includes a suite of text processing libraries for classification, tokenization, tagging, parsing, and other functions. We use Keras API as a sub-module of the TensorFlow library. Scikit-learn library is used for implementing classification algorithms.

6 Data

6.1 Dataset Creation

Data was downloaded via Twitter API using the tweepy library in python. We first set up 'OAuth', a protocol that allows secure authorization to access an API. We have to specify certain app credentials to collect Twitter data (consumer_token, consumer_secret etc.). However, while downloading data, the tweets were fetched using specific keywords for various topics related to sports, technology, news, politics and health. Also, positive and negative emojis were kept in the search filter so that tweets contain emojis too. Tweet data, once downloaded, contains many attributes in addition to text, such as tweet ID, location, URLs, count of likes, retweets etc. From these, we will use the 'full text' attribute in a normal tweet and 'retweet_status.full_text' in case of a retweet to fetch a tweet's complete text. Hashtags also are present as extended entities in tweet attributes. Finally, the data is obtained for further analysis as pandas data frame object in python. After downloading tweet data, suitable preprocessing such as auto-annotation as in Sect. 4.2 could easily be performed on this data frame now.

6.2 Processing Data Features

Raw data collected from Twitter contains noise and characters that may hinder while performing analysis by ML algorithms on textual data. It may include spaces, special characters, punctuations, and emoticons and hashtags (e.g. #SomeHashtag). Thus, we need to apply a certain level of preprocessing to make the data fit for the training of algorithms. Steps for preprocessing Twitter data must consider some specific features such as user mentions (@user_name), '#' character with hashtags, URL links, retweets etc. Also, special characters, punctuations and stopwords are removed, and words are lemmatized. Hashtags are processed to get root words in proper format before including them in the tweet. Emojis are converted to their actual meaning in text form [14]. Lemmatization and stopword removal help in removing plurals, conjugations and unnecessary words. We use the NLTK library in python for recognized stopwords in the English language. We use NLTK word tokenize method for tokenizing the data.

6.3 Data Storage

Python code in Colab fetches and stores data in online cloud storage provided by Google Drive. Our data is present as a python object called dataframe and is stored in a byte-stream using the pickle module in python. This process is called pickling and is very useful, especially in machine learning applications.

7 Experiments and Evaluation

Sentiment analysis and topic popularity analysis was performed on the dataset. The dataset contains 1000 positive and 1000 negative tweets. The creation of dataset was made significantly faster by the automatic annotation technique based on emoticons. This would allow us to make a different dataset again with more tweets later in our analyses. We have performed analyses on different datasets. This includes dataset created as above using Twitter API, 1.6 million tweets dataset available on Kaggle, CMRD and IMDB dataset.

Classification algorithms used for our study include Naïve Bayes (NB), K-nearest neighbour (KNN), Decision Tree (DT), Random Forest (RF) and Support Vector Machine (SVM). Along with this, neural networks such as CNN and LSTM and their hybrid were also analysed on our dataset.

The following steps describe the procedure used for building the classifier and evaluating results:

1. Dataset is loaded and split as test and train data (Described in Sect. 6.1)
2. Apply preprocessing on training data (Described in Sect. 6.2)
3. Include important features for training the model.
4. Use text vectorization and/or Glove word embedding
5. Build the classification model and save it.
6. Preprocess test data and evaluate outcomes using the saved model

Data frame created while loading data helps us to apply preprocessing operation and training the model. The model here could be either a classifier model or a neural network. For neural network models like CNN, LSTM or BiLSTM, we have also used predefined word embedding from glove data, which gives us vector representation of words [20]. These are used as weights that defining the embedding layer in such models (Table 3).

Table 3. An instance of dataset attributes is shown in the table below:

SN	Text	Tr gt	Preprocessed	Hashrem Text	Hash Included sentence	Hash and Smileys	HasSml_one-Words
1	@Hamzaaffan3 @kashyap9991 @ANI Okay fine. Next time don't blame the Government and Modi too #modi ☺.	1	okay fine. next time don't blame the government and modi too #modi ☺.	okay fine. next time don't blame the government and modi too ☺.	okay fine. next time don' t blame the government and modi too ☺ modi.	okay fine. next time don't blame the government and modi too Slightly Smiling Face modi.	okay fine. next time don't blame the government and modi too smiling modi .

We can observe the accuracies for the 2000 and 9000 tweets dataset over various classifier algorithms and neural networks, CNN, LSTM and BiLSTM and their hybrid (Figs. 2, 3 and 5). When we input data that contains hashtags and emojis included as normal text with emoji text as generalized words (smile/sad for positive/negative emoji), we observe the highest accuracies and f-score, about 97–98% for specific algorithms (see Fig. 5). However, such data is assumed to have been suffering from overfitting since annotation of tweets was done on the basis of emoticon data. However, considering data with just hashtags and text denoting emoji meaning, SVM showed the highest score of about 0.96, followed by Random Forest (see Fig. 5).

A detailed analysis was performed on some specific features to analyse relevancy while classifying for sentiment polarity. These features are (1) Hash Included, which refers to using hashtag terms in the tweet, (2) HashEmo, which refers to using hashtag terms along with text form of Emoji as discussed in Sect. 6.2 and (3) HashEmoGen, which refers to hashtag and emojis generalized, i.e., using a more generalized form of emoji text by using reduced text to represent positive and negative emojis. It is observed that there is a significant improvement when using HashEmo, i.e., emoji data with hashtags. Further improvement was seen when using HashEmoGen. This can be seen in Fig. 2(a) and (b) over accuracy and f measure in both 2000 and 9000 tweet datasets. Accuracy and f measure was calculated by considering mean over all algorithms that are listed in Fig. 2.

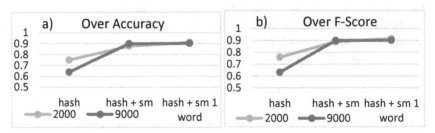

Fig. 2. Comparison of specific features: Hash, Hash + Emoji and Hash Emoji Gen over 2000 and 9000 tweets dataset. (a) is calculated over accuracy while (b) shows plot over f-score

Detailed analysis was performed over various classification algorithms and neural networks. Figure 3 gives us an overview of our findings. Here, we have considered four evaluation metrics, namely, Accuracy, Precision, Recall and F-Measure. Where accuracy gives the ratio of correctly predicted results to total observations, precision is also a measure of correctness of a system giving a ratio of true positives by total results that were predicted as positive. Recall gives us a measure of predicted true positive observations to total observations of the positive class. Broadly Accuracy and F-measure are also sufficient in our case since classes are not imbalanced. Though all are important measures for a model's evaluation, F-Measure singly could account for both precision and recall. It can be seen that CNN-BiLSTM combined model gave high scores (>96%) of overall evaluation metrics considered here. This was followed by SVM, which also gave a consistent and high evaluation score – above 96% for accuracy, above 0.96 for precision and f-measure and slightly less (0.959) as recall score. Certain models such as CNN only and CNN-LSTM combined gave good results over recall and precision (94–97 approx) metrics, respectively. MNB performed quite well with about 90% accuracy but got the least scores when compared to other algorithms in Fig. 3. In this figure, some algorithms like KNN and NB with low scores were not included for better visualization. However, their values can be seen in the table in Fig. 3.

We also evaluated the performance of neural network models used over some standard datasets available online. This included CMRD [15], IMDB [16] and Sentiment 1.6 M tweets [17] data available online. CNN-BiLSTM consistently scored higher for all evaluation metrics, especially over IMDB with about 87% and 88% accuracy and f measure, respectively. A similar pattern was observed for the CNN-LSTM model but with some deviation for recall and precision scores in CMRD and IMDB datasets. Its recall score of 63.68% and 83.20% was lowest among other algorithms over CMRD and IMDB datasets. Only CNN and only LSTM model got the lowest accuracy and f-score for CMRD and IMDB datasets, respectively (Fig. 4).

Features are added logically for increasing the effect of adding new features clearly on the accuracy. For instance, stopwords and lemmatization are preferred at the start, after which the effect of bigram, hashtags and emoticons is analysed. Here, 'hashtag included as text' indicates that text is extracted from hashtags and included in a tweet.

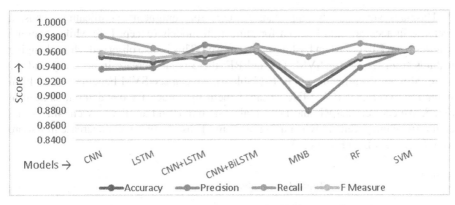

Fig. 3. Performance of Various Models (9000 tweets dataset)

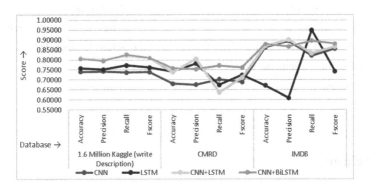

Fig. 4. Illustration of model performance over other datasets

Clubbed words in hashtags are treated as separate words (e.g. #VotingMatters becomes 'Voting Matters'. Porter stemmer library is used for stemming the words to their root form.

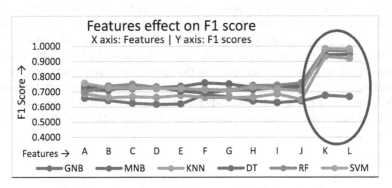

Fig. 5. F1 score for various features with different classifiers

Hashtags are certainly important features for sentiment classification. Thus, a decrease in accuracy can be seen for almost all models (except DT) while changing features from F to G - when hashtags are removed. This is also observed while changing features from G to J for SVM, DT and RF; there is an increase in accuracy and f-1 score. Moving from features G to H (Table 4), we can see SVM benefits greatly from stemming, whereas NB & MNB do not benefit that much for our dataset, as seen in Fig. 5. We can see that RF and DT are almost similar and are the only classifiers that do not show improvement when using bigrams. Thus, we observe a decrease in f-score from feature set D to E in Fig. 5.

Table 4. Features description for Fig. 5

A	Pre-process - URL removal & lower case	G	F + hashtag removed
B	A + stopwords removed	H	G + stemming
C	B + lemmatization	I	F + hash tags as text + stemming
D	C + universal lemmatization *	J	F + hash tag included
E	D + bigram + stemming	K	J + emoji as text
F	D + bigram	L	K + stemming

8 Conclusion

In this paper, we designed and tested various techniques for sentiment analysis and analysed some novel methods for improving accuracy over existing techniques. The framework used efficiently generates a large dataset with auto-annotated class values.

The main components discussed in our approach were the use of CNN – BiLSTM for sentiment analyses, using modified TF-IDF values and Glove vector embedding. Various classification algorithms like SVM, NB, KNN, and Decision Tree were analysed in our study. In addition to accuracy, performance metrics such as f-measure, precision, and recall were also calculated for the comparative analysis. The analysis was performed over various features and models over different datasets, distinctively illustrating their effects. Even more inclusively, our study utilizes hashtags with emoji data which proved to be significant in improving the accuracy of sentiment polarity. We also observed that the CNN-BiLSTM model gives higher accuracy and steady performance over different performance metrics closely followed by SVM. Other results and observations comprehensively indicate the performance of different ML models under specific situations.

References

1. Internet Live Stats. 2019, Twitter Statistics. http://www.internetlivestats.com. Accessed 10 June 2021
2. Hossain, N., Bhuiyan, M.R., Tumpa, Z.N., Hossain, S.A.: Sentiment analysis of restaurant reviews using combined CNN-LSTM. In: 2020 11th International Conference on Computing, Communication and Networking Technologies (ICCCNT). pp. 1–5). IEEE, July 2015
3. The Number of tweets per day in 2020|David Sayce. https://www.dsayce.com/social-media/tweets-day/. Accessed 10 June 2021
4. 10 Twitter Statistics Every Marketer Should Know in 2021 [Infographic]. https://www.oberlo.in/blog/twitter-statistics. Accessed 10 June 2021
5. Sayed, A.A., Elgeldawi, E., Zaki, A.M., Galal, A.R.: Sentiment analysis for arabic reviews using machine learning classification algorithms. In: 2020 International Conference on Innovative Trends in Communication and Computer Engineering (ITCE), pp. 56–63. IEEE, February.2020
6. Li, J., Gao, Y., Gao, X., Shi, Y., Chen, G.: SENTI2POP: sentiment-aware topic popularity prediction on social media. In: 2019 IEEE International Conference on Data Mining (ICDM), Beijing, China, pp. 1174–1179 (2019). https://doi.org/10.1109/ICDM.2019.00143
7. Alfina, I., Sigmawaty, D., Nurhidayati, F., Hidayanto, A.N.: Utilizing hashtags for sentiment analysis of tweets in the political domain. In: Proceedings of the 9th International Conference on Machine Learning and Computing, pp. 43–47. February 2017
8. Almatrafi, O., Parack, S., Chavan, B.: Application of location-based sentiment analysis using Twitter for identifying trends towards Indian general elections 2014. In: Proceedings of the 9th International Conference on Ubiquitous Information Management and Communication, pp. 1–5, January 2015
9. Joshi, K., Bharathi, N., Rao, J.: Stock trend prediction using news sentiment analysis. Int. J. Comput. Sci. Inf. Technol. **8**. 67–76 (2016). https://doi.org/10.5121/ijcsit.2016.8306
10. Gao, T., et al.: DancingLines: an analytical scheme to depict cross-platform event popularity. In: Hartmann, S., Ma, H., Hameurlain, A., Pernul, G., Wagner, R.R. (eds.) DEXA 2018. LNCS, vol. 11029, pp. 283–299. Springer, Cham (2018). https://doi.org/10.1007/978-3-319-98809-2_18
11. Rathod, T., Barot, M.: Trend Analysis on Twitter for Predicting Public Opinion on Ongoing Events. Int. J. Comput. Appl **180**, 13–17 (2018)
12. Ortis, A., Farinella, G.M. and Battiato, S.: An overview on image sentiment analysis: methods, datasets and current challenges. In: ICETE, vol. 1, pp. 296–306 (2019)
13. Abrol, S., Khan, L.: Twinner: understanding news queries with geo-content using twitter. In: Proceedings of the 6th Workshop on Geographic Information Retrieval, p. 10. ACM, February 2010
14. Emojipedia. https://emojipedia.org/people/. Accessed 12 Apr 2017
15. Movie Review Data. https://www.cs.cornell.edu/people/pabo/movie-review-data/. Accessed 7 May 2017
16. Large Movie Review Dataset. https://ai.stanford.edu/~amaas/data/sentiment/. Accessed 7 May 2017
17. Sentiment140 dataset with 1.6 million tweets. https://www.kaggle.com/kazanova/sentiment140. Accessed 7 May 2017
18. Abdullah, M., Hadzikadicy, M., Shaikhz, S.: Sedat: sentiment and emotion detection in Arabic text using CNN-LSTM deep learning. In: ICMLA, pp. 835–840 (2018)
19. Kharde, V., Sonawane, P.: Sentiment analysis of twitter data: a survey of techniques. arXiv preprint arXiv:1601.06971 (2016)

20. GloVe: Global Vectors for Word Representation. https://nlp.stanford.edu/projects/glove/. Accessed 24 Apr 2017
21. Word Embedding: Basics. https://medium.com/@hari4om/word-embedding-d816f643140. Accessed 24 Apr 2017
22. Maas, A., Daly, R.E., Pham, P.T., Huang, D., Ng, A.Y., Potts, C.: Learning word vectors for sentiment analysis. In: Proceedings of the 49th Annual Meeting of the Association for Computational Linguistics: Human Language Technologies, pp. 142–150 June 2011
23. Yaqub, U., Sharma, N., Pabreja, R., Chun, S.A., Atluri, V., Vaidya, J.: Location-based sentiment analyses and visualization of Twitter election data. Digit. Govt. Res. Pract. 1(2), 1–19 (2020)

Mining Spatio-Temporal Sequential Patterns Using MapReduce Approach

Sumalatha Saleti$^{(\boxtimes)}$ ⓘ, P. RadhaKrishna, and D. JaswanthReddy

SRM University AP, Vijayawada, Andhra Pradesh, India
sumalatha.s@srmap.edu.in

Abstract. Spatio-temporal sequential pattern mining (STSPM) plays an important role in many applications such as mobile health, criminology, social media, solar events, transportation, etc. Most of the current studies assume the data is located in a centralized database on which a single machine performs mining. Thus, the existing centralized algorithms are not suitable for the big data environment, where data cannot be handled by a single machine. In this paper, our main aim is to find out the Spatio-temporal sequential patterns from the event data set using a distributed framework suitable for mining big data. We proposed two distributed algorithms, namely, MR-STBFM (MapReduce based spatio-temporal breadth first miner), and MR-SPTreeSTBFM (MapReduce based sequential pattern tree spatio-temporal breadth first miner). These are the distributed algorithms for mining spatio-temporal sequential patterns using Hadoop MapReduce framework. A spatio-temporal tree structure is used in MR-SPTreeSTBFM for reducing the candidate generation cost. This is an extension to the proposed MR-STBFM algorithm. The tree structure significantly improves the performance of the proposed approach. Also, the top-most significant pattern approach has been proposed to mine the top-most significant sequential patterns. Experiments are conducted to evaluate the performance of the proposed algorithms on the Boston crime dataset.

Keywords: Big data · Data mining · MapReduce framework · Sequential pattern mining · Spatio temporal sequences

1 Introduction

Spatio-temporal data deals with the space and position that evolve with the time [1–7]. In spatio-temporal data, data has one or more time dimensions. They arise when data are collected across time as well as space. The movement of an object can be explained by spatial locations sampled at consecutive timestamps. For example, the buses will be moving along the series of streets, people will be travelling from one place to another, but here the routes are not predefined, so we are likely to find the patterns. Sequential pattern mining [8–11] is a data mining task for analyzing sequential data and discovering the sequential patterns. From spatio-temporal sequential patterns, we try to analyze/predict the

© Springer Nature Switzerland AG 2022
K. K. Patel et al. (Eds.): icSoftComp 2021, CCIS 1572, pp. 153–166, 2022.
https://doi.org/10.1007/978-3-031-05767-0_13

past/future movement of the object. Crime incident data set is considered in this paper, which will be having event instances and event types. A neighbourhood space is considered by giving the values like radius 'R' and time 'T'. We try to find out the spatio-temporal sequential patterns with these parameter settings. A spatio-temporal sequential pattern is defined as the sequence of event types, where one event type leads to another event type [4]. For example, the spread of COVID-19 virus from people belonging to one country to another. The virus is spread by human beings in one region to another human being. For instance, Wuhan→Delhi→Calcutta is a sequential pattern that indicates spreading of disease from one place to another.

Spatio-temporal data is being generated in various domains such as transportation, climate science, neuroscience, health care, crime analysis, social media, etc. Integrating spatial and temporal data introduces new challenges for spatio-temporal data mining research in various domains of scientific and commercial importance. Occurrence of one spatio-temporal event will trigger other events which form a sequence of events. Finding the interesting patterns from the sequence of spatio-temporal events is a challenging issue as it involves a large number of candidate patterns. The existing approaches to mining spatio-temporal data assumes the data to be located in a centralized system. However, huge data that is being generated every day across various geographic locations makes it difficult to analyze. Motivated from the aforesaid issues, we aim to develop distributed algorithms using the MapReduce framework that can answer the problem of mining spatio-temporal sequential patterns. The MapReduce framework is developed by Google [12]. It mainly consists of two phases, map and reduce. Two objects that deal with the map and reduce phase are mapper and reducer, respectively. All the communication between the mapper and reducer takes place via ⟨key, value⟩ pairs. The database is distributed to multiple mappers by splitting the input database into disjoint data blocks. The initial processing is done by the mappers which returns the result to an implicit shuffle and sort phase. Here, the intermediate ⟨key, value⟩ pairs are aggregated and sorted. Further, the values connected with the same key are returned to the same reducer and the final set of ⟨key, value⟩ pairs will be emitted to the Hadoop distributed file system (HDFS).

The following are the main contributions of the current work:

1. Proposed a MapReduce algorithm, namely, MR-STBFM for mining spatio-temporal sequential patterns.
2. Extended the MR-STBFM algorithm to reduce the candidate generation cost by using a tree structure. Based on the tree structure, a novel algorithm, MR-SPTreeSTBFM has been proposed that can efficiently handle mining spatio-temporal sequential patterns from big data.
3. Proposed two strategies for mining Top-K patterns efficiently instead of mining all the patterns above the threshold. The proposed algorithms are evaluated on the real dataset. Also, we evaluated the Top-k mining strategies on both the proposed algorithms.

The remaining sections of the paper are organized in the following manner: Sect. 2 gives an overview on the existing literature on spatio-temporal sequential pattern mining. The preliminaries of the current work that helps in defining the problem are given in Sect. 3. Later, the details of the proposed algorithms have been described in Sect. 4. The experimental results are given in Sect. 5. The conclusion and future remarks are presented in Sect. 6.

2   Related Work

Finding the frequent sequential patterns using time without spatial dimension is formulated in [10,13,14]. In [15], a new approach for solving the problem of sequential pattern mining with quadratic time has been discussed and it improves the speed of calculation and data analysis. The problem of discovering frequent sequences from spatio-temporal data is firstly introduced in [5]. The authors proposed the DFS-MINE algorithm that is based on depth-first search. However, it scans the database multiple times to find the frequency of output sequences. Huang et al. [4] proposed two novel algorithms, namely, STS-Miner and Slicing-STS-Miner. These algorithms efficiently mine the spatio-temporal data inorder to discover the sequential patterns. The algorithm of mining spatio-temporal sequences is extended to provide the solution for cascading spatio-temporal patterns in [3].

In [16], a breadth first miner algorithm is used to find the significant sequential patterns and also top-k sequential pattern denoting relations between event types in the dataset. In [17], the authors proposed an efficient algorithm to discover closed spatio-temporal sequential patterns from event data. The CST miner algorithm is used for identifying the closed patterns fastly. Recently, STBFM (Spatio-temporal breadth first miner) [18] was proposed to search for significant sequential patterns from spatio-temporal event instances. Also, the authors extended the algorithm to reduce the time complexity by introducing a SPTree (sequential pattern tree) structure. This led to the SPTreeSTBFM algorithm. Moreover, Top-K pattern mining is proposed that can mine only the topmost significant patterns.

Nowadays there is a lot of anticipation in finding out the sequential patterns from spatio-temporal event data. All the above-mentioned approaches for mining the sequential patterns from spatio-temporal event data assume that the data is processed in a centralized machine. But the real time data is often huge and cannot be handled by the centralized algorithms. There is a need to design distributed algorithms that can deal with big data. Recently, sequential pattern mining is efficiently handled using MapReduce algorithms [19,20]. However, to the best of our understanding, there is no work done on spatio-temporal sequential pattern mining using the MapReduce approach. Hence, in this paper, we aim to find out the most significant spatio-temporal sequential patterns using a MapReduce approach.

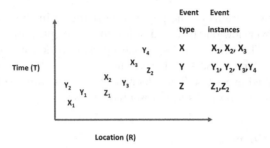

Fig. 1. Spatio-temporal dataset.

3   Preliminaries and Problem Statement

Definition 1 (Neighbourhood space): Let N be the neighbourhood space of instances used in the dataset and it can be any kind of shape which contains the event instances and event types. The neighbourhood space is defined by using two terms i.e., radius R and time interval T. Let us consider an example dataset D which is a set of event instances, $D = \{x1, x2, x3, y1, y2, y3, y4, z1, z2\}$, and F be the set of event types, $F = \{X, Y, Z\}$. The sample dataset is given in Fig. 1.

Definition 2 (Neighbourhood of instances with respect to the event type): For a given instance e, the neighbourhood is defined as $NF(e) = \{e|p \in D(F) \cap$ $\text{distance}(p.location, e.location) \leq R \cap (p.time - e.time) \in [0, T]\}$, where $NF(e)$ is neighbourhood of event instances, R is radius, $D(F)$ is event types in the dataset. For example, neighbourhood of instance $x1$ for the given event type Y is $NY(x1) = \{y1, y2\}$. This indicates that the instance $x1$ satisfies the thresholds R and T for the instances $y1$ and $y2$. (please refer to Fig. 1)

Definition 3 (Spatio-temporal sequential pattern): A spatio-temporal sequential pattern is a sequence of event types, where each event type is joined with another event type. A sequential pattern of length m is denoted as $S[1] \rightarrow S[2] \rightarrow \ldots \rightarrow S[m]$. For instance, $X \rightarrow Y \rightarrow Z$ is the possible sequential pattern of length-3 for the sample dataset given above.

Definition 4 (Set of event instances): A sequential pattern $S = S[1] \rightarrow S[2] \rightarrow \ldots \rightarrow S[n]$ will be having the event instances participating in that particular sequential pattern i.e., $I(S[1]) \rightarrow I(S[2]) \rightarrow \ldots \rightarrow I(S[n])$. For example, let us consider there is a pattern $S : X \rightarrow Y$, if event instances of $X = \{x1, x2, x3\}$ and $Y = \{y1, y2, y3, y4\}$, then $I(S[1]) = \{x1, x2, x3\}$ and $I(S[2]) = \{y1, y2, y3, y4\}$ are the set of possible event instances. Similarly, for $Y \rightarrow Z$, $I(S[1]) = \{y3, y4\}$ and $I(S[2]) = \{z1, z2\}$.

Definition 5 (Participation Ratio): For a sequence S, the participation ratio is defined as the ratio of the number of distinct instances in the neighbourhood space divided by the number of instances in the dataset. The participation ratio for i^{th} element of S is denoted as $PR(S, i)$ and is defined as $PR(S, i) = | I(S, i) |$

$/ \mid D(S, i) \mid$. For the sample dataset, $PR(Y \rightarrow Z, 1) = 2/4 = 0.5$ and $PR(Y \rightarrow Z, 2) = 2/2 = 1$.

Definition 6 (Participation index): Participation index is defined as the minimum from the participation ratios of all the event types of S and it is denoted as $PI(S) = \min(\{PR(S, i), \text{ where } i = 1, 2, 3, 4, \ldots n\})$. For the sequence $S = Y \rightarrow Z$, $PI = \min(\{PR(S, 1), PR(S, 2)\}) = \min(0.5, 1) = 0.5$.

Definition 7 (Topmost spatio-temporal sequential pattern): A spatio-temporal sequential pattern is called topmost K^{th} pattern, if and only if there are $K - 1$ patterns in the top set whose participation index is equal or greater than the participation index of that particular sequential pattern.

Problem Statement: Given a dataset of event instances residing in the Hadoop distributed file system, spatial distance threshold, time window threshold and participation index threshold, the problem of spatio-temporal sequential pattern mining is to mine all the frequent spatio-temporal sequential patterns.

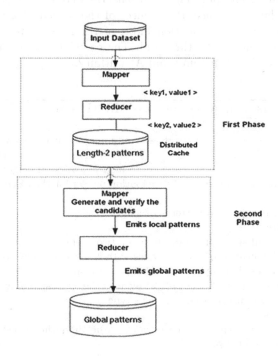

Fig. 2. Proposed MapReduce model.

4   Distributed Mining of Spatio-Temporal Sequential Patterns

4.1   MR-STBFM

It is a two-phase approach, the first MapReduce phase (Algorithm 1) generates all possible spatio-temporal sequential patterns of length 2. Later, the second MapReduce phase (Algorithm 2) generates the output sequences of length $\geq$ 2. The workflow of the proposed model is given in Fig. 2. Mapper of the first phase receives the event instance from its input split and outputs event type as key and neighbourhood instances and set of event instances of event type as values. These are represented as $\langle key1, value1 \rangle$ in Fig. 2. Reducer of the first phase finds all the frequent length-2 patterns and stores them in the distributed cache file. The two event types in length-2 pattern are emitted as key and neighbourhood instances and set of event instances of event type as values. These are represented as $\langle key2, value2 \rangle$ in Fig. 2. All the patterns of length>2 will be generated by the second phase mapper. However, these are referred to as local patterns which are frequent in an input split. The local patterns may not be globally frequent over the entire dataset. Hence, the reducer of the second phase calculates the participation ratio and participation index of the local patterns globally and emits only if they satisfy the threshold. The complete description of the two phases is presented in Algorithm 1 and Algorithm 2.

Algorithm 1. First MapReduce phase of MR-STBFM

Input:

    D ▷ a dataset containing event types and their instances

    F ▷ a set of event types

    R, T ▷ a specification of neighbourhood spaces

    θ ▷ a significance threshold (Participation index) for the discovered sequences.

Output:

    Spatio-temporal sequential patterns of length 2.

    Mapper method:

1: Find the time and space proximity between two event types.
2: Find the neighbourhood of each instance with respect to event type.
3: Find the instances that support neighbourhood instances.
4: Emit the event types of length-2 sequence, the corresponding neighbourhood instances and total instances of an event type.

    Reducer Method:

5: Emit the two event types, the corresponding neighbourhood instances and total instances of an event type.
6: Calculate the participation ratio and participation index.
7: Store the length-2 sequential pattern in a distributed cache file only if the participation index of the pattern satisfies the participation index threshold θ.

Algorithm 2. Second MapReduce phase of MR-STBFM

Input:

    Output from first MapReduce phase

Output:

    All Spatio Temporal sequential patterns of length k, where $k > 2$

    Mapper method:

1: Retrieve the patterns of length 2 and store them in H.
2: Create a hash data structure L, to store locally generated patterns.
3: Let E be the entry retrieved from L.
4: Interpret E and split into event types, neighbourhood instances and total instances of event type.
5: If $A \rightarrow B$ exists in H, where H is a list of spatio-temporal patterns of length 2, then call the function *Generate_sequence* (Algorithm 3) to find the candidate sequences.

    Reducer method:

6: Reducer receives local patterns and their neighborhood instances and total instances from each mapper.
7: Participation ratio and participation index are calculated, and the pattern $A \rightarrow B$ is emitted as output if the participation index of $A \rightarrow B$ satisfies the threshold θ.

Algorithm 3. *Generate_sequence*

Input:

    S ▷ Spatio temporal pattern of length ≥ 1

Output:

    S_k ▷ New sequence

    Function *Generate_sequence(S)*

1: The sequence S is split into two, such that the last event type is B and the remaining is A.
2: Find the participation ratio and participation index of $A \rightarrow B$.
3: If B can be extended, then the neighbourhood event type of B is retrieved from H.
4: The participation ratio of new event type C and participation index of the new sequence $A \rightarrow B \rightarrow C$ is found.
5: If the participation index of the new sequence satisfies the threshold then output the new sequence.
6: New sequence, $S_k \leftarrow (S \rightarrow C)$
7: Call the function *Generate_sequence(S_k)*.

4.2 MR-SPTreeSTBFM

MR-SPTreeSTBFM is an extension to MR-STBFM algorithm. It is also a two-phase MapReduce algorithm. The main difference exists in the generation of candidate sequences. The mapper method of MR-SPTreeSTBFM invokes Algorithm 4 at step 5 of Algorithm 2.

Algorithm 4. Sequence generation using SPTree strategy

1: Initially the candidate pattern set is null.
2: **for** every sequence S_i belongs to set L_{m-1} **do**
3:     **for** every sequence S_j belongs to children of second parent S_i **do**
4:         A new sequence is formed of length $m - 1$.
5:         Instances of the patterns of length $m - 1$ are also equal.
6:         To get the instance $S(m)$, calculate the neighbourhood of instances of $S_i(m-1)$ and $S_j(m - 1)$.
7:         The first parent of S is S_i and second parent of S is S_j.
8:         Add the sequence S to children S_i and add S to C_m.
9:         $Verify\_candidate\_patterns(C_m)//$invoke Algorithm 5
10:     **end for**
11: **end for**

4.3 Top-K Significant Pattern

This section presents two algorithms for finding topmost K patterns. Algorithm 6 can be integrated into the second MapReduce phase of MR-STBFM, whereas Algorithm 7 can be integrated into the second MapReduce phase of MR-SPTreeSTBFM. Topmost significant sequential pattern is the Kth top sequential pattern in the top set only if there are $K - 1$ patterns whose participation index (PI) is equal or greater than the $PI(S)$. Initially we will calculate the participation index of sequence S (Line 3 of Algorithm 6). If PI is greater than the threshold value, then we will check for the top set whose number of patterns are less than $K - 1$ patterns and the sequence S is included into L_m and top set (Lines 4 to 6). If the number of patterns is equal to $K - 1$ patterns then include the sequence S to L_m and the top set, also change the threshold value to the participation index of $Top(K)$ (Lines 7 to 10). If the number of patterns in the top set is equal to K, then include the candidate pattern S to L_m and top (Line 12). Now, if the participation index of S is greater than the threshold value, then the threshold value is to be changed to the value of the actual participation index of top-K pattern (Line 13). Now, the patterns whose participation index is less than the new threshold value will be deleted from L_m and top (Line 15). Algorithm 7 represents the process of verifying the topmost candidate patterns using SPTree strategy. It is integrated into the second phase of MR-SPTreeSTBFM to generate the top K significant patterns.

The top spatio-temporal sequential patterns whose participation index greater than 0.9 are given below: $other\_larcency \rightarrow arrgravated\_assault \rightarrow manslaug(PI = 1)$
$simple\_assault \rightarrow other\_larcency \rightarrow arson(PI = 0.89)$
$drug\_charges \rightarrow simple\_assault \rightarrow gambling\_offense(PI = 0.83)$
The pattern "$other\_larcency- > arrgravated\_assault \rightarrow manslaug$" shows occurrence of three crime event types, namely, $other\_larcency$, $arrgravated\_assault$ and $manslaug$ in the same order as specified.

Algorithm 5. $Verify\_candidate\_patterns(C_m)$

1: Initially the set L_m is null.
2: **for** every sequence S belongs to set C_m **do**
3:     Calculate the participation index of S.
4:     Check whether the participation index is greater than the user defined threshold.
5:     **if** step-4 is true **then**
6:         emit S.
7:     **else**
8:         remove the sequence from children (first parent(s)).
9:     **end if**
10: **end for**

Algorithm 6. Verify topmost candidate pattern

1: Initially the set L_m is null.
2: **for** every sequence S belongs to set C_m **do**
3:     Calculate the participation index of S
4:     **if** the participation index is greater than the user defined threshold **then**
5:         **if** the *top* set is having the number of patterns less than $K - 1$ **then**
6:             Add the sequence S to set L_m and to the *top* set.
7:         **else**
8:             **if** the number of patterns in the *top* set is equal to the $K - 1$ patterns **then**
9:                 Add the sequence S to set L_m and to the *top* set.
10:                 Set the threshold value to the PI of the topmost K pattern.
11:             **else**
12:                 Add S to L_m and to the *top* set.
13:                 **if** the participation index is greater than the user defined threshold **then**
14:                     the threshold value is changed to the PI of the topmost pattern.
15:                     Remove the sequences from the top set and L_m whose PI is less than the new threshold value.
16:                 **end if**
17:             **end if**
18:         **end if**
19:     **end if**
20: **end for**
21: emit L_m.

5   Experimental Evaluation

To assess the performance of proposed algorithms, we have used crime related dataset [21] for Boston city which has 40544 crime related incidents and 26 crime event types. Initially, we set the radius 'R' and time 'T' before experimenting. In the experiments, we have compared two algorithms i.e., MR-STBFM and MR-SPTreeSTBFM with their execution time and number of sequential patterns discovered. The number of patterns generated for R = 200 m and T = 10 days

---

Algorithm 7. Verify topmost candidate pattern using SPTree strategy

---
1: Initially the set L_m is null.
2: **for** every sequence S belongs to set C_m **do**
3:     Calculate the participation index of S
4:     **if** the participation index is greater than the threshold value **then**
5:         **if** the *top* set is having the number of patterns less than $K - 1$ **then**
6:             Add the sequence S to set L_m and to the *top* set.
7:         **else**
8:             **if** the number of patterns in the top set is equal to the $K - 1$ patterns
    then
9:                 Add the sequence S to set L_m and to the *top* set.
10:                 Set the threshold value to the PI of the topmost K pattern.
11:             **else**
12:                 Add S to L_m and to the *top* set.
13:                 **if** the participation index is greater than the threshold value **then**
14:                     the threshold value is changed to the PI of the topmost pattern.
15:                     Remove the sequences from the top set and L_m whose PI is less
    than the new threshold value.
16:                     Remove the sequence S from the children of the first parent of S.
17:                 **end if**
18:             **end if**
19:         **end if**
20:     **else**
21:         remove the sequence S from children of the first parent of S.
22:     **end if**
23: **end for**
24: emit L_m.

---

for various values of participation index is shown in Fig. 3a. For PI $= 0.09$, the number of patterns generated are 4598, and for PI $= 0.085$, the value is 6233, and the value tends to increase higher for lower values of PI. It is 5,93,785 for PI $= 0.05$. It is observed that the number of patterns tends to increase more for PI < 0.06. The execution time of MR-STBFM and MR-SPTreeSTBFM for the above considered R and T values is given in Fig. 3b. It is noticed that the execution time is high for lower values of PI. Especially, in this case, the execution time of both the algorithms increased high for PI < 0.06. From the results, it is observed that MR-SPTreeSTBFM performs better than MR-STBFM with respect to execution time. The reason is that the tree structure reduces the candidate generation cost thereby improving the performance. The number of patterns generated, and the execution time of the proposed algorithms are shown in Fig. 3c and Fig. 3d. This experiment is conducted for R $= 300$ m and T $= 8$ days. It is observed that the number of patterns increased exponentially for PI < 0.2 and this also affects the execution time as shown in Fig. 3d, i.e., the rise in the curve for PI < 0.2. The difference in the execution time between MR-SPTreeSTBFM and MR-STBFM can also be clearly seen for PI < 0.2. The number of patterns generated for PI $= 0.055$ is 3,982 whereas for PI $= 0.015$, it

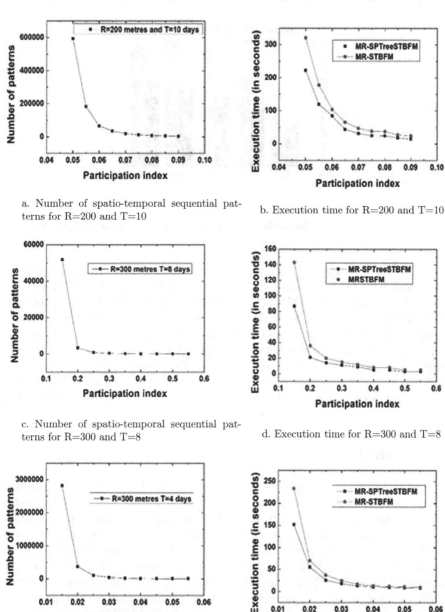

a. Number of spatio-temporal sequential patterns for R=200 and T=10

b. Execution time for R=200 and T=10

c. Number of spatio-temporal sequential patterns for R=300 and T=8

d. Execution time for R=300 and T=8

e. Number of spatio-temporal sequential patterns for R=300 and T=4

f. Execution time for R=300 and T=4

Fig. 3. Performance of proposed algorithms.

is 2,819,490. For this experiment, we have set R = 300 m and T = 4 days. The results are given in Fig. 3e. The execution time for this experiment is shown in

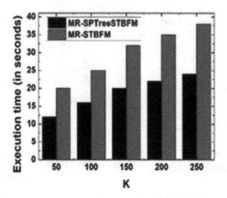

Fig. 4. Execution time for finding Top-K spatio-temporal sequential patterns.

Fig. 3f. The performance of both the algorithms is similar for PI values between 0.055 and 0.03. The difference is clearly known for lower values of PI, i.e., PI < 0.02. On an average, MR-SPTreeSTBFM achieves 21% improvement than the MR-STBFM with respect to the execution time. An experiment is conducted to know the impact of K value on the execution time of the algorithms. The Top-K patterns have been generated by setting k = 50 to k = 250. During this experiment, we set R = 500 m and T = 7 days. The results of this experiment are shown in Fig. 4. It is noticed that MR-SPTreeSTBFM outperforms MR-STBFM for all the values of N, especially for higher values of N, the performance gain is high.

6   Conclusion

In this paper, we proposed MR-STBFM, a new MapReduce algorithm to find the spatio-temporal sequential patterns. In this algorithm, the significance of a sequential pattern is found with the aid of participation ratio and participation index. The significance of a sequential pattern is measured by comparing the participation index of a pattern with the given threshold value. For reducing the generation cost of the candidate patterns, we have used SP-Tree structure and proposed MR-SPTreeSTBFM algorithm. Also, Top-k mining strategy is introduced and integrated with the two proposed algorithms to discover the topmost significant patterns. On an average, MR-SPTreeSTBFM achieves 21% improvement than the MR-STBFM with respect to the execution time. The proposed MapReduce algorithms are the distributed versions of the existing STBFM and SPTreeSTBFM algorithms and can efficiently handle the big data. But, the proposed algorithms cannot discover concise representation of the spatio-temporal patterns. As a future work, we would like to extend the proposed MapReduce algorithms to generate the concise representation [17] of the output set instead of mining all the patterns that satisfy the user specified threshold. Also, the proposed algorithms can be run on multiple datasets to prove their scalability.

References

1. Atluri, G., Karpatne, A., Kumar, V.: Spatio-temporal data mining: a survey of problems and methods. ACM Comput. Surv. **51**(4), 83:1–83:41 (2018)
2. Li, Z.: Spatiotemporal pattern mining: algorithms and applications. In: Aggarwal, C.C., Han, J. (eds.) Frequent Pattern Mining, pp. 283–306. Springer, Cham (2014). https://doi.org/10.1007/978-3-319-07821-2_12
3. Mohan, P., Shekhar, S., Shine, J.A., Rogers, J.P.: Cascading spatio-temporal pattern discovery. IEEE Trans. Knowl. Data Eng. **24**(11), 1977–1992 (2012)
4. Huang, Y., Zhang, L., Zhang, P.: A framework for mining sequential patterns from spatio-temporal event data sets. IEEE Trans. Knowl. Data Eng. **20**(4), 433–448 (2008)
5. Tsoukatos, I.I., Gunopulos, D.: Efficient mining of spatiotemporal patterns. In: Jensen, C.S., Schneider, M., Seeger, B., Tsotras, V.J. (eds.) SSTD 2001. LNCS, vol. 2121, pp. 425–442. Springer, Heidelberg (2001). https://doi.org/10.1007/3-540-47724-1_22
6. Cao, H., Mamoulis, N., Cheung, D.W.: Mining frequent spatio-temporal sequential patterns. In: Proceedings of Fifth International Conference on Data Mining, Houston, TX, USA, pp. 82–89. IEEE (2005)
7. Aydin, B., Angryk, R.A.: Spatiotemporal event sequence mining from evolving regions. In: Proceedings of 23rd International Conference on Pattern Recognition, Cancun, Mexico, pp. 4172–4177. IEEE (2016)
8. Mabroukeh, N.R., Ezeife, C.I.: A taxonomy of sequential pattern mining algorithms. ACM Comput. Surv. **43**(1), 3:1–3:41 (2010)
9. Pei, J., et al.: Mining sequential patterns by pattern-growth: the prefixspan approach. IEEE Trans. Knowl. Data Eng. **16**(11), 1424–1440 (2004)
10. Agrawal, R., Srikant, R.: Mining sequential patterns. In: Proceedings of the Eleventh International Conference on Data Engineering, Taipei, Taiwan, pp. 3–14. IEEE (1995)
11. Mooney, C.H., Roddick, J.F.: Sequential pattern mining - approaches and algorithms. ACM Comput. Surv. **45**, 19:1–19:39 (2013)
12. Dean, J., Ghemawat, S.: MapReduce: simplified data processing on large clusters. ACM Commun. **51**(1), 107–113 (2008)
13. Mannila, H., Toivonen, H., Verkamo, A.I.: Discovering frequent episodes in sequences. In: Proceedings of the First International Conference on Knowledge Discovery and Data Mining, Montreal, Canada, pp. 210–215 (1995)
14. Mannila, H., Toivonen, H., Verkamo, A.I.: Discovery of frequent episodes in event sequences. Data Min. Knowl. Discov. **1**, 259–287 (1997). https://doi.org/10.1023/A:1009748302351
15. Nguyen, T.-T., Nguyen, P.-K.: A new approach for problem of sequential pattern mining. In: Nguyen, N.-T., Hoang, K., Jędrzejowicz, P. (eds.) ICCCI 2012. LNCS (LNAI), vol. 7653, pp. 51–60. Springer, Heidelberg (2012). https://doi.org/10.1007/978-3-642-34630-9_6
16. Maciag, P.S.: Efficient discovery of top-K sequential patterns in event-based spatio-temporal data. In: Proceedings of the 2018 Federated Conference on Computer Science and Information Systems, Poznań, Poland, pp. 47–56 (2018)
17. Maciag, P.S., Kryszkiewicz, M., Bembenik, R.: Discovery of closed spatio-temporal sequential patterns from event data. Procedia Comput. Sci. **159**, 707–716 (2019)
18. Maciag, P.S., Bembenik, R.: A novel breadth-first strategy algorithm for discovering sequential patterns from spatio-temporal data. In: Proceedings of the

Eighth International Conference on Pattern Recognition Applications and Methods, Prague, Czech Republic, pp. 459–466 (2019)

19. Saleti, S., Subramanyam, R.B.V.: A novel mapreduce algorithm for distributed mining of sequential patterns using co-occurrence information. Appl. Intell. **49**(1), 150–171 (2018). https://doi.org/10.1007/s10489-018-1259-2

20. Chen, C.-C., Shuai, H.-H., Chen, M.-S.: Distributed and scalable sequential pattern mining through stream processing. Knowl. Inf. Syst. **53**(2), 365–390 (2017). https://doi.org/10.1007/s10115-017-1037-1

21. Boston-Police-Department: Boston police department: crime incident reports (2014)

A Split-Then-Join Lightweight Hybrid Majority Vote Classifier

Moses L. Gadebe[1](✉) 📵, Sunday O. Ojo[2] 📵, and Okuthe P. Kogeda[3] 📵

[1] Tshwane University of Technology, Pretoria, Private Bag X680, Pretoria 0001, South Africa
gadebeml@tut.ac.za
[2] Durban University of Technology, 41 Sultan Road, Greyville, Durban 4001, South Africa
[3] University of the Free State, P. O. Box 339, Bloemfontein 9300, South Africa

Abstract. Classification of human activities using smallest dataset is achievable with tree-oriented (C4.5, Random Forest, Bagging) algorithms. However, the KNN and Gaussian Naïve Bayes (GNB) achieve higher accuracy only with largest dataset. Of interest KNN is challenged with minor feature problem, where two similar features are predictable far from each other because of limited number of classification features. In this paper the split-then-join combiner strategy is employed to split classification features into first and secondary (KNN and GNB) classifier based on integral conditionality function. Therefore, top K prediction voting list of both classifier are joined for final voting. We simulated our combined algorithm and compared it with other classification algorithms (Support Vector Machine, C4.5, K NN, and Naïve Bayes, Random Forest) using R programming language with Caret, Rweka and e1071 libraries using 3 selected datasets with 27 combined human activities. The result of the study indicates that our combined classifier is effective and reliable than its predecessor Naïve Bayes and KNN. The results of study shows that our proposed algorithm is compatible with C4.5, Boosted Trees and Random Forest and other ensemble algorithms with accuracy and precision reaching 100% in most of 27 human activities.

Keywords: Split-then-join · Ensemble · KNN · Gaussian Naïve Bayes · Lightweight algorithm

1 Introduction

Human Activity Recognition is focused in classifying human activities based on sensor signals from accelerometer, gyroscope and GPS. A smartphone comes with such number of sensors and permits continuous monitoring of numerous physiological signals. However, due to limited storage, processing power and limited memory most researchers in Human Activity Recognition (HAR) use smartphone accelerometer to monitor patients to provide smart healthcare [1, 2]. Moreover, classification algorithms in HAR plays crucial role to classify human activities using sensors data. The classification algorithms also helps researchers to understand human behavior and their environment [1, 2]. However, classification algorithms differs in terms dataset requirements which have impact on storage, memory and processing power of devices. Naïve Bayes, KNN and Support

© Springer Nature Switzerland AG 2022
K. K. Patel et al. (Eds.): icSoftComp 2021, CCIS 1572, pp. 167–180, 2022.
https://doi.org/10.1007/978-3-031-05767-0_14

Vector Machines to accurately make predictions requires training dataset with more classification features [2, 3, 7]. On the other hand, algorithms such as C4.5, Random Forest and Bagging performs better with small dataset [8–13]. The evolution of ensemble algorithms, plays a crucial role in joining algorithms to improve their prediction accuracy. Researchers in [14] proposed number of combing strategies to join a bunch of algorithms, with the aim to increase their prediction. Most of ensemble techniques use combiner strategies (combiner, arbiter and hybrid strategies) proposed in [14]. Most researchers in [9, 12, 13] and [15] employed ensemble algorithms based on voting schemes using tree-based algorithms, because tree-based algorithms produce higher accuracy than simpler algorithms (Naïve Bayes and KNN). Hence, it is prudent to close the gap between tree-based algorithms and simpler algorithms. Ensemble algorithms based on combiner strategies gives the possibilities to improve prediction accuracy of simpler algorithms. In this paper split-then-join approach is proposed based on [14] strategies and integral conditionality function to join voting lists of first and secondary classifiers (Gaussian Naïve Bayes and KNN). The aim of this paper is to augment classification features to increase classification accuracy of data hungry algorithms. The split-then-join approach is simulated in R programming language for comparison using three benchmarking datasets. The remainder of our paper is fashioned as follows: In Sect. 2, related work in HAR is presented. The methodology and experimentation are presented in Sects. 3 and 4 respectively. Experimental results are presented in Sect. 5. Finally, Sect. 6 presents conclusion and future work.

2   Related Work

Bootstrap Aggregating known as Bagging was invented by Breiman [8]. The technique combines different numbers of tree decision (ID3, C4.5) to amalgamate various outputs thereby calculating averages from different decision trees. The employed trees have equal weights and each subsets of training dataset are chosen randomly. Therefore, each incorporated decision tree is taken as subset, then majority vote is computed on each averaged decision tree similar to [14] combiner strategy. Boosting, differently to Bagging uses different subsets of training dataset by reweighting in each iteration such that a single learning model is weighted to construct a final strong classifier. If input dataset with N classification features such that (n_i, k_i) $i = 1, \ldots, N$ where n_i is a feature vector and k_i is a labelled class $k\_i \in 1, \ldots, T$, then iterations are performed with weaker classifier $f(n)$. However, in Bagging, N features are indiscriminately selected with replacement in T iterations from training features [15, 16]. Therefore, weaker classifier is applied on the indiscriminately selected features and the resulting model is stored. But, Boosting focuses on creating stronger classifier compared to Bagging, by giving more influence on successful stronger models. After T iterations, the prediction is made using weighted voting of the predictions for each successful classifier. Random Forest was also introduced by Breiman in 2001 aimed at reducing the danger of over-fitting in constructing ensemble/combined models [8, 15]. One variant of Boosting is AdaBoost, it creates a linear combination of decision trees [15]. Random Forest classifiers consist of a collection of decision trees, because more classifiers are more likely to be well-classified and likely to give appropriate weight to most relevant features [8]. A Random

Forest creates a tree-like structure given as $\{h(n, \Theta k), k = 1, \ldots\}$ and $\{\Theta k\}$ of individually distributed vectors, where each tree casts a unit vote for the most popular class at input n. That is, the prediction of class in training dataset is obtained by majority vote over the predictions of individual trees. Most research work in HAR, uses Random Forest and C4.5 as benchmarking algorithms because they were found to be reliable with smallest dataset [9, 11, 12].

Hence, researchers in [12] employed a group of classifiers to classify human activities using dataset collected from 20 participants using researcher's ASUS ZenFone 5 smartphone. During data collection participants used a smartphone inside their front pockets whilst performing human activities. The smartphone captured tri-axial values for each recorded human activities at frequency rate of 0.5 and 120 instances were collected per minute. Researchers in [12] extracted and annotated time domain features similar to [9] in order to train and test AdaBoost, C4.5, Support Vector Machines (SVM) and Random Forest using WEKA. Simulated algorithms (AdaBoost, C4.5, SVM and Random Forest) reported 98%, 96%, 95% and 93% in accuracy respectively [11]. Authors in [15, 16, 19] proposed ensemble techniques employing tree-oriented algorithms, because they were found to be reliable with smallest dataset and reported accuracy above 98%. However, researchers in [20] proposed a different ensemble model to join a group of expert Naïve Bayes algorithms and average the experts using weighted majority vote strategy. In their [20] work, the technique averages probabilities of each employed algorithm and observe each experts and true classes in a given sample in the dataset. Once, the conditional probabilities were learned their technique employed weighted voting to classifies each unknown sample. They [20] analyzed all expert responses which were jointly averaged to collective classify each sample. The final prediction of their technique is based on individuals experts. They [20] simulated their technique using 3 datasets (MFeat, Optodigit and Pendigit) in comparison to bagging and boosting algorithms. The results of their study, showed slight increment in accuracy close to 97% in all 3 datasets.

A closely related study is reported in [13], they proposed an ensemble technique to join Naïve Bayes Tree, KNN and C4.5 algorithms. The model follows a similar combiner strategy proposed in [14]. In the same way as in [20] the technique of [13] averages voting probabilities of three algorithms, then in each learner a posteriori probability is generated and a class with maximum posteriori is taken as voting hypothesis. In their simulations 10-fold cross validation was conducted on each of 28 datasets using WEKA. The results of their study revealed that their voting ensemble techniques outperformed individual simpler classifiers (Naïve Bayes and KNN) excluding C4.5 classifier. The result presented in [13] are comparable with our reported results in [22]. In this article, we expand our ensemble algorithm using split-then-join approach to join GNB and KNN in marriage of convenience using integral conditionality to address minor features challenges presented in [21–23].

3   Methodology

In this article, split-then-join approach presented in Fig. 1 is used to join KNN and Gaussian Naïve Bayes based on integral conditionality function to improve their prediction accuracy with reduced training dataset [22].

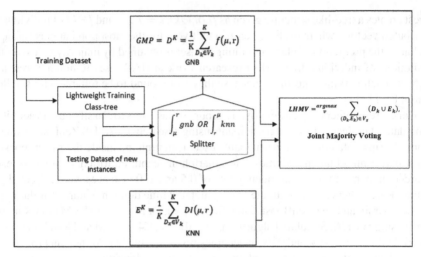

Fig. 1. Split-then-join combiner strategy

Our technique is aimed to address K neighborhood minor features problem, where two records that are closer to each other are recognized far from each other due to limited data point [7, 22]. Different to studies in [13, 14] and [20], we used bottom-up segmentation to select best classification features and thus compress dataset into reduced class-tree. Then, the reduced class-tree is therefore split to upper and lower classifier using our unique integral conditionality rule to accommodate GNB and KNN probability different structures. The split-then-join combiner strategies is presented in two stages as follows:

a. Stage One: Lightweight Training Class-Tree
In [22], we introduced lightweight class-tree to normalize and improve training complexity of lazy classifiers inspired by C4.5 [25]. The technique was borrowed from the C4.5 concept of best attribute selection. Input training dataset with multiple classes $C = \{c_1, c_2, ..., c_N\}$ represented as $TrainerSet^{N \times M}$ class-tree is reduced using R reduction transformers based on our data segmentation technique defined by Eq. (1) and (2) [22]:

$$Data\ Segmentation^{R \times S} = \int_S^R \frac{1}{S} \sum_{x=1}^N p_x \tag{1}$$

where N is rows and M is features from input $TrainerSet^{N \times M}$ and R is compressed subgroups each with S rows given as $R = \{r_1, r_2, ..., r_s\}$ of all mean averages using Eq. (2) [22]:

$$mean = \frac{1}{S} \sum_{x=1}^S i_x \tag{2}$$

Our data segmentation technique is expressed in Algorithm 1 to prepare reduced training class-tree before actual classification.

Algorithm 1: Compressed Lightweight Class-Tree Algorithm

---

1. ***Required*** *: Multi-featured training dataset* $\textbf{TrainerSet}^{N \times M}$ *as* T
2. ***procedure*** *compressDataset()*
3. *Set R to a value to reduce training dataset predicators*
4. ***Output:*** *reduced training class-tree* $\textbf{reducedTree}^{R \times M}$ *to empty*
5. *Set N to number of rows in class K*
6. **for each** *class* $K \in T$ **do**

7. $S \leftarrow \left|\frac{N}{R}\right|$ //split the group into R groups
8. **cols** $\leftarrow$ set to number of attributes in predicator I
9. **Set row** = 1 //the rows of reduced number of groups
10. **for** j=1 to **cols**//for each predicator compute sum and average
11. **for** i = 1 to S **do** //number of reduced splitting group
12. $sum \leftarrow sum + (I_{i,j})$ // Sum of each predicator in element j, i
13. *end-for*
14. $mean_{row,j} \leftarrow \frac{sum}{S}$ //compute mean based on equation 2
15. $row \leftarrow row + 1$ //accumulates count of reduced tree rows
16. $ReducedTree_{row} \leftarrow mean$ // Store each mean reduced group
17. *end-for*
18. *end-for*
19. *end*

The Algorithm expects a training dataset of any dimension in line 1 with reduction transformation value of R size in line 2. Then, the algorithm computes number of S rows for each R group in line 6. Thereafter, the algorithm sums all input predicator i_s in each subgroup R and computes the mean average from line 5 to 14.

b. Lightweight Hybrid Majority Vote Classifier

In this stage, reduced class-tree of best mean features $\phi = \{\mu_1, \mu_2, \ldots, \mu_r\}$ for each class c_r is used as input training dataset. Our proposed LHMV presented in [22] uses reduced training dataset to train our classifier based on union of convenience. In this paper integral conditionality is employed to increase classification feature to minimize minor-features problems reported in [7, 22, 23, 25]. Such that voting results in the first and second classifiers prediction results are stored in D^k and E^k vectors and then joint into union of convenience based on majority voting principle defined by Eq. (3) [22]:

$$LHMV = \overset{argmax}{v} \sum_{(D_k,E_k) \in V_z} (D_k \cup E_k), \tag{3}$$

where K is top potential neighbors in D_k and E_k voting lists of predicted classes labels in a joint vote list V_z. The combination of GNB and KNN is joint probability function $P(C \leq RT) \cup P(C > RT)$ of classification tree C with best mean $\phi = \{\mu_1, \mu_2, \ldots, \mu_r\}$

to classify real-time instances $RT = \{r_1, r_2, r_3, r_j\}$ that factorizes as splitting integral function defined by Eq. (4):

$$P(C|RT) = \left(\sum_{\mu_r \leq r_j} \sum_{\mu_r > r_j} VCF(\mu_r, r_j)DUVF(\mu_r, r_j) \right) \tag{4}$$

where, $VCF(\mu_r, r_j)$ is Vote Cast Function (VCF) known as probability approximation condition $\mu_r \leq r_j$, whereas $DUVF(\mu_r, r_j)$ is the KNN Distance Unit Vote Function within supplementary integral condition $\mu_r > r_j$ to accommodate all missed near-neighbors not within the first VCF conditionality. Our combiner strategy splits the input training dataset into first and secondary classifier then later joins their top K neighbors.

First Classifier: Improved Gaussian Naïve Bayes
As the first classier the GNB is improved and implemented for all best mean averages $\phi = \{\mu_1, \mu_2, \ldots, \mu_r\}$ in each class category c_r that are within first integral condition $VCF(\mu_r, r_j)$ as proper distribution function defined by Eq. (5):

$$VCF(\mu_r, r_j) = \int_{\mu}^{r} p(\mu|r) \tag{5}$$

Expanded to Eq. (6):

$$VCF(\mu_r, r_j) = \begin{Bmatrix} 1 : \mu_r \leq r_j \\ 0 : otherwise \end{Bmatrix} \tag{6}$$

Provided that the integral condition $VCF(\mu_r, r_j)$ is a proper Cumulative Distribution Function (CDF) of all best mean averages μ_r within probability limit function $f(\mu_r, r_j) :\to [0, 1] \, by f (\mu_r \leq r_j)$, which is defined by Eq. (7):

$$P(I \leq RT) = f(\mu_r \leq r_j) = \begin{cases} 1: \mu_r \leq i < r_j \\ 0 : \mu_r > r_j \end{cases} \tag{7}$$

As a result, the probability approximation function holds when the area under $VCF(\mu_r, r_j)$ is in CDF neighborhood. Therefore, each best mean average μ_r in class category c_r is approximated relative to real-time instances r_j as $p(\mu_r|r_j)$ products indicated in Table 1.

The vector D^K accumulates all probability products $p(\mu_r|r_j)$ per class category c_r defined by Eq. (8):

$$D^k = \prod_{\mu_r \leq r_j} p(\mu_r|r_j) \tag{8}$$

where $p(\mu_r|r_j)$ is probability approximation of all μ_r ordered as $\{\mu_1 \leq \mu_2 \leq \mu_3 \leq \cdots \leq \mu_r\}$ in relation to r_j within first integral condition $f(\mu_r \leq r_j)$ per class category c_r.

Table 1. Joint vote cast function of Gaussian Nearest Neighbours

$VCF(c_r, r_j)$	r_1	r_2	r_3	r_j	D^K
c_1	$p(\mu_1\|r_1)$	$p(\mu_2\|r_2)$	$p(\mu_3\|r_3)$	$p(\mu_r\|r_j)$	$(D_1) = \prod_{\mu_r \leq r_j} p(\mu_r\|r_j)$
c_2	$p(\mu_1\|r_1)$	$p(\mu_2\|r_2)$	$p(\mu_3\|r_3)$	$p(\mu_r\|r_j)$	$(D_2) = \prod_{\mu_r \leq r_j} p(\mu_r\|r_j)$
c_3	$p(\mu_1\|r_1)$	$p(\mu_2\|r_2)$	$p(\mu_3\|r_3)$	$p(\mu_r\|r_j)$	$(D_3) = \prod_{\mu_r \leq r_j} p(\mu_r\|r_j)$
c_k	$p(\mu_1\|r_1)$	$p(\mu_2\|r_2)$	$p(\mu_3\|r_3)$	$p(\mu_r\|r_j)$	$(D_k) = \prod_{\mu_r \leq r_j} p(\mu_r\|r_j)$

Therefore, the standard deviation σ under the area of CDF is integral standard normal distribution, defined by equation (9) [22]:

$$GNB = f(\mu_r, r_j) = \int_{\mu_r}^{r} \frac{1}{\sqrt{2\pi\sigma^2}} e^{-\frac{(r-\mu_r)^2}{2\sigma^2}} \qquad (9)$$

where r is a real-time instance and μ_r is mean average from reduced dataset and σ is a standard deviation computed using Eq. (10):

$$Standard\ deviation - \sigma = \sqrt{\frac{1}{k-1}\sum_{r=1}^{k}(r_j - \mu_r)^2} \qquad (10)$$

We modified and transformed the GNB into Gaussian Majority Probability (GMP) defined by Eq. (11):

$$GMP = D^K = \frac{1}{K}\sum_{D_k \in V_k}^{K} f(\mu_r, r_j) \qquad (11)$$

where $f(\mu_r, r_j)$ is integral conditional function of all probability approximations $p(\mu_r|r_j)$ as majority vote list D^k within $f(\mu_r \leq r_j)$, whereas K is number of nearest neighbors in voting list V_k. Figure 2 portrays possible neighbors given A, B and C classes where their mean predicators $\phi = \{\mu_1, \mu_2, \ldots, \mu_r\}$ are within probability limit $f(\mu_r \leq r_i)$.

The data points circled in green in class C are within first limit function $f(\mu_r \leq r_j)$ and are accumulated in D^k voting list. However, those circled in yellow are missed neighbors outside probability limit $f(\mu_r \leq r_j)$. Most data points circled in blue are near-missed neighbors in class B, except one feature circled in red. When looking closely in Fig. 2, class B has 5 potential neighbors which could have resulted to a majority class in

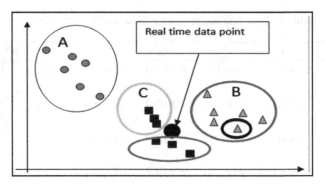

Fig. 2. Closest neighbor data point

relation to real time r_j; but it is not the case due to $f(\mu_r \leq r_j)$ condition, thus resulting into to minor feature problem because eligible features above $\mu_r > r_j$ are discarded.

Second Classifier: Improved Supplementary KNN Classier
In the second classifier all eligible near-missed predicators with smallest distance closer to real-time data point r_j are preserved by employing secondary KNN Distance Unit Vote Function (DUVF) defined by Eq. (12):

$$\text{DUVF}(\mu_r > r_i) = DI(\mu_r, r_j) \tag{12}$$

where DI is smallest similarity distance of all best mean $\phi = \{\mu_1, \mu_2, \ldots, \mu_r\}$ in relation to real time instances $RT = \{r_1, r_2, r_3, r_j\}$ satisfying $f(\mu_r > r_i)$ condition. We modified the distance DI to include integral limit function, such that only best mean μ_r greater than real-time r_j are accommodated using Eq. (13) (see Table 2)

$$DI(I > r_i) = \sqrt{\sum_{x=1}^{k} \int_{r_x}^{\mu_x} (\mu_x - r_x)^2} \tag{13}$$

Table 2. KNN distance vote cast unit

$DI(c_r > r_i)$	r_1	r_2	r_3	r_j	Smallest Distance E^K
c_1	$DI(\mu_1 - r_1)^2$	$DI(\mu_2 - r_2)^2$	$DI(\mu_3 - r_3)^2$	$DI(\mu_r - r_j)^2$	$(E_1) = \sqrt{\sum_{x=1}^{k} \int_{r_x}^{\mu_x} (\mu_x - r_x)^2}$
c_2	$DI(\mu_1 - r_1)^2$	$DI(\mu_2 - r_2)^2$	$DI(\mu_3 - r_3)^2$	$DI(\mu_r - r_j)^2$	$(E_2) = \sqrt{\sum_{x=1}^{k} \int_{r_x}^{\mu_x} (\mu_x - r_x)^2}$
c_i	$DI(\mu_1 - r_1)^2$	$D2(\mu_2 - r_2)^2$	$DI(\mu_3 - r_3)^2$	$DI(\mu_r - r_j)^2$	$(E_k) = \sqrt{\sum_{x=1}^{k} \int_{r_x}^{\mu_x} (\mu_x - r_x)^2}$

Consequently, producing a vector E^K of all top K smallest neighbors within the secondary integral $f(\mu_r > r_i)$ condition to accumulate all smallest nearest neighbors based on DI distance of Eq. (13) for a specific class c_r. Finally, both majority vote lists D^K and E^K in Eq. (11) and (13) are joint as majority vote of convenience $JV^K = D^K \cup E^K$ known as LHMV defined by Eq. (14):

$$LHMV = JV^K =_v^{argmax} \sum_{(D_k,E_k)\in V_z} (D_k \cup E_k), \qquad (14)$$

where, V_z is a majority vote vector with predicted class categories parallel to D_k in conjunction with voting list E_k. Thus, the Eq. (14) computes a majority class from D_k and E_k voting list, such that a class category with more votes is assigned to new instance as class label. The LHMV Eq. (14) is expanded into Algorithm 2.

Algorithm 2: Lightweight Hybrid Majority Vote Algorithm

1. ***Require*** *: TrainerSet$^{N \times M}$ training dataset each with combined*
2. ***Input*** *:*
3. *Set RT$^{Row \times J}$ ← extractRealTimeFeature(acce.x, acce.y, acce.z)//accelerometer X,Y,Z*
4. *Set R to value to resize training dataset rows*
5. *reducedTree$^{R \times M}$ ← compressDataset(TrainerSet$^{K \times N}$, R)//based on Algorithm 1*
6. ***Output*** *: initialise majority vote JV^K to empty*
7. *Set K to top nearest neighbours to control majority vote*
8.   **for** each testRow ∈ RT$^{Row \times J}$ **do** // select each row real time $\{r_1, r_2, r_3, r_j\}$
9. *select each r_j from testRow // select each attribute in testRow*
10.   **for** each ϕ in $C_k \in Ð$ **do** // for each human activity in reduced training set
11. **for** $\mu_r, \in \phi$ **do** //for each multi-featured attribute in class category
12. **if** $\mu_r \le r_j$ **then** // first integration limit for Gaussian Normal Distribution
13. σ ← *stadardDeviation(μ_r, r_j)// compute using equation .10*
14. $D_k \leftarrow D_k \times \dfrac{1}{\sqrt{2\pi\sigma^2}} e^{-\frac{(r_j-\mu_r)^2}{2\sigma^2}}$ // Gaussian probability approximation Naïve Bayes
    equation.8
15. ***end-if***
16. **if** $\mu_r > r_j$ **then** // Second integration supplementary condition
17. $E_k \leftarrow E_k + \left(\sqrt{(\mu_r - r_j)^2} \right)$ //Smallest distance probability using KNN equation
18. ***end-if***
19. ***end-for***
20. ***end-for***
21. *LHMV = JV^K = $_v^{argmax}$ $\Sigma_{(D_k,E_k)\in V_z}(D_k \cup E_k)$, // joint votes list of $D_k \cup E_k$*
22. ***end-for***

Our LHMV algorithms require training dataset and test dataset to classify human activities. Dataset can be divided into training and test dataset using specific ratio 0.7:0.30 or 0.6:0.4 or 50:50 or any splitting ratio. In line 5, training *TrainerSet$^{N \times M}$* is reduced to training class-tree *reducedTree$^{R \times M}$* using Algorithm 1 as preloaded training dataset. Therefore, each row of real time test vector is extracted from dataset RT^J in line 8. Then, probability approximation is estimated between test attribute r_j and training attribute mean μ_r to compute standard deviation σ for each human activity class category C_r in training dataset, see line 13 and line 14 of Algorithm 2. Thereafter, each probability approximation is accumulated into D_k only if the first limit function $f(\mu_r \leq r_j)$ is met otherwise, smallest distance is computed and accumulated into E_k vector. Thereafter, D_k and E_k majority neighbors are sorted in descending and ascending order respectively and joined into marriage of convenience as JV^K majority vote vector in line 21. Lastly, the first Kth human activity class in each vector D_k and E_k are matched and any human activity category with more votes is assigned to new instance r_j as a predicted class.

4 Experimentation

We selected four HAR datasets consisting of raw continuous random variables with 27 combined human activities classes. All the datasets PAMAP2, WISDM and Dataset-HAR-PUC-Rio-Ugulino (PUCRU) were downloaded from UCI of machine learning repository. Thereafter, we removed unimportant attributes such as subject id, gender, age, BMI, weight and height. We then normalized all datasets by removing missing values and capped features to 302 rows per human activity as listed in Table 4. We converted all datasets into Comma Separated Values (CSV) files to meet requirements of R programming language [26, 28, 29].

Table 3. Publicly available human activity recognition dataset

Source	Dataset	Sensor	Features	Instances	Classes
[4]	PAMAP2	3 Colibri wireless inertial measurement units	9	3624	12 classes by 302 instances
[29]	WISDM	Smartphone and Smartwatch Accelerometer	6	5436	12 classes by 302 instances
[30]	PUCRU	Wearable devices Accelerometer	12	1208	5 classes by 302 instances

We simulated our LHMV algorithm and compared it with SVM, C4.5, KNN, Boosted Tree (BT), Naïve Bayes and Random Forest (RT) in *R* programming language. The comparison is conducted in *R* programming language using Caret, e1071 and Rweka libraries [26, 28]. During simulation we used 60 rows as transformation reductions per human activity to compress every dataset to best mean classification class-tree similar to [23]. On each existing R classifiers we used R default settings. In our simulation setup, we used cross validation approach proposed by Kuhn [26, 28]. We employed K-fold cross validation approach to randomly partition each dataset, one after another, into equal K sub-sets; such that K-1 is used as training and K subset is used as testing on each personalized dataset. The cross-validation algorithm iterates K times, such that all K-fold instances are used as both training and exactly once as testing. The results of cross validation are presented in precision, accuracy, recall and f-measure.

5 Experimental Results

We present the results of comparison (LHMV against KNN, SVM, C4.5, BT, and Random Forest (RF)) using three selected datasets given in Table 3. All algorithms using WISDM dataset (consisting of 6 tri-axial values) achieved accuracy, precision, recall and f-measure of 100% across all human activities as shown in Table 4.

Table 4. Comparison results using WISDM dataset

Measurements	SVM	C4.5	RF	BT	KNN	NB	LHMV
Accuracy	0	100	100	100	100	100	100
Precision	0	100	100	100	100	100	100
Recall	0	100	100	100	100	100	100
F-Measure	0	100	100	100	100	100	100

However when PAMAP2 dataset (consisting 9 raw tri-axial attributes from accelerometer, gyroscope and magneto devices) is used, the Random Forest, C4.5 and Boosted Tree achieved 100% in all human activities in accuracy, precision and recall as shown in Table 5.

Table 5. Comparison results using PAMAP2 dataset

Measurements	SVM	C4.5	RF	BT	KNN	NB	LHMV
Accuracy	97	100	100	100	97	97	90
Precision	97	100	100	100	97	97	86
Recall	97	100	100	100	100	100	98
F-Measure	97	100	100	100	98	98	91

The results are similar to preliminary results of tree-based and ensemble algorithms presented in [10, 12], but with improved accuracy, precision and recall above 97% in KNN and Naïve Bayes. Algorithms SVM and KNN reported nil in all human activities as shown in Table 6 using PUCRU training instances whereas all other algorithms, including our LHMV, reported precision, accuracy, recall and f-measure of 100% in static human activities (sitting and standing).

The improved classification accuracy in simpler algorithms is owed to usage of largest training datasets; a confirmation that Naive Bayes and KNN performs optimally than other sophisticated algorithms using largest dataset [22, 23]. The similarity between Naïve Bayes and LHMV in all the presented results is owed to the implementation of GNB as our first classifier and KNN as second classifier using split-then-join strategy. As observed, failure of one classifier (KNN) as shown in Table 6 does not affect the prediction of another classifier unless both classifiers fails. Overall, we can therefore conclude

Table 6. Comparison results using PUCRU dataset

Measurements	SVM	C4.5	RF	BT	KNN	NB	LHMV
Accuracy	0	100	100	100	0	99	83
Precision	0	100	100	100	0	100	83
Recall	0	100	100	100	0	99	83
F-Measure	0	100	100	100	0	100	83

that our LHMV is reliable, effective using reduced classification features of different sizes and is competitive with other algorithms. Our novel split-then-join strategy is effective and suitable to join simpler algorithms based on integral conditionality as compared other strategies presented in [13, 14, 20]. Moreover, the experimental results are competitive with voting ensemble strategies implemented in [10–13, 20]. The results confirm that ensemble algorithm increases predictions accuracy and outwit simpler algorithms (SVM, KNN and Naïve Bayes) [14, 16].

6 Conclusion and Future Work

The split-then-join approach to join GNB and KNN algorithms is presented in this paper. The split-then-join approach compresses and split the training dataset into first and second classifiers and then join their voting lists in marriage of convenience for final prediction. The results of our simulations showed that our LHMV is effectives and competitive with tree-oriented and other ensemble algorithm yet using small and reduced training dataset. We can conclude that our split-then-join approach as the union of convenience improved classification accuracy, precision, recall and f-measure of KNN and Naïve Bayes classifiers. The results reveals that if one algorithm fails, it does not impact the prediction unless both algorithms fails. In all training datasets, our algorithm reached the accuracy and precision between 80% and 100%. In future, we intend to evaluate time-complexity of our LHMV to determine its viability to be implemented on resources constraint smartphone with reduced small training instances.

References

1. Anguita, D., Ghio, A., Oneto, L., Parra, X., Reyes-Ortiz, J.L.: A public domain dataset for human activity recognition using smartphones. In: ESANN, pp. 437–442 (2013)
2. Zhang M, Sawchuk AA. USC-HAD: a daily activity dataset for ubiquitous activity recognition using wearable sensors. In: Proceedings of the 2012 ACM Conference on Ubiquitous Computing, 5 September 2012, pp. 1036–1043. ACM (2012)
3. Parkka, J., Ermes, M., Korpipaa, P., Mantyjarvi, J., Peltola, J., Korhonen, I.: Activity classification using realistic data from wearable sensors. IEEE Trans. Inf Technol. Biomed. **10**(1), 119–128 (2006)
4. Reiss, A., Stricker, D.: Introducing a new benchmarked dataset for activity monitoring. In: Wearable Computers (ISWC), 2012 16th International Symposium on 18 June 2018, pp. 108–109. IEEE (2012)

5. Su, X., Tong, H., Ji, P.: Activity recognition with smartphone sensors. Tsinghua Sci. Technol. **19**(3), 235–249 (2014)
6. Mannini, A., Intille, S.S., Rosenberger, M., Sabatini, A.M., Haskell, W.: Activity recognition using a single accelerometer placed at the wrist or ankle. Med. Sci. Sports Exerc. **45**(11), 2193 (2013)
7. Kaghyan, S., Sarukhanyan, H.: Activity recognition using K-nearest neighbor algorithm on smartphone with tri-axial accelerometer. Int. J. Inform. Models Anal, **1**,146–156 (2012)
8. Breiman, L.: Random forests. Mach. Learn. **45**(1), 5–32 (2001)
9. Kwapisz, J.R., Weiss, G.M., Moore, S.A.: Activity recognition using cell phone accelerometers. ACM SIGKDD Explor. Newsl. **12**(2), 74–82 (2011)
10. Catal, C., Tufekci, S., Pirmit, E., Kocabag, G.: On the use of ensemble of classifiers for accelerometer-based activity recognition. Appl. Soft Comput. **31**(37), 1018–1022 (2015)
11. Daghistani, T., Alshammari, R.: Improving accelerometer-based activity recognition by using ensemble of classifiers. Int. J. Adv. Comput. Sci. Appl. **7**(5), 128–133 (2016)
12. Gupta, S., Kumar, A.: Human activity recognition through smartphone's tri-axial accelerometer using time domain wave analysis and machine learning. Int. J. Comput. Appl. **127**(18), 22–26 (2015)
13. Gandhi, I., Pandey, M.: Hybrid Ensemble of classifiers using voting. In: 2015 International Conference on Green Computing and Internet of Things (ICGCIoT), pp. 399–404. IEEE, October 2015
14. Chan, P.K., Stolfo, S.J.: Experiments on multistrategy learning by meta-learning. In: Proceedings of the Second International Conference on Information and Knowledge Management, pp. 314–323. December. 1993
15. De Stefano, C., Fontanella, F., Di Freca, A.S. A novel naive bayes voting strategy for combining classifiers. In: 2012 International Conference on Frontiers in Handwriting Recognition, pp. 467–472. IEEE, September 2012
16. Reiss, A., Hendeby, G., Stricker, D.: A competitive approach for human activity recognition on smartphones. In: European Symposium on Artificial Neural Networks, Computational Intelligence and Machine Learning (ESANN 2013), 24–26 April, 2013, Bruges, Belgium, pp. 455–460. ESANN (2013)
17. Reiss, A.: Personalized mobile physical activity monitoring for everyday life. Ph.D. thesis in Computer Science, Technical University of Kaiserslautern (2014)
18. Bao, L., Intille, S.: Activity recognition from user-annotated acceleration data. In: International Conference on Pervasive Computing, pp. 1–7 (2004)
19. Ravi, N., Dandekar, N., Mysore, P., Littman, M.L.: Activity recognition from accelerometer data. In: AAAI 2005 July 9 (vol. 5, No. 2005), pp. 1541–1546) (2005)
20. Lockhart, J.W., Weiss, G.M.: Limitations with activity recognition methodology & data sets. In: Proceedings of the 2014 ACM International Joint Conference on Pervasive and Ubiquitous Computing: Adjunct Publication, 13 September, pp. 747–756. ACM (2014)
21. Gadebe, M.L., Kogeda, O.P., Ojo, S.O.: Personalized real time human activity recognition. In: 2018 5th International Conference on Soft Computing & Machine Intelligence (ISCMI), pp. 147–154. IEEE, November 2008
22. Gadebe, M.L., Kogeda, O.P.: A lightweight hybrid majority vote classifier using Top-k dataset. In: Patel, K.K., Garg, D., Patel, A., Lingras, P. (eds.) Soft Computing and its Engineering Applications. icSoftComp 2020, CCIS, vol. 1374, pp. 182–196. Springer, Singapore (2021). https://doi.org/10.1007/978-981-16-0708-0_16
23. Kose, M., Incel, O.D., Ersoy, C.: Online human activity recognition on smart phones. In: Workshop on Mobile Sensing: From Smartphones and Wearables to Big Data 16 April 2012, vol. 16, No. 2012, pp. 11–15 (2012)
24. Kuhkan, M.: A method to improve the accuracy of k-nearest neighbor algorithm. Int. J. Comput. Eng. Inf. Technol. **8**(6), 90 (2016)

25. Ruggieri, S.: Efficient C4. 5 [classification algorithm]. IEEE Trans. Knowl. Data Eng. **14**(2), 438–444 (2002)
26. Arlot, S.: Celisse, A.: A survey of cross-validation procedures for model selection. Stat. Survey **4**, 40–79 (2010)
27. Kuhn, M.: Building predictive models in R using the caret package. J. Stat. Softw. **28**(5), 1–26 (2008)
28. Kuhn, M.: A Short Introduction to the caret package. R Found Stat. Comput, pp.1–10 (2015)
29. Weiss, G.M., Yoneda, K., Hayajneh, T.: Smartphone and smartwatch-based biometrics using activities of daily living. IEEE Access **7**, 133190–133202 (2019)
30. Ugulino, W., Cardador, D., Vega, K., Velloso, E., Milidiú, R., Fuks, H.: Wearable computing: accelerometers' data classification of body postures and movements. In: Barros, L.N., Finger, M., Pozo, A.T., Gimenénez-Lugo, G.A., Castilho, M. (eds.) SBIA 2012. LNCS (LNAI), pp. 52–61. Springer, Heidelberg (2012). https://doi.org/10.1007/978-3-642-34459-6_6

Identification of Barriers in Adoption of IoT: Commercial Complexes in India

Nishani Salvi[1(✉)] and Gayatri Doctor[2]

[1] Faculty of Technology, CEPT University, Ahmedabad, India
nishani.salvi@gmail.com
[2] Faculty of Management, CEPT University, Ahmedabad, India
gayatri.doctor@cept.ac.in

Abstract. The Internet of Things (IoT) has been widely emerging in various sectors with a very wide range of sensors and wireless networks. It is used in many fields like the educational sector, health sector, agricultural sector, even in the construction industry. The study focuses on the implementation of IoT in commercial complexes of India as well as the barriers faced by different stakeholders throughout the lifecycle of the project, which were further categorized into three types. The objective was achieved by quantifying the responses collected from the respondents using correlation method. The variables considered for the correlation analysis were age, awareness, usage, and agreement. Also, benefits of using IoT were observed for this study.

Keywords: Internet of Things · Lighting and temperature adjustment · Smart parking · Commercial complexes

1 Introduction

1.1 Introduction and Need for the Study

The Internet of Things (IoT) has been widely emerging in various sectors with a very wide range of sensors and wireless networks. It is used in many fields like the educational sector, health sector, agricultural sector, even in the construction industry. As per, (Yeo et al. 2020) smart cities and smart buildings are the main highlighted areas where IoT technologies have been used widely. The main aim of introducing IoT technologies in our daily life is to improve the efficiency of work and to improve the quality of life. The need is to reduce direct human interaction as little as possible (Kamel and Memari 2019).

A smart building is basically the one using technology to optimize the performance of a building. All the components of the building are interlinked with each other to effectively regulate the building. The features of smart building are controlling air quality, lighting, comfort, security, etc. smart buildings deliver its occupants comfort and make them productive by creating a comfortable workspace by using sensors (Building Effeciency Initiative, 2011; True Occupancy 2019). From the study of various research

© Springer Nature Switzerland AG 2022
K. K. Patel et al. (Eds.): icSoftComp 2021, CCIS 1572, pp. 181–193, 2022.
https://doi.org/10.1007/978-3-031-05767-0_15

papers and responses obtained from the respondents it was known that almost 80% of the users were unaware about the benefits of IoT. Various application of IoT has been explored in the construction industry – On-site safety monitoring using wearables, Maintenance of fuel and power consumption, Productivity monitoring, Machine control, Site monitoring, Fleet Management, etc. (Commercial Design 2020; Construction Tech 2020; Ecoideaz 2020; HiDecor 2020; HIoTron 2020).

India is at an emerging stage in IoT technology hence the benefits and barriers faced by project stakeholders are not yet completely known. Also, awareness among the users is also yet to be estimated. Thus, there arises a need to study the advantages and drawbacks of IoT in buildings and to identify the awareness among its users. The objective is to understand how IoT technology can be used in commercial complexes.

- To identify which different technologies are used.
- To identify the stakeholders and barriers faced in implementation.
- To identify the benefits of using IoT in buildings.
- To identify the barriers in the implementation of the technology used.
- To identify the awareness among the users.

2 Literature Review

Literature review was carried out to understand the current usage of internet of things in various buildings.

2.1 Smart Buildings

The modern definition of a smart building is a building equipped with all necessary smart devices like lighting, heating, electric fixtures which can be controlled from anywhere with the help of a smartphone or a computer. Usually, smart buildings are assumed to be just automated buildings. But looking at the commercial definition, the building is programmed to monitor the environment and the actions performed by the users, behaviour patterns and to predict the future states of the building (Batov 2015).

García et al. (2017) talk about how a balance is required between the environment and natural resources to provide a level of comfort to the users of the building. The motive is to educate the users and to motivate them for effectively using resources. Also, how old buildings lack to provide comfort to its users and energy-efficient resources. The background of various cities worldwide is discussed here. And a proposed system is discussed along with the layers and organization involved in it. It is also discussed that how smart cities, smart homes, smart buildings are still in the inception phase and is an emerging idea. It is also mentioned that smart city is one of the popular emerging concepts. The objective of this is to reduce the difficulties faced due to urbanization. A definition of Smart City is stated by the author in this study which goes like: "The motive of using smart technologies in the cities is to make everything connected including health, safety, transportation, education, etc." (García et al. 2017).

Batov (2015) mentions the benefits of a smart building by reviewing different art projects. The potential benefits of a smart building are shown in the figure.

By inhabitants comfort the author means that the building understands the behaviors of its inhabitants and tries to provide comfort according to their habits. By adopting the technologies, the building will reduce its energy consumption and help in cost-cutting to the users as well. Since everything in the building is automated it will eventually reduce the time in daily routine chores. Safety is a very important part of any building, adopting smart technologies will help in detecting fire, leads or any hazardous incidents and will warn its occupants. A smart building will also take care of the occupants' health and provide appropriate air conditioning, lighting, temperature and other health care parameters (Batov 2015). Jia et al. (2019) does not directly discuss smart buildings instead he discusses what are the benefits and challenges of using IoT in buildings in a broader perspective. The paper is divided into 4 primary aspects, which are:

- To provide an aggregate knowledge of the IoT domain to the researchers, and industry professionals.
- To focus on the current scenario and future of IoT in the building industry.
- To improve the functionality of the buildings by discussing the latest technologies and recent development in IoT.
- To understand the benefits and challenges of using IoT in buildings for its entire lifespan.

2.2 IoT in Residential Buildings

IoT in residential buildings constitutes smart homes. Smart homes incorporate IoT in healthcare, safety, security, comfort, and energy consumption. Smart homes provide a better quality of life by creating an automated environment. It optimizes the comfort of the users by observing their actions and gives them a better living environment (Alam et al. 2012). The author (Chan et al. 2008) here states that the concept of smart homes is to improve the lifestyle of the elderly and disabled. The major targets of introducing technology are to monitor mobility and physiological parameters, improve comfort, and deliver therapy. By comfort the author means the ease in daily activities using IoT technology, wearable devices and sensors help in monitoring the health of the users and security in homes can be monitored by installing sensors at homes (Alam et al. 2012). Many assistive systems are available: electric wheelchairs, stairlifts, etc. The benefits of using IoT technologies in homes are stated below:

- It benefits the users who are staying alone and need assistance during emergencies.
- It is beneficial to the elderly and disabled people.
- It is beneficial to the people living in remote areas or urban areas with an inadequate amount of health service.
- It is beneficial to people who are suffering from chronic diseases and need continuous help and monitoring.
- It is beneficial to the people who are involved in the health care sector and are practising virtual visits.

The disadvantages are there is insufficient comprehension of user needs and improper demands of products and services used in smart homes are not properly explained to the users and hence the industry dominated the users. The main issue is a breach of privacy and hence older people are hesitant to share their details. Mistrust among the users and the producers results in a smaller number of users. It can be concluded that using IoT technologies in homes can help in taking proper care of the elderly and disabled people. Trust between the agencies is the most important factor in this industry. Also, the privacy and confidentiality of the users must be preserved (Chan et al. 2009).

2.3   IoT in Office Buildings

When internet of things is involved with an office building, it can be said as a smart office. A smart office basically ensures appropriate and effective use of technologies and infrastructure. Nowadays, with the advancement of technology the offices are automated. Also, there is a need of having a transparent environment. By including technology in the office, the systems become transparent and allow to share the information more openly (Bhuyar and Ansari 2016). The objective of having a smart office is to provide comfort to its occupants, improve their efficiency, improve the operations of building system, decrease energy consumption, and improve the life cycle of its utilities. A huge amount of energy consumption is done by office buildings because of their architecture. The IoT can be applied to the entire building, i.e., on lighting, temperature, air conditioning, ventilation, and security. For security purposes, offices use biometric or fingerprint-based identification as it is the safest form of identification. Since there is personal information of the users stored in the system, security becomes the most important factor, and it should remain safe. The biggest disadvantage of connecting every system in the office on one platform is that it is connected to the internet or Wi-Fi and when it shuts down the entire system shuts down, which may lead to losing data if it is not saved (Rafsanjani and Ghahramani 2019). Hence, by using the internet of things the offices will see a reduction in energy consumption, ease in document management, increased productivity and efficiency in the occupants or users.

2.4   Research Gap

Internet of Things (IoT) is an emerging technology which is being used in various applications and sectors. Some benefits of IoT have been discussed and mentioned in books, papers, and articles. As the adoption of IoT is emerging in India the benefits and barriers faced by designers, clients, PMC, stakeholders in the implementation of these technologies in buildings are not known upto its full extent, hence the study aims to investigate the barriers faced by the designers, clients, Project Management Consultants, and stakeholders during implementation, construction, and post construction phase. Also, to identify awareness among the users by comparing various factors like: age, awareness, usage and agreement rating, and to determine the benefits of using IoT in buildings (Fig. 1).

3 Research Methodology

3.1 Introduction

The main objective of the research was to get familiar with the process and to achieve new perception, to test a hypothesis, to determine the characteristics of a group or an individual. The research methodology consisted of four main stages, namely literature review, data collection, data analysis, and conclusion. The stages are shown in the Fig. 2.

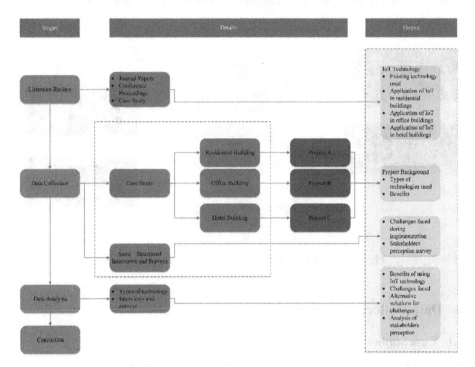

Fig. 1. Research methodology flow chart

Phase 1:

- *Literature Review*: To determine the various technologies like 'smart homes', 'smart offices', etc. used in buildings various papers and conference proceedings were referred. The literature review was carried out for IoT technologies in residential, office and hotel buildings to understand its implementation in different sectors.

Phase 2:

- *Data Collection*: For data collection, three methods were used: A case study for understanding the types of technologies and their benefits; Semi-structured Interviews to understand the challenges faced in the implementation of such technologies and a Questionnaire survey to know about the mindset of the stakeholders regarding the implementation of IoT technologies used in the buildings.

Phase 3:

- *Data Analysis:* Collected data in the form of case studies, interviews, and surveys have been analyzed to provide solutions for commonly faced problems.

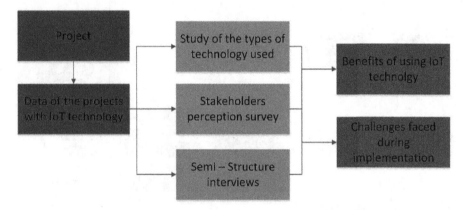

Fig. 2. Data analysis methodology for commercial complexes

4   Data Collection and Analysis

4.1   Case Studies

A total of four case studies had been selected for this research work. The case studies considered are for commercial complexes in India. The case studies had been selected where IoT technologies are used. The first case study selected was Shapath-V, which is in Ahmedabad. It is a 19-storey building and first green building in the state. The features of the building are sensor based pressurized staircase, which prevent fire from expanding to other floors, treated fresh air supply, heat recovery wheel which prevents rising of temperature in the building, CO_2 sensor which monitor CO_2 levels in the build and all the other services of the building are controlled by one system only. The second case study which is Intel SRR3 located in Bangalore and is a 9-storey building with a lead platinum certificate. The features of the building are power saving mode, fuel cell system which connects natural gas into electricity. Then there is smart ceiling water saving feature and video wall. The next case study selected was volteo which is situated in Hyderabad it is an IOT and IT services company and the idea was to build a formal and casual workspace for its employees. The main aim of the company was to showcase their own devices and sensors in their own office buildings so the main features of the building are color-coded zones smart wall lighting in temperature adjustment sensors for energy consumption and water level monitoring so there was a smart wall which was built right in the center of the office which showed surrounding temperature, amount of energy consumed and water level consumption. The last case study was Siemens which is located in in Mumbai, after the Covid pandemic the Siemens office decided to make the office smart and because of that they developed an app which had a death check in and availability feature which also had a call for elevator the app showed your allocated desk where you could change lighting and temperature adjustment according to your preference, there was a service or repair request and sanitation request too (Table 1).

Table 1. Details of the case studies

Case study no.	Category	Name of the project	Location
1	Commercial complexes	Shapath V	Ahmedabad
2		Intel SRR3	Bangalore
3		Volteo	Hyderabad
4		Siemens	Mumbai

4.2 Identification of Barriers

After taking interviews with the stakeholders and vendors different types of barriers were identified, which were further categorized into: Technical, Financial and Miscellaneous barriers. The causes for the barriers were also identified and is mentioned below:

1. Technical Barriers

   a. *Data Security:* The consumers were hesitant to upload their data on the cloud due to security reasons. Due to an increase in the number of hijacks and security breach the consumers refrain from adding or uploading their data online.
   b. *Compatibility:* With the advancement in technology, old sensors installed at the offices are not compatible with new ones. This creates a problem in linking the old sensors with the new sensors.
   c. *Interoperability:* The difference between two different types of systems can cause a problem with their operations. It is difficult to maintain the operations and connections between two systems.
   d. *Poor Design of Sensors:* Vendors to gain profit, often compromise with the quality of sensors, which results in poor performance of sensors as well as of the building.
   e. *Unpredictable Actions of Sensors*: Sensors are based on technology, hence there are chances that there can be a technical glitch in the system which may cause the sensors to behave in an unwanted manner.

2. Financial Barriers

    a. *Investment:* Not all firms are ready to invest in these technologies, due to the lack of awareness among the consumers. And the sensors are expensive and are not widely used in many cities in India.

    b. *Maintenance:* There is an extra cost for the maintenance of these sensors. In case the sensors require any repair the cost of it is very high, this is the reason for fewer people investing in them.

    c. *High Price of Sensors:* Since these technologies are not widely used, the sensors are expensive.

    d. *Limiting the Sensors outside the Office areas:* The builders do not provide these technologies inside the offices, which becomes an extra cost for the consumer to install these sensors and technology.

    e. *Updating the Systems:* Often vendors tell the consumers that the devices and sensors do not require any updates, but as and when new technologies come up, updates are required, and the vendors then charge extra for updating the devices and systems.

3. Miscellaneous Barriers

    a. *Consumer Awareness:* Due to a lack of awareness amongst the consumers, the technologies are not yet known and very well used.

    b. *Existence of the Vendor:* There is always a question to the consumer about the existence of vendors in the next 10 years. Due to the rapid advancement, there are chances that the vendor no longer exists in the market.

    c. *High Expectations of Consumer:* Consumers have very high expectations from the technology, with a minimum budget. The vendor cannot meet the requirements of the consumer for their price which becomes a challenge to the vendor.

    d. *Authentication of Vendors:* Consumers are worried about the authentication of vendors since the technologies are less known in the industry.

    e. *Slow Adoption of Technology:* Due to the lack of awareness, the growth of these technologies is not increasing in the way it is required.

    f. *Absence of IoT Guidelines:* Since these technologies and sensors are not yet known to the people, there are no strict rules and guidelines to be followed.

    g. *Compromise in the Quality of Sensors:* There are no strict rules and regulations for the use of the sensors and technologies, the vendors often tend to compromise with the quality of sensors.

    h. *Approval Process:* There are different approval for the installation of sensors, which is a hassle to the consumers, as well as it takes very long for the confirmation.

    i. *Change in Demand:* With the advancement in technology, there is always a different requirement for different consumers. The vendors find it hard to keep up with the demands of the consumers.

    j. *Obsolete sensors:* New technologies and sensors are coming up in the market every day which makes the old technologies and sensors obsolete.

4.3 Data Analysis

Data analysis was carried out for the questionnaire survey which helped in identifying the awareness among the people. A total of 104 surveys were carried out on different categories of people. Some were working in IoT or IT firms, while others were not working with any technology in their workspace. Based on the survey correlation was identified between four different variables. The hypothesis formulated for this study has been tested using an appropriate statistical method and is discussed in this part. Pearson's correlation (r) was used for this study to understand the linear relationship between the variables were awareness, age, usage, and agreement.

H_{01}: *There is no correlation between age and awareness regarding internet of things.*

Correlation value obtained from the analysis showed that there is no significant relationship between the variables awareness and age. The R-value obtained was $(-0.330 < -0.5)$ hence, the mentioned hypothesis was accepted. This showed that awareness does not increase or decrease with increase or decrease in age.

H_{02}: *There is no correlation between awareness regarding internet of things and usage of the technologies.*

Correlation value obtained from the analysis showed that there is no significant relationship between the variable's awareness and usage. The R-value obtained was $(0.102 < 0.5)$ hence, the mentioned hypothesis was accepted. This showed that usage of technology does not increase with an increase in awareness.

H_{03}: *There is no correlation between awareness regarding internet of things and agreement on investing in the technologies.*

Correlation value obtained from the analysis showed that there is no significant relationship between the variable's awareness and agreement. The R-value obtained was $(0.141 < 0.5)$ hence, the mentioned hypothesis was accepted. This showed that agreement on investing in technologies does not increase with an increase in awareness.

H_{04}: *There is no correlation between agreement on investing and usage of the technologies.*

After identifying the Pearson's coefficient, the R-value obtained was $(0.368 < 0.5)$. The results showed that the null hypothesis was accepted as there was no significant relationship between the variables agreement and usage. This showed that agreement on investing in technologies does not increase the usage of it. The direction of the variables is shown in the figure. A summary table of tests performed along with the interpretation in shown in the Table 2.

4.4 Analysis for Semi-structured Interviews

The interviews were carried out to identify the barriers faced by the stakeholders and vendors involved in the project. The barriers were further categorized into three groups. The table and graph shown the number of times similar kind of barriers were faced by the stakeholders and vendors. The Fig. 3 shows the technical barriers faced by the stakeholders and vendors. The highest frequency of barrier faced was data security, and interoperability, while compatibility and unpredictable actions of sensors were less comparatively. The financial barriers faced by the stakeholders and vendors. Investment, maintenance, and the high price of sensors were the major barriers faced by the

Table 2. Summary Table of R-value

Variable	R-Value	Result	Interpretation
Awareness and Age	-0.330	Null Hypothesis Accepted	No significant relation
Awareness and Usage	0.102	Null Hypothesis Accepted	No significant relation
Awareness and Agreement	0.141	Null Hypothesis Accepted	No significant relation
Agreement and Usage	0.368	Null Hypothesis Accepted	No significant relation

stakeholders and vendors. The Figs. 3 and 4 shows miscellaneous barriers faced by the stakeholders and vendors. Consumer awareness was a major concern and barrier faced by the stakeholders. While compromise in quality of sensors, approval process, high expectations and existence vendors were also few major barriers faced.

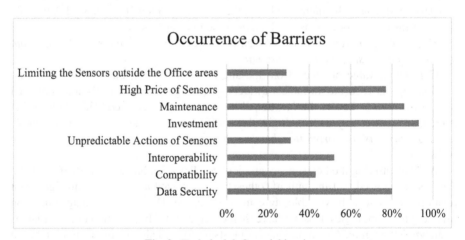

Fig. 3. Technical & financial barriers

4.5 Benefits of Using IoT

After the identification of awareness among the people, the benefits of using IoT in commercial complexes were identified with different stakeholders. The benefits identified are discussed below:

1. Reduction in energy consumption: Including IoT in commercial complexes will reduce energy consumption by 10%–30%. Also, it will help in financial savings significantly and can make up to green building standards.

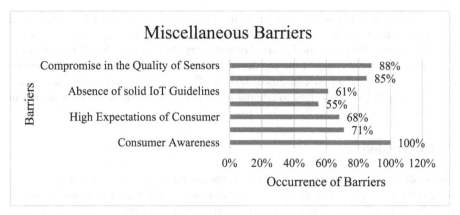

Fig. 4. Miscellaneous barriers

2. Improving efficiency of the building: HVAC and lighting is one of the most common steps used for improving the efficiency of the building. Since, all the rooms or offices in a complex has a different requirement, installing sensors can monitor and adjust the requirements of the room accordingly making it more efficient. The lighting and temperatures adjust automatically based on the requirements of the rooms.

3. Maintenance of the Building: Maintenance of all the services like HVAC, electrical, plumbing, firefighting, etc. is supposed to be done on regular basis. Including IoT for these services can help a lot in timely maintenance. The sensors can detect any abnormal behaviour of these services and can prevent it from any further more damage. By this, a lot more money can be saved from the maintenance of these services.

4. Increase in Productivity of Residents: Since the sensors installed in the offices monitor the requirements of their residents, the main aim is to give a comfortable experience to their residents. The Health and safety of their residents is the focus and to implement it effectively.

5. Efficient use of Resources: Insights identified from the building has been used in the planning of proper usage of resources.

6. Proper Utilization and Optimization of Space: Installing sensors helps in the management and utilization of space efficiently. The sensors help in monitoring the usage pattern of its residents and can organize the places accordingly. The sensors keep a schedule of the usage pattern and can tell the availability of the space too.

7. Safety and Security: Safety and security is the most commonly used domain in commercial complexes. Sensors have been installed to monitor any unusual activity or fatality. They send an alert in case of any unusual activity.

8. Lighting control: The usage of lighting is maximum in any building and also it is neglected more often. This tends to a lot of wastage of electricity. Smart lighting has been used to avoid wastage of electricity and the lights can be programmed based on the usage of the rooms. Timers have been set for the lights to turn on and off.

9. Temperature Control: Air conditioning is a major part of any building, but the use of it is done sensibly. Usually, the air conditioning is turned on even when a room

is not in use, this results in wastage of energy. Sensors installed for temperature control helps the admin to take necessary actions when the usage is less or more.

10. Plug Load: Plugs are used every day and everywhere, even when the devices are not in using the energy from the devices is being used. Since, it is not practically possible to unplug all connected devices of a complex, sensors installed in the building helps the devices to automatically cut off the load when they are not in use.

5 Conclusion

The study aimed at finding the barriers to the adoption of IoT in commercial complexes and identifying various benefits of implementing it. Also, awareness was identified regarding the IoT among the people. By the identification of barriers, the stakeholders can come with measures to eliminate these barriers and provide better service to the users. In this study, four case studies of commercial complexes were identified to understand different types of technologies used. Furthermore, the different stakeholders and vendors were identified who were involved in the project at different stages of its construction. Semi-structured interviews were carried out with identified stakeholders and vendors to identify various barriers observed in the implementation of IoT. After the identification of barriers, a people's perception survey was carried out to identify awareness amongst them. The barriers identified were categorized into three sections, i.e., technical barrier, financial barrier, and miscellaneous barrier. The technical barrier included Data Security, Compatibility, Interoperability, Unpredictable Actions of Sensors. The financial barrier included Investment, Maintenance, High Price of Sensors, Limiting the Sensors outside the Office areas. And lastly, miscellaneous barriers included Consumer Awareness, Existence of the Vendor, High Expectations of Consumer, Authentication of Vendors, Absence of solid IoT Guidelines, Approval Process, Compromise in the Quality of Sensors. For technical barriers, the highest frequency of barrier faced was data security, and interoperability, while compatibility and unpredictable actions of sensors were less comparatively. For financial barriers Investment, maintenance, and the high price of sensors were the major barriers faced by the stakeholders and vendors. For miscellaneous barriers, Consumer awareness was a major concern and barrier faced by the stakeholders. While compromise in quality of sensors, approval process, high expectations and existence vendors were also few major barriers faced. People's perception survey involved in correlation analysis of four different variables. The variables identified for correlation were awareness, age, usage, and agreement. Correlation analysis aimed at finding a significant relationship between awareness and age, awareness and usage, awareness and agreement, agreement, and usage. The results showed that the null hypothesis was accepted as there was no significant relationship between the variables awareness, agreement, and usage.

References

Alam, M.R., Reaz, M.B.I., Ali, M.A.M.: A review of smart homes - past, present, and future. IEEE Trans. Syst. Man Cybern. Part C Appl. Rev. **42**(6), 1190–1203 (2012). https://doi.org/10.1109/TSMCC.2012.2189204

Batov, E.I.: The distinctive features of " smart" buildings. Proc. Eng. **111**(TFoCE), 103–107 (2015). https://doi.org/10.1016/j.proeng.2015.07.061

Bhuyar, R., Ansari, S.: Design and implementation of smart office automation system. Int. J. Comput. App. **151**(3), 37–42 (2016). https://doi.org/10.5120/ijca2016911716

Building Effeciency Initiative. (2011). What is a Smart Building? https://buildingefficiencyinitia tive.org/articles/what-smart-building

Chan, M., Campo, E., Estève, D., Fourniols, J.Y.: Smart homes - current features and future perspectives. Maturitas **64**(2), 90–97 (2009). https://doi.org/10.1016/j.maturitas.2009.07.014

Chan, M., Estève, D., Escriba, C., Campo, E.: A review of smart homes-Present state and future challenges. Comput. Methods Programs Biomed. **91**(1), 55–81 (2008). https://doi.org/10.1016/j.cmpb.2008.02.001

Commercial Design: Revolution in smart buildings (2020). https://www.commercialdesignindia.com/3847-revolution-in-smart-buildings

Construction Tech: IoT in Construction, The Big Picture (2020). https://unearthlabs.com/blog/con struction-tech/iot-in-construction/

Ecoideaz: Making Buildings Smarter in India: The IoT way (2020). https://www.ecoideaz.com/expert-corner/making-buildings-smarter-in-india-the-iot-way

García, O., Chamoso, P., Prieto, J., Rodríguez, S., De La Prieta, F.: A serious game to reduce consumption in smart buildings. Commun. Comput. Inf. Sci. **722**, 481–493 (2017). https://doi.org/10.1007/978-3-319-60285-1_41

HiDecor. IoT in Workspace (2020). https://hidecor.in/iot-in-workspace/

HIoTron. IoT in Construction: Analysis of Innovations and Use Cases (2020). https://www.hio tron.com/iot-in-construction/

Jia, M., Komeily, A., Wang, Y., Srinivasan, R.S.: Adopting internet of things for the development of smart buildings : a review of enabling technologies and applications. Autom. Constr. **101**(July 2018), 111–126 (2019). https://doi.org/10.1016/j.autcon.2019.01.023

Kamel, E., Memari, A.M.: State-of-the-art review of energy smart homes. J. Archit. Eng. **25**(1), 03118001 (2019). https://doi.org/10.1061/(asce)ae.1943-5568.0000337

Rafsanjani, H.N., Ghahramani, A.: Extracting occupants' energy-use patterns from Wi-Fi networks in office buildings. J. Build. Eng. **26**(April), 100864 (2019). https://doi.org/10.1016/j.jobe.2019.100864

True Occupancy. What is a Smart Building? (2019). https://www.trueoccupancy.com/blog/what-is-a-smart-building

Yeo, C.J., Yu, J.H., Kang, Y.: Quantifying the effectiveness of IoT technologies for accident prevention. J. Manage. Eng. **36**(5), 04020054 (2020). https://doi.org/10.1061/(ASCE)ME.1943-5479.0000825

A Dynamically Adapting Framework for Stock Price Prediction

Shruti Mittal ⓘ and C. K. Nagpal$^{(\boxtimes)}$ ⓘ

JC Bose University of Science and Technology, YMCA, Faridabad 121006, Haryana, India
nagpalckumar@rediffmail.com

Abstract. The available research work in the area of stock price prediction is focused on generic prediction which doesn't involve stock specific issues. The fact remains that a stock may have quite different on-going trend than the general stock market. Thus a price prediction mechanism which is tailor made for the stock under consideration would be a welcome affair. This paper proposes such a tailor made mechanism that involves the initial creation of candidate predictions, selection of the best pair and subjecting this pair to further reduction of error using back propagation learning. The results obtained are quite precise and scalable for the extended period.

Keywords: Predictive analytics · Stock price prediction · Supervised learning · Statistical learning · Regression analysis · Long short-term memory (LSTM) · Time series analysis

1 Introduction

Recent past has seen a flood of research papers relating to stock price predictions [7–9, 12–15]. The reasons for this flood include: ease of data availability in electronic form, advances in machine learning, availability of open source APIs for these algorithms. The hypothesis behind all these works is that the machine learning will identify hidden patterns in data, capture trend and there will be a prediction breakthrough [10, 18]. Thus research community has tried to explore different strategies based upon deep learning networks [19–21], different types of signal/activation functions [17, 24], incorporation of impact of news [23] and social media sentiments etc. [22]. But the success is still far off. The reason for the same is that the share market is not totally governed by mathematical or logical patterns but it is also driven by other factors such as human manipulations, greed, social & economic sentiments and announcement of govt. policies etc. All these factors can affect the different stocks in different manners which may be good/ better for one stock and bad/worse for the other.

We, therefore, are of the view that it is inane to look for a generic, static and universally applicable prediction strategy that shall be appropriate for all the stocks in all the stock markets across the globe. Every stock in the stock market has its own individuality and therefore requires its own distinctive treatment. Finding such treatments for each stock in every stock market is a voluminous and tedious task. If one can design a stock specific

© Springer Nature Switzerland AG 2022
K. K. Patel et al. (Eds.): icSoftComp 2021, CCIS 1572, pp. 194–208, 2022.
https://doi.org/10.1007/978-3-031-05767-0_16

customized strategy keeping in view the individuality of the stock, the results will be more accurate. This paper is an effort in this direction. The work involves selection of a pair of stock specific prediction strategy from a set of created options. This selected pair is subjected to error correction learning for the further reduction of error thereby minimizing the prediction error.

Organization of the Paper. The paper contains 5 sections. Section 2 talks about literature survey. Section 3 talks about the details of the proposed mechanism. Section 4 provides the implementation details. Section 5 provides detailed results and analysis thereof. Section 6 concludes the paper in the light of results.

2 Literature Survey

As discussed above, a significant work is in progress in stock price prediction domain, most of which is based upon machine learning algorithms. In this section we take a look on some of the available recent literature in the area of stock price prediction.

Jingyi Shen et al. [1] proposed a model based on feature engineering and deep learning for predicting stock price. Their work includes pre-processing of the stock market dataset, utilization of multiple feature engineering techniques combined with a customized deep learning based system for stock market price trend prediction. The work proposed has performed well when compared with other machine learning models.

Kara et al. [2] in their work proposed two models based on ANN & SVM and compared their performance on Istanbul Stock Exchange. Ten technical indicators were selected as inputs of the proposed models. The experimental results showed that ANN performed better than SVM.

Qiu and Song et al. [3] in their work proposed a model based on artificial neural network optimized using genetic algorithm to predict the direction of the Japanese stock market. The work categorizes the technical indicators of the stock market into two major input categories and predict the direction of the daily stock market index. The optimization of the ANN with the help of Genetic Algorithm has enhanced the performance of the system.

Guresen et al. [4] analyzed the effectiveness of various neural networks like multilayer perceptron (MLP), dynamic artificial neural network (DAN2) and the hybrid neural networks which use generalized autoregressive conditional heteroscedasticity (GARCH) to extract new input variables on daily exchange rate values of NASDAQ Stock Exchange index. It was observed in their work that classical ANN model MLP outperforms DAN2 and GARCH-MLP.

Fischer et al. [5] in their work equated the performance of various deep learning techniques with long short-term memory for financial market predictions. They analyzed the performance of various LSTM model variants built upon daily close price, daily volume weighted average prices, weekly close price and Weekly volume weighted average prices and concluded that LSTM outperforms other deep learning networks and that it effectively extracts the useful information from the noisy financial time series data.

Krauss et al. [6] in their paper implemented and analyzed the effectiveness of deep neural networks (DNN), gradient-boosted-trees (GBT), random forests (RAF), and several ensembles of these methods for all stocks in S&P 500 from 1992 to 2015. Daily one-day-ahead trading signals were generated based on various trading strategies like daily returns, standard deviation, daily directional accuracy at the portfolio level and daily trading frequency and various trading strategies were built before investing.

McNally S. et al. [10] have compared the performance accuracy of various deep learning algorithms (optimized RNN and LSTM) with ARIMA model for Bitcoin Price Index. Their major observation is that the deep learning algorithms have better accuracy than ARIMA model and also that LSTM takes longer time to train than RNN.

Seo M. et al. [17] Proposed an ANN based model containing multiple hidden layers combined with Google Domestic Trends (GDT). They concluded that with more number of hidden layers and varying activation functions, accuracy of model can be improved. This, in our view, may not be always true. They have claimed that the performance of the hybrid model with GDTs outperformed GARCH model and the model without GDTs.

Pang X et al. [18], introduces a stock vector similar to word vector in deep learning and proposed deep LSTM based neural network with embedded layer and LSTM with automatic encoder with vectorized data for stock market prediction.

After going through the above literature it can be seen that most of the strategies are generic, with one or more machine learning algorithms [25–27] at their base. These algorithms [27, 28] can capture the common trend prevalent amongst all the stocks leaving behind the individuality of the stock as an outlier matter. Moreover, none of them has incorporated a provision for the human manipulations or political or social impact which an individual stock may be undergoing.

The distinct features of the proposed work are as follows:

- Stock Specific Strategy instead of the general one.
- Generation of multiple predictions instead of relying upon the single one.
- Extraction of statistical extensions from the predicted base set.
- Improving the input data quality through candidate selection with best proximity.
- Fine tuning of the predictions using error reduction through back propagation.

We suggest that the most important thing in the stock market is the on-going trend for the stock under consideration. This line of actions may or may not be in-line with the stock sector and the general stock market trend. This paper provides a strategy to generate a stock price prediction mechanism which is tailor-made for the stock under consideration.

3   Proposed Framework

3.1   Framework Description

The proposed mechanism involves taking up of the past price data for the stock under consideration from 1st July 2016 to 30 June 2019 on daily basis [16, 29–31, 32]. The data so obtained included Stock Price, Sector Index and Nifty 50 index at the close of the day. This data was used for the supervised learning with past n days data as input and the stock close price of n + 1th day as the output using sliding window mechanism as shown in the Fig. 1.

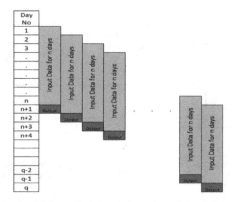

Fig. 1. Sliding window of size n for price prediction

A primary set of predictions Sp is created using a sliding window of 5, 10, 20 and 30 days with LSTM on Python platform. Sp is subjected to the statistical extensions using max, min and mean operations leading to the creation an additional set of prediction called mix bag predictions Sm. Union of Sp and Sm is done to create comprehensive set of predictions Sc.

All these predictions were compared with the actual price of the stock in future and coefficient of correlation (CO) and mean absolute error (MAE) is computed. Using this information, two top candidate predictions C1 and C2 are chosen. The basis for the Choice of C1 and C2 is MAE and CO is used to ensure the minimum correlation between the actual price and the candidate prediction. If CO is less than 0.95 the candidate prediction is not used.

Now these candidate predictions also have some error though quite small compared to others. Next task is to reduce the error in candidate predictions to the minimum level in order to bring it closest to the actual price of the stock. To accomplish this, back propagation algorithm was applied with two candidate predictions as input and the actual close price as the target. This task was performed on WEKA tool using Linear Regression. The results so obtained were tested and it was found that there has been a marked improvement in the prediction accuracy. Figure 2 shows the complete diagram of the design and implementation phase of the experiment conducted by us. Now the details of candidate predictions (c1 and c2) and the back propagation networks (ANN id) details were recorded for the stock under consideration in the data base.

The task is repeated for a set of stocks and for each stock C1, C2 and ANN ID is recorded as shown in Table 1. The details of the Table1 were used for making the predictions about the stock in the prediction phase as shown in Fig. 3.

Initially the work was performed for next day prediction but the quality of the results encouraged us to extend the experiment for generating predictions for next week, next fortnight and next month. So in general the proposed work uses n days of stock price, stock index and sector index as input and the stock price on (n + k)th day as the output with k = 1, 7, 15, 30. The results so obtained have been discussed in the next section.

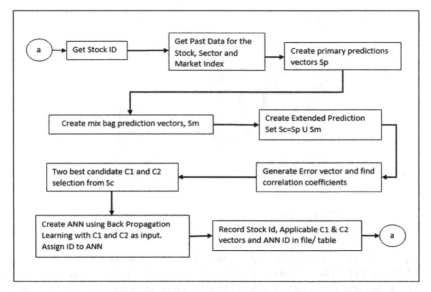

Fig. 2. Design and implementation phase to be repeated periodically for continuous adaption

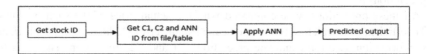

Fig. 3. Prediction phase

Table 1. Recording of ANN ID

Stock Id	C1	C2	ANN ID
101	Pred 5	Pred Max	A_1
102	Pred10	Pred 30	A_2
103	Pred 30	Pred Avg	A_3
.			
.			
N	Pred 20	Pred 30	A_N

3.2 Framework Exemplified

We demonstrate the working of the proposed mechanism with the help of the Bajaj Auto Stock. The data was collected for the past three years (from 1st July 2016 to 30 June

2019) [16, 29–31, 32] and comprehensive set of prediction of the stock price are created for the test period from 1st July 2019 to 6th November, 2020 using the above mentioned mechanism as shown in Table 2. Here Pred5, Pred10, Pred20 and Pred30 indicate the predictions obtained on the basis of past 5, 10, 20, 30 days data. Also we have statistically extended set of predictions:

PredMin = min (Pred5, Pred10, Pred20, Pred30).

PredMax = max (Pred5, Pred10, Pred20, Pred30).

PredAvg = average (Pred5, Pred10, Pred20, Pred30).

Table 2. Primary and statistically extended predictions for Bajaj Auto

Date	Close price	Pred 5	Pred 10	Pred 20	Pred 30	Pred-Min	Pred-Max	Pred-Avg
7/1/2019	2910.7	2904.18	2922.27	2913.55	2912.27	2904.18	2922.27	2913.07
7/2/2019	2886.25	2886.18	2893.69	2900.07	2906.18	2886.18	2906.18	2896.53
7/3/2019	2895.35	2894.43	2908.39	2904.13	2907.72	2894.43	2908.39	2903.67
7/4/2019	2895.55	2894.88	2905.16	2904.95	2908.21	2894.88	2908.21	2903.3
7/5/2019	2838.2	2837.35	2852.05	2853.78	2858.79	2837.35	2858.79	2850.49
7/8/2019	2779.9	2775.3	2802.36	2795.26	2798.6	2775.3	2802.36	2792.88
7/9/2019	2784.75	2779.55	2808.05	2791.46	2792.5	2779.55	2808.05	2792.89
7/10/2019	2741.9	2739.95	2769.01	2755.99	2760.9	2739.95	2769.01	2756.46
7/11/2019	2711.65	2708.03	2747.74	2723.25	2727.99	2708.03	2747.74	2726.75
7/12/2019	2721.75	2718.77	2759.56	2727.71	2730.99	2718.77	2759.56	2734.26
7/15/2019	2714.25	2715.27	2754.02	2724.67	2729.86	2715.27	2754.02	2730.96
7/16/2019	2732.25	2734.73	2772.76	2739.94	2744.94	2734.73	2772.76	2748.09
7/17/2019	2696.45	2699.18	2738.6	2713.39	2720	2699.18	2738.6	2717.79
7/18/2019	2632.85	2630.59	2681.44	2653.49	2658.91	2630.59	2681.44	2656.11
7/19/2019	2557.55	2550.11	2611.8	2578.24	2581.02	2550.11	2611.8	2580.29

Table 3 shows the error between actual and predicted price. Here AE-5, AE-10, AE-20, AE-30, AE-MIN, AE-MAX and AE-AVG show the absolute error for the predictions Pred5, Pred10, Pred20, Pred30, PredMin, PredMax and PredAvg respectively. Table 4 shows the Mean Absolute Error (MAE) for all the predictions in S_c for Bajaj Auto and some other stocks. On the basis of lowest values of MAE-20 and MAE-30, Pred20 and Pred30 become the candidate predictions for the purpose of further error reduction. On the basis of similar argument, Pred20 and PredMin become the candidate predictions for Eicher Motors. Table 5 shows the correlation between Close Price Vector and the various Price Prediction Vectors. For example, CO-5 indicates the value of correlation coefficient between the close price and Pred5. Table 6 shows the candidate predictions C1 and C2 for the different stocks along with their ANN-ID.

Table 3. Error between different predictions for Bajaj Auto

Date	Close price	AE-5	AE-10	AE-20	AE-30	AE-MIN	AE-MAX	AE-AVG
7/1/2019	2910.7	6.52	11.57	2.85	1.57	6.52	11.57	2.37
7/2/2019	2886.25	0.07	7.44	13.82	19.93	0.07	19.93	10.28
7/3/2019	2895.35	0.92	13.04	8.78	12.37	0.92	13.04	8.32
7/4/2019	2895.55	0.67	9.61	9.4	12.66	0.67	12.66	7.75
7/5/2019	2838.2	0.85	13.85	15.58	20.59	0.85	20.59	12.29
7/8/2019	2779.9	4.6	22.46	15.36	18.7	4.6	22.46	12.98
7/9/2019	2784.75	5.2	23.3	6.71	7.75	5.2	23.3	8.14
7/10/2019	2741.9	1.95	27.11	14.09	19	1.95	27.11	14.56
7/11/2019	2711.65	3.62	36.09	11.6	16.34	3.62	36.09	15.1
7/12/2019	2721.75	2.98	37.81	5.96	9.24	2.98	37.81	12.51
7/15/2019	2714.25	1.02	39.77	10.42	15.61	1.02	39.77	16.71
7/16/2019	2732.25	2.48	40.51	7.69	12.69	2.48	40.51	15.84
7/17/2019	2696.45	2.73	42.15	16.94	23.55	2.73	42.15	21.34
7/18/2019	2632.85	2.26	48.59	20.64	26.06	2.26	48.59	23.26
7/19/2019	2557.55	7.44	54.25	20.69	23.47	7.44	54.25	22.74

Table 4. Mean absolute error of different automobile companies for next day prediction

Sector	Stock	MAE-05	MAE-10	MAE-20	MAE-30	MAE-MIN	MAE-MAX	MAE-AVG
Automobile	Bajaj Auto	21.9	37.4	12.4	14.1	14.3	36.2	18
	Eicher Motors	72.7	99	39.3	54.9	39.4	103.8	46.1
	Heromotocorp	26.4	86.9	118.9	49.9	48	121.5	44.1
	Mahindra	11.4	20.3	16.4	17.5	19.3	20.4	6.1
	Maruti	76.7	68.1	189.4	172.4	185.3	173.5	78.9
	Tata Motors	8	22.3	23.5	14.8	8	24.7	13.2

Table 5. Correlation of different automobile companies for next day prediction

Sector	Stock	CO-05	CO-10	CO-20	CO-30	CO-Min	CO-Max	CO-Avg
Automobile	Bajaj Auto	0.985	0.997	0.999	0.998	0.998	0.991	0.998
	Eicher Motors	0.994	0.996	0.998	0.996	0.998	0.995	0.998
	Heromotocorp	0.997	0.989	0.988	0.993	0.995	0.988	0.997
	Mahindra	0.997	0.999	0.994	0.994	0.996	0.999	0.998
	Maruti	0.994	0.997	0.984	0.977	0.987	0.986	0.992
	Tata Motors	0.993	0.998	0.991	0.985	0.993	0.993	0.995

Table 6. Candidate selection of automobile sector for next day

Stock Id	C1	C2	ANN ID
Bajaj Auto	Pred20	Pred 30	**BA-1**
Eicher Motors	Pred20	Pred-min	EM-1
Heromotocorp	Pred-5	Pred-ave	HM-1
Mahindra	Pred-5	Pred-ave	MH-1
Maruti	Pred-5	Pred-10	MR-1
Tata Motors	Pred-5	Pred-min	TM-1

After describing the working of the proposed framework through illustration let us take up the actual experiment.

4 The Experiment and the Results

To assess the credibility of the proposed mechanism, the past price data was taken from the period 1/7/2019 to 6/11/2020. This data was collected from various websites [16, 29–31, 32]. This data was divided into two parts: training part (from 1/7/2019 to 4/8/2020) and the test part (from 5/8/20 to 6/11/2020). The system was trained for the 16 Nifty 50 stocks and their performance was evaluated. Table 7 shows the mean absolute error (MAE) for the candidate predictions C1 and C2 and MAE-ATR obtained after the training process for the daily prediction during the testing phase. The last column of the table shows the reduction in error with mean of candidate1 and candidate2 prediction as the base. It can be seen that after applying the ANN, the MAE gets reduced significantly and error reduction upto 77% have been witnessed making the predictions quite accurate. At some places, there is an increase in the error, which is due to due the spikes in the stock prices.

Table 7. Reduction in error for daily price prediction after backpropagation learning

Sector	Stock	MAE-C1	MAE-C2	MAE-ATR	%Error reduction
Automobile	Bajaj Auto	9.61	9.3	3.77	60.13
	Eicher Motors	27.53	27.6	11.33	58.90
	Heromotocorp	15.74	19.71	17.83	−0.59
	Mahindra	10.81	5.07	4.81	39.42
	Maruti	79.62	36.11	30.45	47.38
	Tata Motors	8.49	8.49	2.6	69.38

(continued)

Table 7. (*continued*)

Sector	Stock	MAE-C1	MAE-C2	MAE-ATR	%Error reduction
FMCG	Asian Paints	76.78	76.78	107.5	−40.01
	Britannia	101.22	120.39	59.19	46.58
	Hindustan Unilever	54.53	40.11	22.97	51.46
	Nestle	1366.54	1366.54	309.06	77.38
	Titan	13.01	17.95	12.79	17.38
IT	HCL	32.95	43.11	30.38	20.12
	Infosys	79.69	79.69	63.57	20.23
	TCS	104.64	116.12	92.83	15.90
	Tech Mahindra	12.47	12.47	9.35	25.02
	Wipro	8.49	6.23	6.7	8.97

Table 7 shows the average reduction in MAE after applying the process. To illustrate the convergence process we are augmenting the convergence graphs in Figs. 4, 5, 6, 7, 8, 9, 10, 11 and 12 indicating the reduction in the prediction error on the daily basis for 9 Nifty 50 companies taking 3 companies each from IT, FMCG and Automobile sectors for the period 7/10/2020 to 6/11/2020.

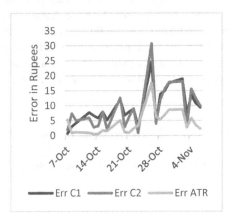

Fig. 4 Error reduction in Bajaj Auto

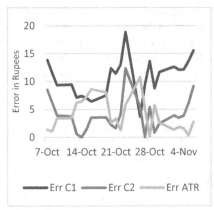

Fig. 5. Error reduction in Mahindra & Mahindra

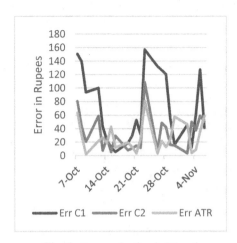

Fig. 6. Error reduction in Maruti

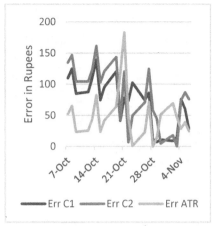

Fig. 7. Error reduction in Britannia

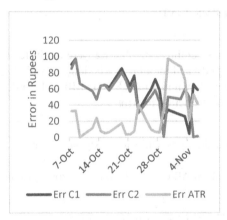

Fig. 8. Error reduction in Hindustan Unilever Limited

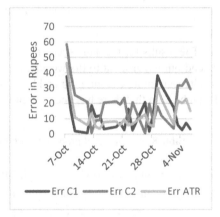

Fig. 9. Error reduction in Titan

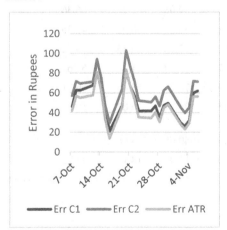

Fig. 10. Error reduction in HCL Technologies

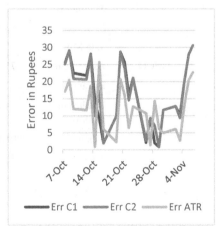

Fig. 11. Error reduction in Tech Mahindra

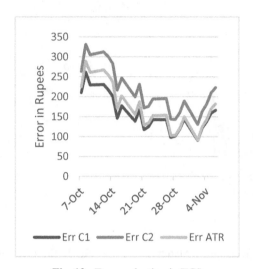

Fig. 12. Error reduction in TCS

The results for the daily predictions were so enthusiastic that we were motivated to extend our experiment to the weekly, fortnightly and monthly predictions as well. The results were quite encouraging and have been shown in Tables 8, 9 and 10 in the order. It can be seen that there has been a significant reduction in error.

Table 8. Extension of proposed work from daily predictions to weekly predictions

Period	Sector	Stock	MAE-C1	MAE-C2	MAE-ATR	%Error reduction
Weekly	Automobile	Bajaj Auto	61.64	61.46	61.71	−0.26
		Eicher Motors	63.27	63.48	62.19	1.87
		Heromotocorp	14.4	16.56	8.38	45.87
		Mahindra	7.09	9.08	2.72	66.36
		Maruti	80.81	36.65	30.92	47.35
		Tata Motors	5.55	3.48	2.55	43.52
	FMCG	Asian Paints	75.07	75.07	71.2	5.16
		Britannia	167.74	178.23	58.45	66.21
		Hindustan Unilever	54.53	74.84	26.03	59.76
		Nestle	2001.95	1940.65	244.68	87.59
		Titan	35.99	17.95	14.1	47.72

(continued)

Table 8. (*continued*)

Period	Sector	Stock	MAE-C1	MAE-C2	MAE-ATR	%Error reduction
	IT	HCL	30.32	30.32	27.64	8.84
		Infosys	112.28	106.86	54.87	49.92
		TCS	48.05	48.05	51.73	−7.66
		Tech Mahindra	10.72	9.16	10.41	−4.73
		Wipro	3.69	4.68	2.63	37.16

Table 9. Extension of proposed work from daily predictions to fortnightly predictions

Period	SECTOR	Stock	MAE-C1	MAE-C2	MAE-ATR	%Error reduction
Fortnightly	Automobile	Bajaj Auto	9.61	9.3	3.79	59.92
		Eicher Motors	63.27	59.94	63.02	−2.30
		Heromotocorp	15.98	20	12.48	30.63
		Mahindra	16.73	14.55	15.54	0.64
		Maruti	80.81	36.65	30.98	47.25
		Tata Motors	5.99	8.12	6.88	2.48
	FMCG	Asian Paints	75.15	75.15	106.45	−41.65
		Britannia	151.45	101.79	29.47	76.73
		Hindustan Unilever	33.83	33.83	30.46	9.96
		Nestle	1366.13	1366.13	309.3	77.36
		Titan	12.99	17.89	12.42	19.56
	IT	HCL	30.32	30.32	27.72	8.58
		Infosys	112.28	106.86	54.49	50.27
		TCS	48.05	48.05	49.87	−3.79
		Tech Mahindra	10.72	9.16	10.21	−2.72
		Wipro	3.69	4.68	5.02	−19.95

Table 10. Extension of proposed work from daily predictions to monthly predictions

Period	SECTOR	Stock	MAE-C1	MAE-C2	MAE-ATR	%Error reduction
Monthly	Automobile	Bajaj Auto	4.19	4.01	3.87	5.61
		Eicher Motors	16.4	17.45	8.3	50.96
		Heromotocorp	28.48	29.06	25.34	11.92
		Mahindra	5.56	11.1	5.73	31.21
		Maruti	69.04	39.48	29.99	44.73
		Tata Motors	4.38	4	4.56	−8.83
	FMCG	Asian Paints	64.73	64.73	92.95	−43.60
		Britannia	94.34	93.22	30.78	67.18
		Hindustan Unilever	35.16	35.16	35.07	0.26
		Nestle	1322.57	1322.57	347.78	73.70
		Titan	13.17	17.41	11.45	25.11
	IT	HCL	65.57	63.16	60.95	5.31
		Infosys	109.28	109.28	56.97	47.87
		TCS	50.17	50.17	49.85	0.64
		Tech Mahindra	10.4	9.02	10.28	−5.87
		Wipro	3.75	4.92	5.76	−32.87

5   Conclusion

This paper presents a stock price prediction strategy which is both simplistic and effective. The hallmark of the strategy is that it is not general but stock specific. The localization of the precdiction mechanism helps in generating the quite accurate results that may not be feasible in a generic proposal. The prediction accuracy has not only been quite high but is scalable as well, which is indicated by the weekly, fortnightly and monthly predictions. Though the paper shows results for three sectors of the Nifty50, for the purpose of illustration but in actual practice, it was carried out for many other sectors as well and the results were quite encouraging. We hope that the strategy will be useful for the software developers and the end user investers.

The results confirm efficacy of the proposed mechanism. Some critics may object that repeating the similar customized process for each stock is not worthwhile. In this regard, it is stated that with the availability of parallel processing techniques, huge memory spaces, and superfast processor, this is quite possible. With money being the most motivating factor, all these efforts are really worth.

References

1. Shen, J., Omair Shafiq, M.: Short-term stock market price trend prediction using a comprehensive deep learning system. J. Big Data **7**(1), 1–33 (2020). https://doi.org/10.1186/s40537-020-00333-6
2. Kara, Y., AcarBoyacioglu, M., Baykan, Ö.K.: Predicting direction of stock price index movement using artificial neural networks and support vector machines: the sample of the Istanbul stock exchange. Expert Syst. Appl. **38**(5), 5311–5319 (2011)
3. Qiu, M., Song, Y.: Predicting the direction of stock market index movement using an optimized artificial neural network model. PLoS ONE **11**(5), e0155133 (2016)
4. Guresen, E., Kayakutlu, G., Daim, T.U.: Using artificial neural network models in stock market index prediction. Expert Syst. Appl. **38**(8), 89–97 (2011)
5. Fischer, T., Krauss, C.: Deep learning with long short-term memory networks for financial market predictions. Eur. J. Oper. Res. **270**(2), 654–669 (2018). https://doi.org/10.1016/j.ejor.2017.11.054
6. Krauss, C., Do, X., Huck, N.: Deep neural networks, gradient-boosted trees, random forests: statistical arbitrage on the S&P 500. Eur. J. Oper. Res. Elsevier **259**(2), 689–702 (2017)
7. Krauss, C.: Statistical arbitrage pairs trading strategies: Review and outlook. J. Econ. Surv. **31**, 513–545 (2016)
8. Atsalakis, G.S., Valavanis, K.P.: Surveying stock market forecasting techniques–part II: soft computing methods. Expert Syst. Appl. **36**(3), 5932–5941 (2009)
9. Sermpinis, G., Theofilatos, K., Karathanasopoulos, A., Georgopoulos, E.F., Dunis, C.: Forecasting foreign exchange rates with adaptive neural networks using radial-basis functions and particle swarm optimization. Eur. J. Oper. Res. **225**(3), 528–540 (2013)
10. McNally, S., Roche, J., Caton, S.: Predicting the price of bitcoin using machine learning. In: Proceedings—26th Euromicro International Conference on Parallel, Distributed, and Network-Based Processing, PDP 2018, pp. 339–43 (2018)
11. Weng, B., Lu, L., Wang, X., Megahed, F.M., Martinez, W.: Predicting short-term stock prices using ensemble methods and online data sources. Expert Syst. Appl. **112**, 258–273 (2018)
12. Kumar, D., Murugan, S.: Performance analysis of Indian stock market index using neural network time series model. In: International Conference on Pattern Recognition, Informatics and Mobile Engineering, pp 72–78. IEEE (2013)
13. Adebiyi, A.A., Ayo, C.K., Adebiyi, M.O., Otokiti, S.O.: Stock price prediction using neural network with hybridized market indicators. J. Emerg. Trends Comput. Inf. Sci. **3**(1), 1–9 (2012)
14. Si, Y.W., Yin, J.: OBST-based segmentation approach to financial time series. Eng. Appl. Artif. Intell. **26**(10), 2581–2596 (2013)
15. Metcalfe, A.V., Cowpertwait, P.S.: Introductory Time Series with R, pp. 2–5. Springer, Berlin (2009). https://doi.org/10.1007/978-0-387-88698-5
16. National Stock Exchange of India. https://www.nseindia.com/. Accessed 23 Nov 2020
17. Seo, M., Lee, S., Kim, G.: Forecasting the volatility of stock market index using the hybrid models with google domestic trends. Fluct. Noise. Lett. **18**(01), 1–17 (2019)
18. Pang, X., Zhou, Y., Wang, P., Lin, W., Chang, V.: An innovative neural network approach for stock market prediction. J. Supercomput. **74**, 1–21 (2018)
19. Fischer, T., Krauss, C.: Deep learning with long short-term memory networks for financial market predictions. Eur. J. Oper. Res. **270**(2), 654–669 (2018)
20. Senapati, M.R., Das, S., Mishra, S.: A novel model for stock price prediction using hybrid neural network. J. Inst. Eng. (India) Ser. B. **99**(6), 555–563 (2018)
21. Rather, A.M., Agarwal, A., Sastry, V.N.: Recurrent neuralnetwork and a hybrid model for prediction of stock returns. Expert Syst. Appl. **42**(6), 3234–3241 (2015)

22. Khan, W., Ghazanfar, M.A., Azam, M.A., Karami, A., Alyoubi, K.H., Alfakeeh, A.S.: Stock market prediction using machine learning classifiers and social media, news J. Ambient. Intell. Humaniz. Comput. 1 24 (2020).https://doi.org/10.1007/s12652-020-01839-w

23. Xu, J., Murata, T.: Stock market trend prediction with sentiment analysis based on LSTM neural network (2019)

24. Kim, T., Kim, H.Y.: Forecasting stock prices with a feature fusion LSTM-CNN model using different representations of the same data. PLoS ONE 14(2), e0212320 (2019)

25. Wang, J., Wang, J., Fang, W., Niu, H.: Financial time series prediction using elman recurrent random neural networks. Comput. Intell. Neurosci. **2016**, 1–14 (2016)

26. Pang, X., Zhou, Y., Wang, P., Lin, W., Chang, V.: An innovative neural network approach for stock market prediction. J. Supercomput. **76**, 2098–2118 (2020)

27. Yang, F., Chen, Z., Li, J., Tang, L.: A novel hybrid stock selection method with stock prediction. Appl. Soft Comput. **80**, 820–831 (2019)

28. Adebiyi, A.A., Adewumi, A.O., Ayo, C.K.: Comparison of ARIMA and artifcial neural networks models for stock price prediction. J. Appl. Math. **2014**, 1–7 (2014)

29. Bao, W., Yue, J., Rao, Y.: A deep learning framework for financial time series using stacked autoencoders and long-short term memory. PLoS ONE (2017). https://doi.org/10.1371/journal.pone.0180944

30. Yahoo finance. https://finance.yahoo.com/. Accessed 23 Dec 2020

31. Rediff money. https://money.rediff.com/index.html/. Accessed 23 Dec 2020

32. Moneycontrol. https://www.moneycontrol.com/. Accessed 23 Dec 2020

Evaluating Binary Classifiers with Word Embedding Techniques for Public Grievances

Khushboo Shah(✉) , Hardik Joshi , and Hiren Joshi

Gujarat University, Ahmedabad-9, India
khushbooshah@gujaratuniversity.ac.in

Abstract. Public grievance happens when the administration fails to meet the expectations of people. For any government, it is preeminent need to settle people's complaints. The government offers great deal of services in almost all areas of life to its citizens. Indian Government provides various e-platforms such as websites and mobile applications for Indians to register their complaints for any of the services but due to lack of awareness of these platforms and the lengthy registration process, people are turning to easily accessible user-friendly sources like social media to convey their grumble through microblogging. In this experiment we have focused on Grievances posted on Twitter for Indian Railways. Identifying genuine complaints among lacs of social media messages is quite a big challenge. This paper aims to use Machine Learning binary classifiers along with word embedding techniques to perform this classification job that can predict that the given text is a grievance or not. To find out the best classifier with the combination of suitable word embedding techniques, we associated six binary classifiers: Naïve Bays, Support Vector Machine, Logistic Regression, Decision Tree, Random Forest, K-Nearest Neighbor, with three-word embedding techniques: TFIDF, Word2Vec, BERT. Experiments have shown that, in many cases, text classification results are very close to each other, which can make the situation puzzled to pick the right model. Hence, we have used AUC ROC curves and K-Fold cross-validation method to validate the experimental results.

Keywords: AUC score · Binary classifiers · Public grievance · ROC curve · Word embedding techniques

1 Introduction

Complaints are statements about the expectations that have not been met. They are likewise the chances for the organization to reconnect with clients by fixing an assistance or item breakdown. Thusly, "complains are gifts that customers give to the businesses" [1]. Public grievance happens when the administration fails to meet the expectations of people. For any government, it is preeminent need to settle people's complaints. Indian Government provides various e-platforms such as websites and mobile applications for Indians to register their complaints for any provided services but due to lack of awareness of these platforms and the lengthy registration process, people are turning to easily accessible user-friendly sources like social media to convey their grumble through

© Springer Nature Switzerland AG 2022
K. K. Patel et al. (Eds.): icSoftComp 2021, CCIS 1572, pp. 209–221, 2022.
https://doi.org/10.1007/978-3-031-05767-0_17

microblogging. Identifying genuine complaints among lacs of social media messages is quite a big challenge. To deal with this large amount of e-content, Artificial Intelligence (AI) comes into the picture. These electronic data contain valuable information, and to get new insights from it, Natural Language Processing (NLP) and Machine Learning (ML) are employed in conjunction, which are the branches of AI. ML algorithms are used to create intelligent products and services using this e-data, depending on the type of data and task carried out [2]. Classification is a crucial data mining method with a wide range of applications to classify the various types of data used in pretty much every field of our life. Classification is utilized to categorize the item according to its features with respect to the predefined set of classes [3]. Binary Classification has been comprehensively studied and various classifiers are used in predictive modeling for the task. In this type of classification, the data is classified by assigning two class labels like 'yes' and 'no' [4]. Spam email filtering is a well-known example of binary classification where the ML algorithm classifies emails into two classes – 'spam' or 'not spam' [5]. Real-world data comes with noise, inconsistency, and inadequacy, which can obscure useful patterns and lead to poor performance of the model hence data cleaning becomes essential [6]. After manual classification with 'Yes' and 'No' labels, data pre-processing is the next step towards this journey. Since ML models accept only numbers, a further job is to convert the cleaned data into machine acceptable format which is done by word embedding methods. Many numerical representation techniques are available but choosing the right one would be semantically meaningful as it can have a massive impact on the results of predictive models [7]. In this experiment, we have used TFIDF, Word2Vec, and BERT techniques for word vectorization. The last step is to select the right binary classifier. Numerous ML supervised algorithms have come up for each kind of predictive task [2]. In this paper we have used six binary classifiers: Naïve Bays, Support Vector Machine, Logistic Regression, Decision Tree, Random Forest, K-Nearest Neighbors and evaluated them in combination with all three word embedding methods. Due to the wide variety of data, it is hard to pick the appropriate classifier and vectorization method for the given dataset. This paper aims to test six classifiers with three word embedding techniques with diverse ratios of the dataset. In this experiment, to prepare the dataset, tweets were downloaded from the Twitter posted for Indian Railways. The main purpose is to automatically identify that the given tweet is a grievance or not with the help of predictive model. This experiment was done in four tracks with doubling the records every time. We started with 500 then 1000, 2000 and 4000 records in Track 1, Track 2, Track 3 and, Track 4 respectively. As shown in Fig. 1, many-to-many relationships between binary classifiers and word embedding techniques were established during the experiment.

Experiments have shown that, in many cases, text classification results are very close to each other, which can make the situation puzzled to pick the right model. Hence, we have used AUC ROC curves and K-Fold cross validation method to validate the experimental results.

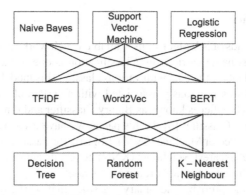

Fig. 1. Each word embedding technique is tested with each binary classifier

2 Government, Public Grievance and Social Media Platforms

Twitter and Facebook are acquiring popularity among people as a citizen grievance management system where they can lodge their complaints. Twitter is the most prominent platform to publish the content via microblogging. Because of the wide reachability and network among users, National Government also uses Twitter to connect with the general public [8]. Because of free-form nature of microblogging and rapid growth of content, automatic identification of the type of tweet is technically a big challenge [9].

3 Word Embedding Techniques and Binary Classifiers

3.1 Word Embedding Techniques

Representation of text into real number vectors is known as Word Embedding. It has been one of the essential methods used to perform NLP [10]. The principle benefit of Word Embedding is, it permits to offer a more effective and meaningful representation of words by keeping up the logical resemblance of them and by forming low dimensional vectors [11]. As a late, a few models have been proposed and capitulated state-of-the-art results in numerous NLP undertakings. Albeit these studies consistently guarantee that their models are superior to the past ones dependent on different evaluation factors [12].

In this experiment, we have used three well-known Word Embedding techniques TFIDF (Term Frequency Inverse Document Frequency), Word2Vec (W2V), and BERT (Bidirectional Encoder Representations from Transformers), which have completely different working methods from each other.

In TFIDF, Word frequency represents the number of occurrences of a given word in the document and Inverse document frequency refers to an overall significance of a word in the document [13]. W2V is capable of capturing the relation of one word with other words, context of a word, semantic and syntactic similarity, and so forth. W2V is perhaps the most prominent method to learn word embeddings with shallow neural networks [14]. BERT has the capability to cope with NLP jobs like supervised text classification for any type of corpus, without human intervention yet deliver great results [15]. This NLP model was aimed to pre-train deep bidirectional representation from unlabeled text and result into fine-tuned using labeled text [16].

3.2 Binary Classifiers

There are different kinds of ML algorithms available which work on a variety of data and types of undertakings being carried out. Classification is one of the major domains in ML where the model labels the data with various classes. This model undergoes training with the data which was previously associated with a certain class. In case there are just two classes, the classifier is known as Binary Classifier and if multiple classes then referred as Multi-class Classifier [3]. In this paper, we have used six binary classifiers: Naive Bayes, Support Vector Machine (SVM), Logistic Regression (LR), Decision Tree, Random Forest and K-Nearest Neighbors (KNN), to test its compatibility with all three word embedding techniques for India Railways Dataset.

Naïve Bayes is the probabilistic binary classifier that works on Bayes' theorem. Unlike other algorithms, it does not use iterative modeling, hence it is linear in time and its core upside is, it works with a lesser amount of data than conventional classifiers [2]. A support vector machine (SVM) is a supervised machine learning algorithm that learns by example to assign labels to objects [17]. It uses classification algorithms for the binary classification problems where it takes labeled training data for each category as input and predicts a category for new text. Logistic Regression is a broadly used basic classification classifier which maps the relationship between dependent and independent variables. The core mathematical idea that underlies logistic regression is the logit - the natural logarithm of an odds ratio [18]. Decision Tree is mainly used to simplify complicated relationships among input variables and target variables. It divides the original input variable into meaningful subsets based on past knowledge of training data [19, 20]. Random Forest Classifier is one of the top successful machine learning algorithms which demonstrated to be an exceptionally well-known and amazing technique in pattern recognition and ML for high-dimensional classification and skewed problems [21]. KNN is a non-parametric classification technique, which is basic yet successful in many cases. For data D record R to be classified, its K nearest neighbors are found and this structures a neighborhood for R [22].

4 K-Fold Cross Validation Technique

The K-fold Cross Validation (KCV) technique is one of the top used methods by researchers for model selection and error assessment of classifiers. The KCV consists in splitting a dataset into k subsets; then, iteratively, some of them are utilized to learn the model, while the others are exploited to assess its performance [12].

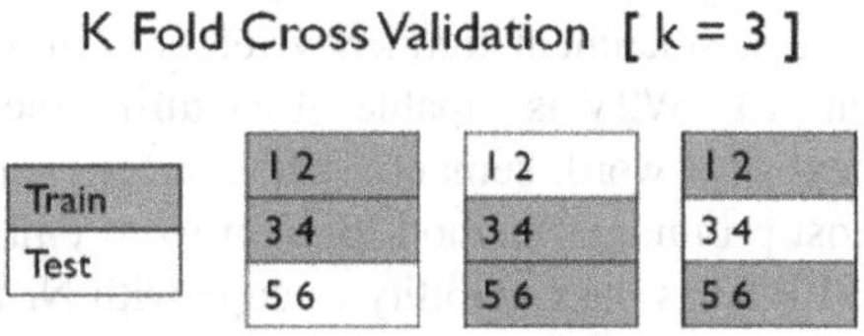

Fig. 2. Total 6 data sample is used. Highlighted boxes show training data and other are testing dataset. As k's value is 3, total three folds created with 2 datasets where each sample goes for training and testing during any of the fold execution.

KCV first shuffles the dataset randomly and then splits it into K groups. For each unique group, it takes the group as a test data set and the remaining groups as a training dataset. After then it fits the model with the training dataset and evaluates it on the test dataset. It keeps the evaluation score and discards the model. This process repeats for each group. The beauty of this model is each observation in the dataset is assigned to an individual group and stays in that group for the duration of the process. Hence, each record is given the opportunity to be used in the hold data set one time and used to train the model k-1 times [12].

5   Experiment Process

We prepared our dataset for the Indian Railways. This data is original tweets posted by the users on Twitter for Indian Railways' Handles. These tweets downloaded in.csv file.

Corpus built for two classes and annotations done manually. If a tweet is a grievance, then it was labeled as 'YES' else 'NO'. Because of the limitation of 140 characters in a single tweet, users use slang, acronyms and multiple languages as well. Due to free typing facility, there is no defined structure of tweet thus grammatical errors and spelling mistakes have been rashly expanded [9]. This noisy data need to be cleaned with the help of Natural Language Processing (NLP) techniques to retrieve the right information [23]. Important information such as Trian No, Coach No and PNR No. extracted from the tweet. After retrieving these data, stopword removal, spelling correction and stemming applied on remining words. As a result Final Words we got which used as an input for the word embedding model (see Fig. 4). To provide machine acceptable input format to the classifier, vectors generated through word embedding techniques.

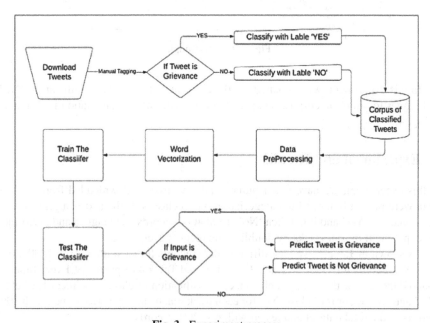

Fig. 3. Experiment process

Numerous elements influence the result of the Machine Learning (ML) model on a given project. The quality and representation of the data are predominant factors. During the model training phase, knowledge extracting becomes complex if data is noisy, irrelevant, redundant, and unreliable. Notably, data preparation and sifting steps take a significant amount of processing time in ML issues. Data pre-processing comprises stages like – data cleaning, data normalization, feature extraction, feature selection, and so forth. And at the end, the final result comes out which is the actual training data set for the ML model [24].

CLASS	TWEET	TIDY_TWEET	TRAIN_NO	PNR_NO	COACH_NO	FINAL_WORDS
YES	@RailMinIndia @sureshpprabhu dhauli worst train.12822 coach C1 ac not wrkng. refund money .why is this coach even there when ac nt working?	dhauli worst train12822 coach c1 ac not wrkng refund money why is this coach even there when ac nt working	12822		c1	dhauli worst train coach c ac wrkng refund money coach even ac nt working
YES	@RailMinIndia my PNR is 2118887897 train number 12983. Very bad experience with badding. Just wrapped in paper not even washed.	my pnr is 2118887897 train number 12983 very bad experience with badding just wrapped in paper not even washed	12983	2118887897		pnr train number bad experience badding wrapped paper even washed
YES	@RailMinIndia Pnr8647770808 very bad situation..all unreserved passengers sitting in resrvation coach causing inconvinience to passengers	pnr8647770808 very bad situationall unreserved passengers sitting in resrvation coach causing inconvinience to passengers		8647770808		pnr bad situationall unreserved passengers sitting resrvation coach causing inconvinience passengers
YES	@RailMinIndia PNR 6253134881 train no 22940. Dirty toilets and bedroll. Pathetic. Please help.	pnr 6253134881 train no 22940 dirty toilets and bedroll pathetic please help	22940	6253134881		pnr train dirty toilets bedroll pathetic please help
YES	@RailMinIndia pnr no. 8147850160 train. 59014 S S2 Coach suspension is not working getting heavy bumps on head please repair it..now..	pnr no 8147850160 train 59014 s s2 coach suspension is not working getting heavy bumps on head please repair itnow	59014	8147850160	s2	pnr train coach suspension working getting heavy bumps head please repair itnow
NO	@RailMinIndia please help one poor villager is no more just, HIRAKUD station SAMBALPUR divi. https://t.co/vlm0trDyQq	please help one poor villager is no more just hirakud station sambalpur divi				please help one poor villager hirakud station sambalpur divi
NO	@RailMinIndia isn't tatkal booking is allowed at every station counter? unable to book at teghra station(TGA)	isnt tatkal booking is allowed at every station counter unable to book at teghra stationtga				isnt tatkal booking allowed every station counter unable book teghra stationtga
NO	@sureshpprabhu @RailMinIndia trying to book tatkal ticket from MFP to NDLS for last 1 week and have failed due to your poor website performa	trying to book tatkal ticket from mfp to ndls for last 1 week and have failed due to your poor website performa				trying book tatkal ticket mfp ndls last week failed due poor website performa
NO	@RailMinIndia Train no: 12017. Pnr: 2646724070. Reserved 2 tckts, 1 wtng. If only chair seats, then why allow waiting in shatabdi trains.	train no 12017 pnr 2646724070 reserved 2 tckts 1 wtng if only chair seats then why allow waiting in shatabdi trains	12017	2646724070		train pnr reserved tckts wtng chair seats allow waiting shatabdi trains

Fig. 4. Corpus sample

In this experiment, we annotated 4000 records and cleaned them with various NLP techniques to build the corpus. Sample of corpus with manual annotation is displayed in Fig. 4.

6   Experiment and Results

In this experiment, we have taken Indian Railways tweets, downloaded from Twitter. Data were manually tagged to categorize into two classes. If the text is a grievance, it was tagged as 'Yes' and if not then 'No'. Total 4000 records were tagged and performed the experiment in four tracks by doubling the records every time.

As shown in Table 1, we took different data records for each track. 500, 1000, 2000 and 4000 records for Track 1, Track 2, Track 3, and Track 4 respectively. Every time we doubled the data with the little unbalanced classification of class. 55% records of class 'Yes' and 45% records of class 'No' for each track. Each classifier was trained with 80% records of the total dataset and tested with rest 20% records.

Table 1. Experiment performed in four tracks

	Track 1	Track 2	Track 3	Track 4
Total records	500	1000	2000	4000
Yes	225	550	1100	2200
No	175	450	900	1800
Train	400	800	1600	3200
Test	100	200	400	800

For each track, vectors were generated through all three word embedding techniques and after that, all six ML classifiers were trained and tested to check the accuracy of grievance classification. NB does not support negative vectors; hence it is not tested with W2V, BERT, and K-Fold.

Track 1: Table 2 shows classification accuracy score out of 1 for 500 records. The combination of BERT with LR provided the highest accuracy of 0.91 while W2V with DT provided the lowest accuracy of 0.66. K-Fold Validation suggests that RF's performance is better than other classifiers with the accuracy score of 0.85.

Table 2. Track 1 with 500 records and highlighted cells represent highest accuracy

	NB	SVM	LR	DT	RF	KNN
TF-IDF	0.81	0.81	0.79	0.78	0.77	0.81
Word2Vec	NA	0.83	0.84	0.66	0.86	0.81
BERT	NA	0.88	**0.91**	0.78	0.88	0.88
2 Fold	NA	0.77	0.79	0.67	0.80	0.75
5 Fold	NA	0.81	0.82	0.69	0.84	0.79
10 Fold	NA	0.80	0.82	0.67	**0.85**	0.79

Track 2: Table 3 shows classification accuracy score out of 1 for 1000 records. The combination of W2V with LR provided the highest accuracy of 0.89 while BERT with DT provided the lowest accuracy of 0.76. K-Fold Validation suggests that LR's performance is better than other classifiers with the accuracy score of 0.86.

Track 3: Table 4 shows classification accuracy score out of 1 for 2000 records. The combination of W2V with RF provided the highest accuracy of 0.83 while BERT with DT provided the lowest accuracy of 0.71. K-Fold Validation suggests that LR and RF both perform better than other classifiers with the accuracy score of 0.81.

Table 3. Track 2 with 1000 records and highlighted cells represent highest accuracy

	NB	SVM	LR	DT	RF	KNN
TF-IDF	0.81	0.82	0.83	0.81	0.85	0.82
Word2Vec	NA	0.83	**0.89**	0.78	0.85	0.82
BERT	NA	0.84	0.87	0.76	0.84	0.84
2 Fold	NA	0.83	0.85	0.72	0.84	0.80
5 Fold	NA	0.83	**0.86**	0.74	0.85	0.80
10 Fold	NA	0.84	**0.86**	0.73	0.85	0.81

Table 4. Track 3 with 2000 records and highlighted cells represent highest accuracy

	NB	SVM	LR	DT	RF	KNN
TF-IDF	0.82	0.82	0.82	0.73	0.78	0.74
Word2Vec	NA	0.8	0.82	0.74	**0.83**	0.79
BERT	NA	0.79	0.81	0.71	0.79	0.79
2 Fold	NA	0.78	0.80	0.71	0.80	0.75
5 Fold	NA	0.79	**0.81**	0.72	**0.81**	0.76
10 Fold	NA	0.79	**0.81**	0.71	**0.81**	0.77

Track 4: Table 5 shows classification accuracy score out of 1 for 4000 records. The combination of W2V with RF provided the highest accuracy of 0.93 while TFIDF with KNN provided the lowest accuracy of 0.81. K-Fold Validation suggests that RF's performance is better than other classifiers with the accuracy score of 0.97.

Table 5. Track 4 with 4000 records and highlighted cells represent highest accuracy

	NB	SVM	LR	DT	RF	KNN
TF-IDF	0.86	0.89	0.87	0.88	0.89	0.81
Word2Vec	NA	0.85	0.84	0.92	**0.93**	0.82
BERT	NA	0.90	0.86	0.92	0.90	0.89
2 Fold	NA	0.83	0.83	0.85	0.90	0.78
5 Fold	NA	0.84	0.84	0.94	0.96	0.83
10 Fold	NA	0.84	0.84	0.96	**0.97**	0.84

6.1 AUC - ROC Curve

In ML, model performance measurement is an essential job. For classification problem, AUC (Area Under Curve) ROC (Receiver Operating Characteristics) curve are most reliable measures [25]. The ROC curve is a graph that visualizes the relation between the true positive rate and the false positive rate for a model under varying decision thresholds [26]. In ROC curve, the predictive capabilities of a variable is summarized by the AUC [27].

Our experiment results show that except Naïve Bayes and SVM, all four classifiers give highest result with different word embedding methods. Hence, we calculated AUC scores for Logistic Regression, Decision Tree, Random Forest and KNN. Compare to all other classifiers, AUC score of Random Forest is the highest one (see Fig. 5). Logistic Regression's score is similar to Random Forest for 500 records but as we increase the records, Random Forest dominates all other classifiers (see Fig. 5).

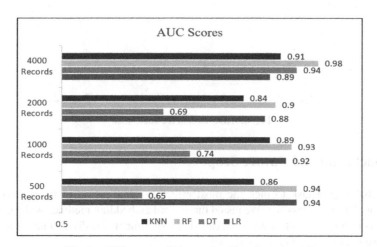

Fig. 5. AUC scores of binary classifiers in each track

ROC curves clearly depict that in each track, Random Forest performs the best while Logistic Regression's result is more near to Random Forest but Decision Tree and KNN's performances are really poor in maximum cases (see Fig. 6).

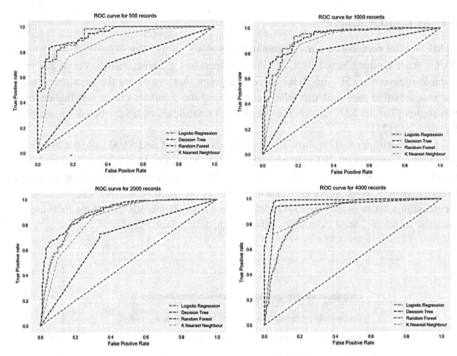

Fig. 6. ROC curves for each track

7   Conclusion and Future Scope

It is a challenging task to pick the most effective combination of word embedding method with binary classifier. We tested three word embedding methods with six binary classifiers on the Indian Railways dataset. The experiment was divided into four tracks where we doubled the data in each track.

From all four tracks only track 1 gave highest accuracy in BERT while in all other tracks W2V performed well. Result of Track 1 shows that LR is the best classifier with 0.91 accuracy but this is not supported by K-Fold and AUC scores where both supporting RF with accuracy 0.85 and 0.94 respectively. In Track 2 experiment results are best for LR that is 0.89 which is also supported by K-Fold which has a score 0.86. But in Track 3 and 4 Experiment results, K-Fold and AUC scores – all shows that RF is the best classifier for this dataset with the scores 0.83, 0.81, 0.90 respectively in Track 3 and 0.93, 097, 0.98 respectively for Track 4 (see Fig. 7).

After preprocessing the downloaded tweets, model does feature selection to perform prediction operations. Prediction of a class is completely based on rich selection of features. Random Forest is best in feature selection as it adds randomness to the classifier, while expanding the trees. Rather than looking for the main essential feature while splitting a node, RF looks for the best feature among a random subset of features. This outcomes in a wide variety that results in a superior model. Hence, Random Forest outperformed in our dataset.

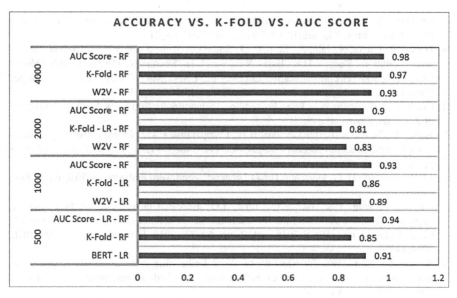

Fig. 7. Overall highest accuracy vs. k-fold scores vs. AUC scores

In future, we can increase the dataset and observe the performance of Random Forest against other classifiers which can help to pick the correct model for this Indian Railways dataset to apply other applications other than the classification.

References

1. Barlow, J., Moller, C.: A Complaint is a Gift. Berrett-Koehler Publishers, Inc., Oakland (2008)
2. Bahel, V., Pillai, S.: A comparative study on various binary classification algorithms and their improved variant for optimal performance. In: IEEE Region of 10 Symposium, June, pp. 5–7 (2020)
3. Patil, T.R., Sherekar, S.S.: Performance analysis of naive bayes and J48 classification algorithm for data classification. Int. J. Comput. Sci. Appl. **6**(2), 256–261 (2013). www.researchpublications.org
4. Ranjitha, K.V.: Classification and optimization scheme for text data using machine learning naïve bayes classifier. In: 2018 IEEE World Symposium on Communication and Engineering, pp. 33–36 (2018)
5. Vyas, T., Prajapati, P., Gadhwal, S.: A survey and evaluation of supervised machine learning techniques for spam e-mail filtering. In: Proceedings of 2015 IEEE International Conference on Electrical, Computer and Communication Technologies, ICECCT 2015 (2015). https://doi.org/10.1109/ICECCT.2015.7226077
6. Zhang, S., Zhang, C., Yang, Q.: Data preparation for data mining. Appl. Artif. Intell. **17**(5–6), 375–381 (2003). https://doi.org/10.1080/713827180
7. Latysheva, N.: Why do we use word embeddings in NLP? towardsdatascience.com (2019). https://towardsdatascience.com/why-do-we-use-embeddings-in-nlp-2f20e1b632d2. Accessed 14 July 2021
8. Agarwal, S., Sureka, A.: Investigating the role of twitter in e-governance by extracting information on citizen complaints and grievances reports. In: Reddy, P.K., Sureka, A.,

Chakravarthy, S., Bhalla, S. (eds.) BDA 2017. LNCS, vol. 10721, pp. 300–310. Springer, Cham (2017). https://doi.org/10.1007/978-3-319-72413-3_21

9. Goyal, M., Gupta, N., Jain, A., Kumari, D.: Smart government e-services for indian railways using Twitter. In: Sharma, D.K., Balas, V.E., Son, L.H., Sharma, R., Cengiz, K. (eds.) Micro-Electronics and Telecommunication Engineering. LNNS, vol. 106, pp. 721–731. Springer, Singapore (2020). https://doi.org/10.1007/978-981-15-2329-8_73

10. Gerth, T.: A Comparison of Word Embedding Techniques for Similarity Analysis. University of Arkansas, Fayetteville (2021)

11. Naili, M., Chaibi, A.H., Ben Ghezala, H.H.: Comparative study of word embedding methods in topic segmentation. Procedia Comput. Sci. **112**, 340–349 (2017). https://doi.org/10.1016/j.procs.2017.08.009

12. I.I. Systems and D. O. Identifier: How to generate good word embedding. IEEE Intell. Syst. (2016)

13. Liu, C.Z., Sheng, Y.X., Wei, Z.Q., Yang, Y.Q.: Research of text classification based on improved TF-IDF algorithm. In: 2018 International Conference on Intelligent Robotics and Control Engineering. IRCE 2018, vol. 2, pp. 69–73 (2018).https://doi.org/10.1109/IRCE.2018.8492945

14. Karani, D.: Word2Vec, Introduction to Word Embedding and. towardsdatascience.com. https://towardsdatascience.com/introduction-to-word-embedding-and-word2vec-652d0c 2060fa. Accessed 15 Jul 2021

15. González-Carvajal, S., Garrido-Merchán, E.C.: Comparing BERT against traditional machine learning text classification. arXiv Prepr. arXiv2005.13012, Ml (2020). http://arxiv.org/abs/2005.13012

16. Rong, X.: word2vec Parameter Learning Explained. arXiv Prepr. arXiv1411.2738, pp. 1–21 (2016). http://arxiv.org/abs/1411.2738

17. Shelke, P.P.P., Korde, A.N.: Support vector machine based word embedding and feature reduction for sentiment analysis-a study. In: Proceedings of 4th International Conference on Computing Methodologies and Communication ICCMC 2020, pp. 176–179 (2020). https://doi.org/10.1109/ICCMC48092.2020.ICCMC-00035

18. Peng, C.Y.J., Lee, K.L., Ingersoll, G.M.: An introduction to logistic regression analysis and reporting. J. Educ. Res. **96**(1), 3–14 (2002). https://doi.org/10.1080/00220670209598786

19. Song, Y.Y., Lu, Y.: Decision tree methods: applications for classification and prediction. Shanghai Arch. Psych. **27**(2), 130–135 (2015). https://doi.org/10.11919/j.issn.1002-0829.215044

20. Patil, S., Kulkarni, U: Accuracy prediction for distributed decision tree using machine learning approach. In: Proceedings of International Conference on Trends Electronic Informatics, ICOEI 2019, vol. 2019-April, ICOEI, pp. 1365–1371 (2019). https://doi.org/10.1109/icoei.2019.8862580

21. Azar, A.T., Elshazly, H.I., Hassanien, A.E., Elkorany, A.M.: A random forest classifier for lymph diseases. Comput. Methods Programs Biomed. **113**(2), 465–473 (2014). https://doi.org/10.1016/j.cmpb.2013.11.004

22. Guo, G., Wang, H., Bell, D., Bi, Y., Greer, K.: KNN model-based approach in classification. In: Meersman, R., Tari, Z., Schmidt, D.C. (eds.) On the Move to Meaningful Internet Systems 2003: CoopIS, DOA, and ODBASE. Lecture Notes in Computer Science, vol. 2888, pp. 986–996. Springer, Heidelberg (2003). https://doi.org/10.1007/978-3-540-39964-3_62

23. Joshi, H.J.H.S.K.: Smart approach to recognize public grievance from microblogs. Towar. Excell. UGC HRDC GU **13**(02), 57–69 (2021)

24. Kotsiantis, S.B., Kanellopoulos, D.: Data preprocessing for supervised leaning. Int. J. **1**(2), 1–7 (2006). https://doi.org/10.1080/02331931003692557

25. Narkhede, S.: Understanding AUC - ROC Curve (2019). https://towardsdatascience.com/understanding-auc-roc-curve-68b2303cc9c5. Accessed Sep 16 2021

26. Brzezinski, D., Stefanowski, J.: Prequential AUC: properties of the area under the ROC curve for data streams with concept drift. Knowl. Inf. Syst. **52**(2), 531–562 (2017). https://doi.org/10.1007/s10115-017-1022-8

27. Muschelli, J.: ROC and AUC with a binary predictor: a potentially misleading metric. J. Classif. **37**(3), 696–708 (2019). https://doi.org/10.1007/s00357-019-09345-1

Database Concentration Method for Efficient Image Retrieval Using Clustering and Image Tag Comparison

Soorya Ram Shimgekar⬭, Preetham Reddy Pathi⬭, and V. Vijayarajan(✉)⬭

School of Computer Science and Engineering, Vellore Institute of Technology,
Vellore, India
vijayarajan.v@vit.ac.in

Abstract. As the world moves towards a digital civilization, we generate more and more data along the way. With this increase in the number
of images on the internet and with social media companies, faster and
scale-able image retrieval algorithms are needed for efficiency. In this
paper, we propose a tag based image retrieval algorithm which also uses
clustering for better efficiency in large data sets. Although there are multiple existing algorithms that use tags most of them use tag completion
but not new tag generation. So, they cannot be used for completely unlabelled data. The proposed algorithm makes use of IBM's(International
Business Machines) VisualRecognitionV3 to generate new tags for completely unlabelled data. We make use of K-Means algorithm to firstly
cluster the images into different clusters based on features extracted
using ImageNet weights of ResNet50 architecture. We then use IBM's
VisualRecognitionV3 to label the folders based on its contents. Similar
steps are also carried out with query image and using word-vectorization
and cosine similarity, top-n most relevant folders are found out. Then
relevant images from these best folders are saved as output based on
similarity of image features, found using ResNet-50 model on ImageNet
Weights.

Keywords: Neural networks · K-means clustering · ResNet-50 ·
Feature extraction · Cosine similarity

1 Introduction

Content Based Image Retrieval (CBIR) is the process of retrieving most relevant
images of a given input image from the database using various computer vision
(CV) methods to collect features from the images [1]. It is also called Query
by Image Content (QBIC). In CBIR the images are retrieved without any cor-
relating text or annotations with the images. Instead, the images are retrieved
depending on the features in the images [2]. Efficient and better performing
CBIR models make a meaningful contribution to our life as it is used daily by
all of us in all the places. We upload a lot of images to the internet from our

© Springer Nature Switzerland AG 2022
K. K. Patel et al. (Eds.): icSoftComp 2021, CCIS 1572, pp. 222–234, 2022.
https://doi.org/10.1007/978-3-031-05767-0_18

personal photos saved in the cloud to the posts we post on Instagram. With this ever-growing database of images it gets harder by the day to retrieve relevant images. Comparing features of input image and all the database images using different similarity measures is normally used in image retrieval [3]. Image retrieval is generally classified into three categories, text-based, content-based, semantic-based which use different features to retrieve the images [4]. In this paper we will be focusing on a hybrid method which uses both Content Based and Text Based Retrieval Methods. Content Based Image Retrieval (CBIR) uses different features in the image and compares the similarity of those features among all the images in the database. The techniques used in CBIR are derived from signal processing and other image processing areas. The features are extracted from the images and saved in the database. When a Query image arrives, its features are extracted and compared with the database image features. Some features used in the CBIR include colour, texture, shape and other salient points in an image.

There is a lot of research being done in the field of CBIR and each of those approaches serve different purposes in case of queries or databases. The objective of this paper is to propose a novel approach where the images are gone through clustering and then, some random images of the cluster are taken for which tags are generated. Those tags are compared with the tags of the query image.

The paper consists of 6 sections starting with an abstract. In our Literature review we have reviewed publications with unique methods for image retrieval and have also surveyed upon various proposals that are required for our proposals. From the reviews, under Sect. 3, a motivation for writing this paper is stated. In the Sect. 4, the proposed system is discussed where a detailed algorithm is also presented along with the various techniques used. In Sect. 5 the results are discussed along with stating the proposals limitation and further improvement. Finally, in Sect. 6 we conclude our paper with our findings and final takeaways.

2 Literature Review

In this section, we reviewed publications with unique methods for image retrieval along with various proposals that are required in our proposal as well.

2.1 Content Based Image Retrieval

The authors in [5] introduced a process which uses a relevance based ranking approach to estimate the relevance of images in a given data set. It uses Relevance based Ranking i.e. images are re-ranked depending on the visual and semantic information. The mAP was observed to be 0.6 on the NUS-WIDE dataset. It is the least precise model of all the proposed methods in the paper. And the precision was observed to decrease dramatically with an increase in depth(n).

The method in [6] is used to fill the missing and noisy tags for given images. It used Tag Matrix Completion (TMC) i.e. The image-tag relation by a tag matrix is represented and the optimal tag matrix consistent is searched. It resulted in an mAP of 0.66 when used on the Corel dataset. As the paper only addresses the datasets with labelled data, datasets without labels go unaddressed.

In the quest for finding a more efficient image retrieval method the authors in [7] introduced a method to classify and test CBIR models using deep learning for image tagging which uses NR+CNN i.e. Adding a "Noise Robust Layer" which decreases the impact of noise on the result. This resulted in an mAP of 0.53 on the very extensive NUS-WIDE-270K dataset. But, in some cases the noise may have a positive effect on the final result and reduction of the impact of noise might hinder the result in such instances.

The deep supervised hashing method mentioned in [8] is used to find binary codes for an efficient image retrieval on big datasets. They use an CNN model that inputs of two images and outputs whether the images are similar or not. This method gives an mAP value of 0.2196 when their "Sigmoid-m-1" model is used and tested on the CIFAR-10 dataset. Although the proposed method.

The authors of the paper [9] proposed a method, similar to the previous paper, which produces binary codes for images which helps with retrieval times. Here, they use minimal loss quantization criteria for their neural network for better results. It results in an mAP of 0.1943 when tested with the CIFAR-10 dataset.

In the pursuit of making an efficient image retrieval methods using binary codes, authors of [10] also propose an image retrieval method based on binary codes derived from images. Here, the loss between the real-valued feature vector and the derived binary vector is reduced but the method fails to improve the mAP on CIFAR-10 as it only produces an mAP of 0.1617.

In the process to improve AlexNet (one of the most popular image classification model), the authors of [11] made it efficient and feasible for image retrieval. The output of this improved AlexNet model is binary codes which are used for retrieving the images. The method posts an mAP of 0.4055 when tested with the CIFAR-10 dataset which is also the dataset used by this paper.

2.2 Clustering

In [12], the author proposes an Image Segmentation methodology which uses the K-Means clustering algorithm. Here, subtractive clustering is used prior to K-Means clustering in order to generate centroid points based on the value of the data points in the dataset. This process which uses both subtractive clustering and K-Means clustering results in the segmentation of image. In [13], the author proposes an Image Segmentation algorithm using fuzzy K-Means clustering algorithm. It concludes that using a fuzzy K-Means clustering algorithm was better than K-Means clustering algorithm in image analysis. Fuzzy Logic gives a basic method to come to an end based on obscure, ambiguous, boisterous, or missing data. In [14], the authors compare multiple clustering algorithms which are used in Image segmentation. Image segmentation is a very important aspect for multiple computer vision applications like self-driving cars where other cars and objects need to be segmented to process from the camera feed. In the paper it was observed that modified fuzzy C-Means algorithms are used the most for image segmentation.

2.3   Object Classification and Tag Generation

In [15], the author has performed a thorough study of the new accomplishments in this field achieved by profound learning strategies. In excess of 300 findings are included for this overview, covering numerous parts of conventional object detection. It was observed that region-based detection techniques performed better when given the freedom of computational cost when compared to unified object detection techniques. In [16], the author has compared different image classification methods which use Convolution Neural Networks (CNNs). The dataset used was ImageNet and the accuracy metric used is error. ResNet from 2015 was observed to have the lowest error rate among the compared models. Looking at the tag completion and generation models, in [17], the author proposes an original idea of a method for image tag completion of incomplete tags based on CNNs. The model takes in the incomplete tag and the image and outputs the completed tag for the image. As new methods with help of RNNs surface, old methods like this are no longer required as most of the image databases are not labeled. As the tag completion method cannot be used in the present case, in [18], the authors propose a tag generation method to generate multiple tags from an image using a combination of Convolution Neural Networks (CNNs) and Recurrent Neural Networks (RNNs).

2.4   Similarity Coefficient

As most of the literature and other works being published electronically, the need for finding similarity between them and classifying them has been ever increasing. In [19], the authors propose a new similarity metric based on the Longest Common Subsequence (LCS) string matching method. Most of the semantic similarity metrics are only applied on large documents to understand the meaning of the text and the metrics cannot be efficiently used for single sentences or short paragraphs. It was observed that the proposed method outperforms most of the then-current approaches. In [20], the authors explore the idea of connecting an Inception V3 network and a ResNet network as a hybrid for better results as ResNets and Inception V3 have nearly the same performance on images. It was concluded that the Inception-ResNet v1 network has a similar computational cost to that of Inception v3 and results in better performance although marginal. In [21], the authors compare the best CNN based models today i.e. Inception and ResNet to show the similarities and differences between them. It was found that the features extracted from the images by both the models are very similar and in some cases the Inception model extracts more intricate features from the images. As both the models were trained on the same dataset it does not come as a surprise that they both extract similar features.

2.5   Feature Extraction

In [22], the author reviews and compares different feature extraction techniques for CBIR in particular which is a very important problem to be optimized upon as the size of the databases are growing in an unprecedented manner. The authors conclude that text based retrieval is very efficient and offers a lot of advantages if implemented correctly. In [23], the authors implement and compare different feature extraction and classification methods including CNN based models. On the CIFAR-10 dataset, CNNs showed 85% accuracy when compared to 37% of the SVM classifier and 46% of the vanilla neural network classifier. This shows how better the CNN models are compared to other models. In [24], the authors proposed CNN based spatial classification of hyperspectral images (HSIs) which is used in various satellite feed related use cases. It was observed that this proposed CNN based model gives better performance than that of other neural network models.

3   Motivation

The main motivation behind our proposed CBIR model is to reduce the time taken to retrieve the image from very large databases. Using just tag similarity and feature similarity on images with relevant tags as the iterative process for retrieving images for a query image, this process is much more efficient and consumes less time when compared to traditional methods which use SIFT, SURF or other Deep Learning based methods. In the data-driven world that we live in today, it is important that we are able to retrieve the images and videos we generate everyday without the user having to wait and the server not having to use a lot of computation. As more and more people come online, create and share more images and videos it's crucial to have a proper retrieval system which is accurate, efficient and consumes less time.

The Image Retrieval system proposed in this paper uses a clustering and tag generation based technique to retrieve the most relevant images. The clustering of the image saves a lot of computation as we do not generate tags for every image in the dataset. Using feature extraction methods we were able to achieve greater accuracy in clustering when compared to clustering the images directly. With the tag comparison based technique, we will save a lot of time in computation as we only need to compare text and not features extracted for all images like in traditional methods. This strikes a good balance between accuracy and efficiency.

4   Proposed System

We have proposed an Image Retrieval method which uses clustering and tag generation to retrieve the most relevant images very efficiently. We will start with extracting features of images as clustering the features gives better results than clustering the images themselves. ResNet-50 with the weights of ImageNet were used here to extract the features of the images. After extracting the features from the images, we cluster them with K-Means Clustering. The clustered images are then stored into folders. We randomly take n(=10) images from each folder and run them through the IBM VisualRecognitionV3 which generates tags for the images. We then concatenate all the unique tags of each folder and name the folder with the generated string. The algorithm of this part where the database concentration happens is given in Algorithm 1.

When a Query image is given, it is run through IBM VisualRecognitionV3 to generate tags. The tags are then added to all the folder names. In order to do any mathematical operations with text, vectorization is needed. We run the folder names through Count Vectorizer to get vectors. All the vectors are tested for cosine similarity with query tag. For all the images in the folders with highest x(=3) similarity, extract features of them using ResNet-50. Now apply cosine similarity between the features of query image and features of top x folders. Retrieve the top desired number of results. The algorithm for query image retrieval is given in Algorithm 2. An overview of the proposed system is given through the Architecture Diagram in Fig. 1.

Algorithm 1: Database Clustering

Result: Clustered Folders with tags as folder names
initialization;
resnet ⟵ ResNet50(weights='imagenet');
for *image in all images in the database* **do**
　　temp⟵ resnet(image);
　　add temp to array;
end
initialize KMeans;
cluster array;
save each cluster as a folder;
for *each folder* **do**
　　load n(=10) random images from folder;
　　for *each image in random images* **do**
　　　　tag ⟵ VisualRecognitionV3(image);
　　　　tag ⟵ SnowBall Stemmer(tag) ;
　　　　add tag to tags ;
　　end
　　string ⟵ unique(tags);
　　folder name ⟵ string ;
end

4.1 Feature Extraction from Images

Although clustering is a very useful unsupervised machine learning algorithm, it can be said that clustering images is not its forte. But, clustering with the

---

Algorithm 2: CBIR for Query Image

Result: Most Relevant Images
initialization;
resnet ⟵ ResNet50(weights='imagenet');
query_tag ⟵ VisualRecognitionV3(query_image) ;
query_tag ⟵ SnowBallStemmer(query_tag) ;
Append query_tag to all folder names;
Clean all the strings;
query_vector ⟵ CountVectorizer(query_tag);
for *i in all folder names* **do**
 | vector ⟵ CountVectorizer(i) ;
 | add vector to vector_list ;
end
for *i in vector_list* **do**
 | similarity ⟵ CosineSimilarity(query_vector,vector_list) ;
 | add similarity to similarity_list;
end
query_features ⟵ **for** *i in folders with top x(=3) similarity* **do**
 | **for** *image in i* **do**
 | | feature ⟵ resnet(image);
 | | add feature to feature_list;
 | **end**
end
for *i in feature_list* **do**
 | similarity ⟵ CosineSimilarity(query_features,i) ;
 | add similarity to result_list;
end
Output images with top 'n' similarity;

---

extracted feature vectors makes it very much usable. In feature extraction, we extract all the different features that particular neural network is using to classify it. In our case, we use the infamous ResNet-50 model for extracting the features from images as we do not have to do any training to get usable results from it. Clustering becomes very easy with feature vectors instead of image matrices. Figure 3 shows the architecture of ResNet-50.

4.2 Clustering

Clustering is the process of grouping similar data points in a dataset. Clustering is an unsupervised learning technique and works properly if implemented with other data processing techniques. In this proposed system, we will be using clustering because we don't need to train it on any data beforehand. K-Means Clustering is a very simple clustering algorithm which has good results without making the system complex to implement. As clustering is a non-supervised algorithm it does not need any kind of training. Which is important here as

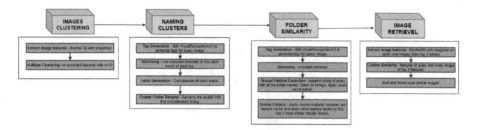

Fig. 1. Architecture of the proposed system

there are no labels to the images at the start. K-Means clustering uses euclidean distance to calculate the distance between the centroid and data points.

4.3   Object Classification and Tag Generation

The next part of the proposed system is tag generation. Generating tags to an image requires classifying the object. Old models used to only give single output such as the name of the object and this is not ideal in our case as, most of the images in reality have multiple objects in them. To deal with this issue we have use IBM's visual recognition v3 API. that returns various details other than the object present in the image. Details such as colours present, shapes are also returned by the API.

4.4   Similarity Coefficient

The process of calculating the similarity coefficient between cluster folder tag and query image tag, a three step process is employed. Firstly, stemming is applied on the tags to preprocess the tag string, which is followed by count vectorization to convert the string into a 1×300 feature vector for each tag string. Finally, cosine similarity is calculated between the cluster tag feature vector and query tag feature vector. Although cosine similarity metric is a string-based algorithm, it has a good track record of being computationally feasible.

Count Vectorization: As text is not supported by machine learning models, we will have to convert the labels into vectors which we can give as input to the model. There are different vectorization techniques. Count Vectorization can also be called a one-hot encoding. For every unique word in the corpus a count is assigned and for every occurrence of that word the count of it is used in making the vector. It is a bag of words (BoW) type of vectorizer meaning, the order of words in the sentence does not matter to the vectorizer. The Fig. 2 shows the basic architecture of a Count Vectorizer. It is only concerned with the presence of the word and not it's placement in the sentence. For applications like our's where we need to just find the similarity between different sentences, a BoW based model makes the most sense and is better for efficiency.

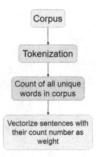

Fig. 2. Count vectorizer

Stemming: Stemming is the process of replacing a word by its root word so that two different forms of the same word are not considered separately. An ML model is easily variable by the data it gets. We discard the prefixes and the suffixes and just leave the root word. Although we can still find similarities between different sentences and it will be accurate without stemming, we will still be gaining efficiency as there are less unique words. SnowBall stemmer has different rules for stemming a word. Although there are multiple stemmers available to use, SnowBall stemmer has been used here as it has a proper middle ground for finding root words. Although Lemmatizer does a similar job, it uses a predefined corpus to find the root word of a given word. So, Lemmatizers consume a lot more computation than stemmers as stemmers just use a set of rules to stem. For this particular case, stemming was used as we don't need the advantages offered by lemmatizing.

Cosine Similarity: It is a similarity metric used to find how similar or different two sentences or documents are. It calculates the cosine of the angle between the two vectors being compared in a multidimensional space. This does not take into account the euclidean distance between the vectors and is not prone to changing when the size of the sentence or document are further. The input here is the vector outputted by the Count Vectorizer and it outputs the similarity value between the sentences or documents. It is immune to the effect of the number of common words which is used by the Euclidean distance.

5   Results and Discussion

5.1   Dataset Used: CIFAR-10

In this paper, we have used the standard CIFAR-10 dataset. It contains 50000 images of 32×32 dimensions and RGB colour format. The dataset consists of 10 classes with 5000 images from each class. The classes in CIFAR-10 include airplane, automobile, bird, cat, deer, dog, frog, horse, ship, truck. This dataset was proposed in the paper [25].

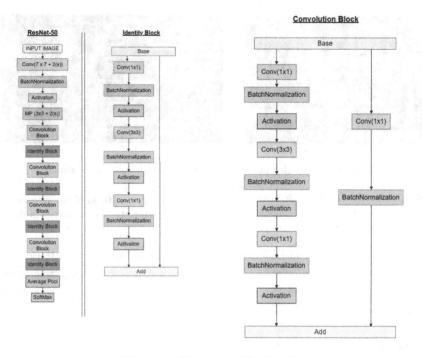

Fig. 3. Architecture of ResNet-50

5.2 Input and Output

In this sub-section, we show the results of "Database Concentration" part of our proposed method. Here, in the first image Fig. 4 we can see that there are many pictures in a single "data" folder which are not classified. Whereas, in the second image we can see that the images are clustered and stored in folders according to the clustering results. It can be seen that the folders are also named with the obtained tag names from the tag generation algorithm. This was tested on a small 100 image dataset on a local machine to test the working of the model. A comparison of the ratio of number of images to retrieve and number of images in the dataset in case of the CIFAR-10 dataset is shown in Table 2.

5.3 Results

The comparison table and a graph plotted with the mAP values for better understanding of the data is available in Table 1. We can see that the proposed algorithm was holding up against the CBIR algorithms compared. Although the proposed method is not the best out there, it's a master of none which balances the time taken with the retrieval accuracy. This can be further improved by using different CBIR methods in the query retrieval step and by using better tag generation models. As for the database concentration aspect, the method only needs to search through 15000 images instead of searching through 50000

Fig. 4. Dataset folder before and after the Database Concentration

Table 1. mAP% between proposed method and other proposals on CIFAR-10 dataset

Method	mAP%
Sigmoid-m-1	21.96
DeepBit	19.46
DeepHashing	16.17
OANIR	40.55
Proposed	**43.82**

Table 2. Ratio of No. of images to retrieve to No. of images in dataset

Before/After	Ratio
Before concentration	5000:50000
After concentration	5000:15000

images in case of CIFAR-10 dataset. Using this concept the ratio of number of images to retrieve to number of images to search of concentrated dataset comes to 1:3 all the way from 1:10 as shown in Table 2. This improves the time required for retrieval of relevant images by a large extent. A line graph visualisation of the results and comparison with other methods has been shown in Table 1.

5.4   Limitations

There are however, certain limitations to this proposed method. As we focused mostly on the method for database concentration, focus on the CBIR model used in the final step did take a hit. This method can be further improved upon by using more sophisticated CBIR methods for the final step of query image retrieval. Using other CBIR focused Deep Learning models like AlexNet and OANIR would have improved the mAP of the proposed model.

6   Conclusion

The paper has proposed a novel content-based image retrieval system which is based on feature extraction, tag generation, and clustering. These types of improvements are needed to help with the efficiency and speed of the retrieval process in large databases. Although the method is not as accurate as some of the other approaches, this is more efficient than other feature comparison methods. As we're just comparing the tags and not the whole feature set, it will be much less computationally intensive. But since the main aim of this paper was to enhance the retrieval process by concentrating the dataset, we are successful in it.

References

1. Mahmoudi, S., Benjelloun, M.: Using global shape descriptors for content medical-based image retrieval. In: Handbook of Research on ICTs and Management Systems for Improving Efficiency in Healthcare and Social Care, pp. 492–502. IGI Global (2013)
2. Schaefer, G.: Effective and efficient browsing of large image databases. In: Handbook of Research on Digital Libraries: Design, Development, and Impact, pp. 142–148. IGI Global (2009)
3. Image Retrieval System. SSLA, An Engineering Solutions Company (2013). https://www.ssla.co.uk/image-retrieval-system/
4. Alkhawlani, M., Elmogy, M., El Bakry, H.: Text-based, content-based, and semantic-based image retrievals: a survey. Int. J. Comput. Inf. Technol. 4(01), 58–66 (2015)
5. Wang, Y., et al.: Joint hypergraph learning for tag-based image retrieval. IEEE Trans. Image Process. 27(9), 4437–4451 (2018)
6. Wu, L., Jin, R., Jain, A.K.: Tag completion for image retrieval. IEEE Trans. Pattern Anal. Mach. Intell. 35(3), 716–727 (2012)
7. Fu, J., Rui, Y.: Advances in deep learning approaches for image tagging. APSIPA Trans. Signal Inf. Process. 6 (2017)
8. Liu, H., et al.: Deep supervised hashing for fast image retrieval. In: Proceedings of the IEEE Conference on Computer Vision and Pattern Recognition, pp. 2064–2072 (2016)
9. Lin, K., et al.: Learning compact binary descriptors with unsupervised deep neural networks. In: Proceedings of the IEEE Conference on Computer Vision and Pattern Recognition, pp. 1183–1192 (2016)
10. Liong, V.E., et al.: Deep hashing for compact binary codes learning. In: Proceedings of the IEEE Conference on Computer Vision and Pattern Recognition, pp. 2475–2483 (2015)
11. Bai, C., et al.: Optimization of deep convolutional neural network for large scale image retrieval. Neurocomputing 303, 60–67 (2018)
12. Dhanachandra, N., Manglem, K., Chanu, Y.J.: Image segmentation using K-means clustering algorithm and subtractive clustering algorithm. Procedia Comput. Sci. 54, 764–771 (2015)
13. Dehariya, V.K., Shrivastava, S.K., Jain, R.C.: Clustering of image data set using k-means and fuzzy k-means algorithms. In: 2010 International Conference on Computational Intelligence and Communication Networks, pp. 386–391. IEEE (2010)
14. Aslam, Y., et al.: A review on various clustering approaches for image segmentation. In: 2020 Fourth International Conference on Inventive Systems and Control (ICISC), pp. 679–685. IEEE (2020)
15. Liu, L., et al.: Deep learning for generic object detection: a survey. Int. J. Comput. Vis. 128(2), 261–318 (2020)

16. Sultana, F., Sufian, A., Dutta, P.: Advancements in image classification using convolutional neural network. In: 2018 Fourth International Conference on Research in Computational Intelligence and Communication Networks (ICRCICN), pp. 122–129. IEEE (2018)
17. Geng, Y., et al.: A novel image tag completion method based on convolutional neural transformation. In: Lintas, A., Rovetta, S., Verschure, P.F.M.J., Villa, A.E.P. (eds.) ICANN 2017. LNCS, vol. 10614, pp. 539–546. Springer, Cham (2017). https://doi.org/10.1007/978-3-319-68612-7_61
18. Wang, J., et al.: CNN-RNN: a unified framework for multi-label image classification. In: Proceedings of the IEEE Conference on Computer Vision and Pattern Recognition, pp. 2285–2294 (2016)
19. Islam, A., Inkpen, D.: Semantic text similarity using corpus-based word similarity and string similarity. ACM Trans. Knowl. Discov. Data (TKDD) 2(2), 1–25 (2008)
20. Szegedy, C., et al.: Inception-v4, inception-ResNet and the impact of residual connections on learning. In: Proceedings of the AAAI Conference on Artificial Intelligence, vol. 31 (2017)
21. McNeely-White, D., Beveridge, J.R., Draper, B.A.: Inception and ResNet features are (almost) equivalent. Cogn. Syst. Res. 59, 312–318 (2020)
22. Chadha, A., Mallik, S., Johar, R.: Comparative study and optimization of feature-extraction techniques for content based image retrieval. arXiv preprint arXiv:1208.6335 (2012)
23. Jogin, M.: Feature extraction using convolution neural networks (CNN) and deep learning. In: 2018 3rd IEEE International Conference on Recent Trends in Electronics, Information & Communication Technology (RTEICT), pp. 2319–2323. IEEE (2018)
24. Yue, Y., et al.: Spectral-spatial classification of hyperspectral images using deep convolutional neural networks. Remote Sens. Lett. 6(6), 468–477 (2015)
25. Krizhevsky, A., Hinton, G., et al.: Learning multiple layers of features from tiny images (2009)

Microstructure Image Classification of Metals Using Texture Features and Machine Learning

Hrishikesh Sabnis[1], J. Angel Arul Jothi[2(✉)] [ID], and A. M. Deva Prasad[3]

[1] Department of Mechanical Engineering, Birla Institute of Technology and Science Pilani Dubai Campus, Dubai International Academic City, Dubai, UAE
f20170023@dubai.bits-pilani.ac.in
[2] Department of Computer Science, Birla Institute of Technology and Science Pilani Dubai Campus, Dubai International Academic City, Dubai, UAE
angeljothi@dubai.bits-pilani.ac.in
[3] Department of Mechanical Engineering, Manipal Academy of Higher Education Dubai Campus, Dubai International Academic City, Dubai, UAE

Abstract. This research was conducted on microstructural images of four different metals such as mild steel, stainless steel, aluminium, and copper. The research aims to classify the microstructure images of the four metals obtained using light-optical microscope using texture features and machine learning techniques. The images were subjected to different steps like image pre-processing, texture feature extraction and classification. Classical machine learning algorithms like Support Vector Machines, Naïve Bayes, K-Nearest Neighbor, Decision Tree, Random Forest, and XGBoost were used. Experimental results show that the XGBoost classifier achieves the highest classification accuracy of 98% using normalized texture features and cross validation using 10-folds.

Keywords: Microstructure images · Machine learning · Support Vector Machine · Naïve Bayes · K-Nearest Neighbor · Decision Tree · Random Forest · XGBoost

1 Introduction

The structure of a material is predominantly classified as: crystal structures and microstructures. Crystal structures refer to the mean position of the atoms inside the unit cell and are completely characterized by the lattice type, which helps describe the appearance of the material on an atomic scale. The structure, as a whole, is made up of atoms, while its lattice consists of points and the system is a set of axes. In other words, this structure is simply an ordered array of atoms, ions or molecules. On the other hand, microstructures refer to the arrangement of phases and defects within a material which tend to affect the physical properties along with the behavior of the materials. These properties are studied to be modified in order to produce specific properties for material processing as they are rich in information relating to morphology. Microstructures help describe the appearance of the material on a nn-cm length scale as they provide

© Springer Nature Switzerland AG 2022
K. K. Patel et al. (Eds.): icSoftComp 2021, CCIS 1572, pp. 235–248, 2022.
https://doi.org/10.1007/978-3-031-05767-0_19

the inner structure of the material which stores the genesis of the material and varies across different materials. Microstructures often help reveal and evaluate critical properties of materials and are hence preferred for the study of material science. Analyzing microstructure images of metals helps manufacturers to determine product reliability. Also, microstructure decides the mechanical properties of metals [1].

Machine learning (ML) is the science of building automated systems that learn from samples (data) and improve through experience. This is one of the rapidly growing fields today that integrates computer science and statistics. ML algorithms have been applied in various fields like medicine, commerce, manufacturing, education and marketing. One of the important reasons of the impact of ML on various fields is due to the large availability of data and computational facilities [2]. ML and image processing algorithms have been used for the study and analysis of microstructure images [3, 4].

Usually, microstructure image analysis of materials is done to determine the strength, ductility, hardness, and corrosion resistance of the materials. However, in this work we aim to develop an automated method to distinguish the microstructure images of metals such as mild steel, stainless steel, aluminium, and copper with the aid of texture features and machine learning. This automated identification of metals using microstructure images can be incorporated into an industrial robot that can aid the robot during a) inspection of materials to distinguish different materials and b) material handling. This work can be further extended to select appropriate materials automatically by the robot for designing and manufacturing a product.

This paper is organized as follows: Sect. 2 details the literature survey. Section 3 describes the dataset. Methodology is provided in Sect. 4. The evaluation metrics and the experiments conducted are explained in Sect. 5. Section 6 gives the results followed by conclusion.

2 Literature Review

This section reviews the previous work on the application of machine learning techniques to microstructure images. Ma et al. used Random Forest (RF) classifier to predict 10 different processing classes from microstructure images of an alloy of uranium and molybdenum. Noise from the images was removed by bilateral filters and the images were segmented using k-means algorithm. Area, spatial, Haralick and Local Binary Pattern (LBP) features were extracted from the images. The method reported a F1 score value of 95.1% [5].

Naik et al. explored the classification of metallurgical phases (ferrite, pearlite, and martensite) of steels using texture features and pixel intensity variations of phases. Gray Level Co-occurrence Matrix (GLCM) features were extracted from the images followed by feature ranking and feature selection. ML classifiers such as K-Nearest Neighbors (KNN), Linear Discriminant Analysis (LDA), Naïve Bayes (NB), and Decision Trees (DT) were used to classify the images [6].

DeCost et al. employed the bag of visual features image representation for creation of generic microstructural signatures. Support Vector Machines (SVM) classifier was used to classify the microstructures into one of the seven groups. Harris-Laplace and Difference of Gaussian (DoG) operators were employed, focusing on the detection of image gradient structures. This method achieved an accuracy of 83% with a standard deviation of 3% [7].

Gola et al. studied the classification of microstructure images of low carbon steel into three classes such as martensite, pearlite, or bainite. Three parameter groups were used for feature extraction, namely morphological parameters, textural features and substructure parameters. A correlation matrix was used to remove highly corelated features. This was followed by a feature selection method to reduce dimensions and model complexity. SVM classifier was used to classify the images which exhibited promising results [8].

Abouleta worked towards the classification of 18 classes of copper and copper alloys microstructures based upon image processing and neural networks classifier. The images were resized and converted to gray scale. Total of 22 first and second-order texture features such as contrast, dissimilarity, homogeneity, similarity, angular second moment, entropy, etc. were extracted. Further a feed forward artificial neural networks (ANN) model was built to classify the microstructure images. The proposed method achieved an average discrimination rate of 97.62% [9].

Chowdhury et al. studied image-based microstructure recognition for dendritic morphologies with the aid of computer vision and machine learning methods. There were two tasks performed during this study: differentiating between micrographs that shows dendritic morphologies and those that do not, and the different cross-sectional views were identified. Multiple features like the texture features, shape features, visual bag of words features and Convolutional Neural Networks (CNNs) features were extracted from the images. The features were classified using SVM, voting, KNN and RF models. It was concluded that the features extracted using CNNs have the highest power to discriminate the microstructure images. The highest classification accuracy of 92% and 97% was reported for task 1 and task 2 respectively [10].

Papa et al. laid emphasis on using computer techniques for automated classification of metallographic images of cast iron by graphite particles characterization. Texture features was extracted using Gabor filter and classified using Optimum-Path Forest (OPF) classifier. The study used three types of cast iron: nodular, gray, and malleable. The OPF classifier demonstrated better accuracy and speed for the specified task [11].

Sarkar et al. explored the classification of ultrahigh carbon steel microstructural images into three classes using ensemble approach. The Rotational Local Tetra Pattern (RLTrP) features were initially extracted from the images. These features were used to train the base classifiers like SVM, KNN, and RF separately. The outcome of the three base classifiers were combined using a novel fuzzy integral measure. The SVM, RF and KNN classifiers achieved an accuracy of 93.52%, 91.11% and 92.59% respectively. However, the proposed ensemble approach achieved an improved accuracy of 96.11% [12].

Baskaran et al. used a two-stage machine learning and image processing approach for classifying and segmenting microstructure images of titanium alloy. In the first stage a custom-made CNN built of three convolutional layers and one fully connected layer was used to classify the microstructure images into one of the three classes namely lamellar, duplex and acicular. The CNN yielded an accuracy of 93%. The second stage used marker-based watershed transform and Histogram of Oriented Gradients (HOG) method to extract morphology features [13].

Gupta et al. studied the automatic recognition of steel types and subsequent identification of the different phases of steel from microstructure images. A large dataset
consisting of ferrite-pearlite and martensite-austenite micrographs were used for this
study which were obtained from a SEM. The LBP features were obtained from the
images which were fed to a RF classifier to predict the steel type. After this the phase
quantification was done using Otsu's thresholding [14].

Thus, the literature review shows that there has been work on microstructure images
using machine learning and image processing techniques for a) classification of different
classes of various alloys (uranium-molybdenum, copper, titanium), b) classification of
different phases of steel c) material classification and d) segmentation of structures from
microstructure images. In contrast to the previous work, this research work focusses on
classifying different metals (materials) using the microstructure images by extracting
texture features and employing several ML classifiers. As previously stated this could
help in industrial automation.

3 Dataset

The dataset for this study was collected manually and consists of 2000 images. Sample
images from the dataset are given in Fig. 1. A vital component of this research was
acquiring proper microstructural images of the metal samples. The following steps were
followed for preparing the metal samples and acquiring the images from them:

- Sheets of a metal of size approximately 30 × 30 cm with thickness of 1 mm was taken.
  From the sheets, 10 samples each of 2 × 2 cm were cut for capturing the images.
- Modelling clay (MC) was used to keep the samples in place. This eases the process
  of grinding, shining and polishing.
- Polyvinyl Chloride (PVC) pipes were cut with diameter 4–5 cm and length 5 cm. The
  MC was placed within the PVC pipe, with the sample of the metal placed onto the
  top surface of the pipe and was fixed with the help of the clay.
- Ten such samples were prepared for each of the metals and kept for 24 h to allow the
  clay to harden and the metal sample to settle.
- Once settled, the surface of the sample was then grinded using sandpapers of P240,
  P600, P800 and P1200. These are micro grit sandpapers which were used in the order
  of very fine abrasives (P240), followed by extra fine abrasives (P600), and finally
  super fine abrasives (P1200).
- Once completed, a metal polishing and surface finishing machine was used to provide
  a smoother surface and remove all the scratches from the top of the surface.
- After polishing, a few drops of etching agent were placed on the surface of the metal
  and dabbed with a tissue to avoid corrosion from taking place.
- The sample was finally placed on the optical microscope and the images were captured
  at 400 × magnification with the help of a software called 'ENVISION'.
- Approximately 50 images of each sample were taken moving the sample under the
  microscope by 0.1 mm after every image. It is ensured that the grains are visible and
  all the edges are captured so that any anomalies can be avoided.
- A total of approximately 500 images were captured for each metal in JPG format.

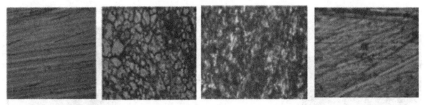

Fig. 1. Sample images of stainless steel, mild steel, copper, and aluminum respectively from the dataset.

4 Methodology

Figure 2 describes the proposed method. After acquiring the images for the study, the next step was to pre-process the images. This was followed by extracting texture features from the images. Then, machine learning classifiers were trained and tested to identify the metals using the extracted features.

Fig. 2. The proposed method

4.1 Image Pre-processing

After acquiring the images, they are subjected to pre-processing to enhance image quality and further prepare them for feature extraction. Denoising and contrast enhancement were the two pre-processing techniques applied to the images. The input images from the dataset were converted to gray scale. The noise in the images were removed by non-local denoising technique. The contrast of the denoised images were enhanced by histogram equalization to obtain the pre-processed images. The steps in pre-processing the images are given by Fig. 3.

4.2 Feature Extraction

After pre-processing, texture features were extracted from the images. Texture is the repeating pixel intensity values that forms a pattern in the images. Texture features capture the spatial distribution of intensity levels in an image [15]. The distribution of the pixel intensities in the microstructure images of each metal were different and this fact was captured using the texture features. Two types of texture features were extracted from the images. They are first-order statistics and second-order statistics features.

First-Order Statistics. The histogram of an image is the basis for the computation of first-order statistics. These features are directly computed from the pixel values of the

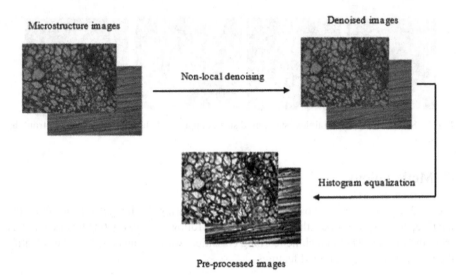

Microstructure images

Denoised images

Non-local denoising

Histogram equalization

Pre-processed images

Fig. 3. Steps involved in pre-processing the images

original image [15]. Let the probability distribution of the pixels at the various gray levels in an image be denoted by Eq. (1).

$$P(I) = \frac{\text{Number of pixels with gray level I}}{\text{Total number of pixels in the image}} \tag{1}$$

For a grayscale image the possible gray levels are denoted as $N = 0, 1, 2,\ldots, 255$. Using the probability histogram of the image given above, other first-order histogram features also known as measures of central moments such as mean (m), standard deviation, skewness and kurtosis can be calculated.

The mean pixel intensity of an image (mean) is given by Eq. (2). Variance evaluates the deviation of the gray level values from the mean of the histogram. Skewness measures how skewed the histogram is with respect to the mean. For a perfectly balanced histogram, the skewness is zero. For unbalanced histograms it is either a positive or a negative quantity. Kurtosis measures the sharpness of the histogram [15, 16].

$$m = \sum_{i=0}^{N-1} i * P(i) \tag{2}$$

The variance, skewness, and kurtosis are given by Eq. (3).

$$cm_k = \sum_{j=0}^{N-1} (i - m)^k * P(i) \tag{3}$$

where N refers to the number of possible gray levels in an image and the value of $k = 2$, 3, 4 for variance, skewness, and kurtosis respectively. In this work, 4 first-order statistics features like mean (m), standard deviation, skewness and kurtosis were extracted from the images.

Second-Order Statistics. The drawback of first-order statistics is that they do not provide the relevant information about the relationship between the gray level values of the pixels in an image. One method to capture the relationship between the gray level values of the pixels in an image is the gray level co-occurrence matrix proposed by Haralick [17]. It is a tabular structure whose dimensions are governed by the maximum gray level value (N) of the pixels in an image. It tabulates how frequently two pixels with a specific value which are separated by a distance (d) and placed in a direction (θ) appear in a particular window. The direction (θ) usually takes four different values such as 0°, 45°, 90° and 135°. The co-occurrence probabilities that are stored in the GLCM are used to compute different statistics which characterizes the texture features of the image. Let G_{ij} denote the $(i, j)^{th}$ entry in the GLCM divided by the sum of the elements of GLCM. In this work, the following 4 second-order statistics or GLCM features were extracted from the given images with values of d = 1 and $\theta = 0°$.

- Energy: This feature is used to detect any disorders in textures and measures the textural uniformity and is given by Eq. (4). In the case of a less homogenous image, the GLCM will have many small entries [18, 19].

$$\text{Energy} = \sum_i \sum_j G_{ij}^2 \tag{4}$$

- Contrast: Measures the local variations present in the image and is given by Eq. (5) [18, 19].

$$\text{Contrast} = \sum_i \sum_j (i - j)^2 G_{ij} \tag{5}$$

- Homogeneity: Measures the homogeneity of an image and is given by Eq. (6). It assumes larger values for smaller gray tone differences in pair elements. Contrast is inversely related to homogeneity. Therefore, if homogeneity decreases the contrast increases and vice versa [18, 19].

$$\text{Homogeneity} = \sum_i \sum_j \frac{G_{ij}}{1 + (i - j)^2} \tag{6}$$

- Correlation: Measures the gray tone dependencies in the image and is given by Eq. (7).

$$\text{Correlation} = \frac{\sum_i \sum_j (i - \mu_r)(j - \mu_c) G_{ij}}{\sigma_r \sigma_c} \tag{7}$$

where μ_r, μ_c, σ_r, σ_c are the mean and standard deviation of the row and the column values of the GLCM respectively [18, 19].

4.3 Material Identification and Classification

The 8 texture features (4 first-order and 4 second-order) extracted from the microstructure images of various metals were used to train and test several supervised machine learning classifiers. During training a classifier is supplied with examples (data points) belonging to different classes and it learns a model. During testing, the model is used to assign class labels to unknown examples. For this research, we used six different classifiers like: Support Vector Machine, K-Nearest Neighbors, Naïve Bayes, Decision Tree, Random Forest and XGBoost. The following paragraphs detail the classifiers used.

Support Vector Machine. Support vector machine is a supervised machine learning technique which aims to classify data points belonging to different classes by developing a hyperplane across a d-dimensional space, where d refers to the number of features used. This hyperplane acts as a decision boundary. The data points which are closer to the hyperplane are called support vectors as they influence the position and orientation of the hyperplane. Changing the support vectors changes the position of the hyperplane. SVM can be of two types depending on the nature of the data: linear SVM and non-linear SVM. If the data points are not separable using a hyperplane in the original feature space a non-linear SVM is used. The concept behind the non-linear SVM is to use a kernel function which maps the original data to a higher dimensional space to make it separable [20, 21]. In this work, we experimented the linear SVM and non-linear SVM with radial basis function (RBF) and sigmoid kernels. The experiments showed the linear SVM to obtain the highest classification accuracy. Hence, in all our experiments we have used linear SVM.

K-Nearest Neighbors. KNN classifiers qualify as a non-parametric, supervised technique. It is a lazy learner as it simply stores the training data rather than learning from it. It performs classification of the test data at the time of testing with the aid of the stored data. KNN computes the distance between the test data and the available training data points. It then assigns the class label to the test data based on the class labels of the K nearest neighbors [20, 21]. In this work, the Euclidean distance was used and all the K neighbors of a test point were assigned equal weights. Different values of K were experimented and a value of 5 shows the highest accuracy for this work.

Naïve Bayes. The Naïve Bayes classifier works relatively fast compared to the other classifier algorithms as it works on Bayes theorem of probability to help predict the class of the test data. NB algorithm works on the assumption of class conditional independence (i.e.) given the class label of a sample, the feature values are independent of each other. The Naïve Bayes classifier computes the conditional probability of a class given a test sample which depends on the product of the prior probability of the class and the conditional probability of the test sample given the class. The conditional probability of all the classes given the test sample is computed and the test sample is assigned the class whose conditional probability is the highest [20, 21].

Decision Tree. Decision tree classifier is a supervised machine learning technique. Its structure represents a tree consisting of a set of nodes and edges. There is a distinct node called the root node which does not have a parent. The internal nodes including the root node represent a test on a feature (attribute) of the dataset while the edges from a node denote the possible values (outcomes) of the feature present inside the node. Each of the leaf nodes represents the outcome (class label). At every internal node the best attribute is identified using a measure of impurity like 'Gini or Entropy' [20, 21]. In this work,

'Gini' was used as the measure of impurity. The tree was grown until all leaves contained samples belonging to the same class or until all leaves contained less than 2 samples.

Random Forest. Random forest is a type of supervised learning algorithm proposed by Breiman in 2001. It consists of an ensemble of decision trees trained using the 'bagging' method. Random forests create subsets of the features and build smaller trees using those subsets which helps prevent overfitting which may occur in a DT model. The number of trees that the algorithm builds, i.e. estimators, help create these subsets. A higher count of estimators usually increases the performance, making the prediction more stable, however slowing down the computation [22, 23]. This work used 100 estimators with 'Gini' impurity measure. As in the case of DT, each tree was grown until all leaves contained samples belonging to the same class or until all leaves contained less than 2 samples.

XGBoost Classifier. XGBoost classifier [24] is a scalable, sequential, decision tree like classifier which uses gradient boosting. It applies an ensemble learning technique which uses the principle of boosting where the error predictions of the previous models are improved by constructing new models. Moreover, the gradient descent method is used to reduce the error when new models are added.

5   Evaluation Metrics and Experiments Conducted

5.1   Metrics Used

The classifiers were evaluated using metrics like precision, recall, F1 score, and accuracy. Let TP, TN, FP, and FN denote true positives, true negatives, false positives and false negatives respectively. Precision evaluates the accuracy for positive predictions by the classifier and is given by Eq. (8). Recall measures the ability of a classifier to find the positive instances and is given by Eq. (9). F1 score is the weighted harmonic mean of precision and recall and is given by Eq. (10). It lies in the range of 1.0 and 0.0. Accuracy is defined as the fraction of correct predictions to the total number of predictions and is given by Eq. (11).

$$\text{Precision} = \frac{TP}{(TP + FP)} \tag{8}$$

$$\text{Recall} = \frac{TP}{(TP + FN)} \tag{9}$$

$$F1score = \frac{2 * (\text{Recall} * \text{Precision})}{(\text{Recall} + \text{Precision})} \tag{10}$$

$$\text{Accuracy} = \frac{TP + TN}{TP + TN + FP + FN} \tag{11}$$

5.2 Experiments

Classifier models require a train set and a test set. The training set is used to train the classifier models while the test set is used to test the classifier models. In this work, the hold out method and the 10-fold cross validation method were used for evaluating the classifier models. For the hold out method, we used an 80:20 ratio where 80% of the dataset was used as train set and the remaining 20% was used to test the models. The texture features extracted were of different ranges. Hence, in order to study the impact of this on the accuracy of classification, we used two approaches like normalized mode and non-normalized mode. For the normalized mode, the feature values were normalized using min-max normalization whereas for the non-normalized mode, the original features extracted were used as such for training and testing the classifier models.

6 Results and Discussions

6.1 Non-normalized Mode

Table 1 and Table 2 provide the results obtained when non-normalized features were used to test the machine learning classifiers. Also, Table 1 shows the results when the hold out method of evaluation was used and Table 2 shows the results when the 10-fold cross validation was used.

It can be seen from Table 1 that for non-normalized features and hold out method, XGBoost and random forest showed the highest accuracy, F1score, precision and recall of 95%. The decision tree classifier attained an accuracy, F1score, precision and recall of 92%. This is followed by KNN classifier which achieved an accuracy of 87% and F1score, precision and recall of 86%. The NB and SVM classifiers achieved an accuracy of 60% and 54% respectively.

Table 1. Results of the various classifiers using non-normalized features and hold out method of evaluation.

Classifier	Accuracy	F1 score	Precision	Recall
XGBoost	0.95	0.95	0.95	0.95
Naïve Bayes	0.6	0.54	0.58	0.58
K-Nearest Neighbors	0.87	0.86	0.86	0.86
Decision Tree	0.92	0.92	0.92	0.92
Random Forest	0.95	0.95	0.95	0.95
SVM	0.54	0.52	0.52	0.52

Similarly, Table 2 shows that the XGBoost classifier achieved the highest accuracy and F1 score of 96% for non-normalized features and 10-fold cross validation. Also, the XGBoost classifier achieved a precision and recall value of 95%. Then, the decision tree and RF classifiers achieved an accuracy, F1score, precision and recall of 94%. The NB and SVM classifiers achieved an accuracy of 60% and 58% respectively. The KNN classifier achieved the lowest accuracy of 47%.

Table 2. Results of the various classifiers using non-normalized features and 10-fold cross validation method of evaluation.

Classifier	Accuracy	F1 score	Precision	Recall
XGBoost	0.96	0.96	0.95	0.95
Naïve Bayes	0.6	0.54	0.58	0.58
K-Nearest Neighbors	0.47	0.47	0.48	0.47
Decision Tree	0.94	0.94	0.94	0.94
Random Forest	0.94	0.94	0.94	0.94
SVM	0.58	0.55	0.58	0.54

Table 3. Results of the various classifiers using normalized features and hold out method of evaluation.

Classifier	Accuracy	F1 score	Precision	Recall
XGBoost	0.95	0.95	0.95	0.95
Naïve Bayes	0.6	0.57	0.6	0.6
K-Nearest Neighbors	0.82	0.81	0.81	0.81
Decision Tree	0.89	0.89	0.89	0.89
Random Forest	0.93	0.93	0.93	0.93
SVM	0.63	0.57	0.59	0.59

Table 4. Results of the various classifiers using normalized features and 10-fold cross validation method of evaluation.

Classifier	Accuracy	F1 score	Precision	Recall
XGBoost	0.98	0.98	0.98	0.98
Naïve Bayes	0.6	0.57	0.6	0.6
K-Nearest Neighbors	0.63	0.62	0.62	0.62
Decision Tree	0.9	0.9	0.9	0.9
Random Forest	0.96	0.95	0.95	0.95
SVM	0.63	0.5	0.45	0.45

6.2 Normalized Mode

Table 3 and Table 4 show the results obtained when normalized features were used to test the machine learning classifiers. Table 3 shows the results when the hold out method of evaluation was used and Table 4 shows the results when the 10-fold cross validation was used.

It can be observed from Table 3 that in the case of normalized features and hold out method, XGBoost classifier achieved the highest accuracy, F1score, precision and recall of 95%. The random forest classifier obtained an accuracy, F1score, precision and recall of 93%. The DT classifier achieved an accuracy, F1score, precision and recall of 89%. The KNN classifier reported an accuracy, F1score, precision and recall of 82%, 81%, 81% and 81% respectively. The SVM and NB classifiers achieved an accuracy of 63% and 60% respectively.

Table 4 shows that for normalized features and 10-fold cross validation the XGBoost classifier achieved the highest accuracy, F1score, precision and recall of 98%. This was followed by the random forest classifier which obtained an accuracy of 96% and F1score, precision and recall of 95%. The decision tree classifier obtained an accuracy, F1score, precision and recall of 90%. The KNN and SVM classifiers achieved an accuracy of 63%. The NB classifier achieved the lowest accuracy of 60%.

Thus, from Tables 1, 2, 3 and 4 the results are found to be almost similar for the normalized and non-normalized features. Same has been observed on changing the technique from the hold out method to 10-fold cross validation. In all the experiments, the XGBoost classifier was found to outperform the other methods. The highest classification accuracy of XGBoost classifier for all the experiments was 98% and was obtained when normalized features and 10-fold cross validation was used. Then, the random forest and decision tree classifiers had performed better when compared with the rest of the classifiers. It could also be noted that the SVM and NB classifiers performed poorly for this dataset for all the experiments. SVM achieved a classification accuracy of 63% for the normalized features whereas it obtained an accuracy of less than 60% using non-normalized features. The NB classifier however reported the same classification accuracy of 60% for all the four experiments. The K-NN classifier performed better in hold out experiments achieving a classification accuracy of around 82% whereas for the 10-fold experiments it reported low classification accuracy values. The train test split could be a reason for this behavior of KNN.

7 Conclusion

This work presented texture features as an effective descriptor to classify microstructure images of mild steel, stainless steel, aluminum, and copper. Different classifier models were trained and experimented to identify the best model. The results are promising and motivate the development of automated systems using texture features and machine learning classifiers to classify microstructure images of metals. In future, the use of a rich set of texture features on more metal samples could be explored. Also, it would be interesting to investigate the use of deep learning models for the identification and classification of microstructure images of metals.

References

1. Clemens, H., Mayer, S., Scheu, C.: Microstructure and properties of engineering materials. In: Peter, S., Andreas, S., Helmut, C., Svea, M. (eds.) Neutrons and Synchrotron Radiation in Engineering Materials Science: From Fundamentals to Applications Wiley, 2nd edn., pp 1–20. Wiley-VCH Verlag GmbH & Co. KGaA (2017)
2. Jordan, M.I., Mitchell, T.M.: Machine learning: trends, perspectives, and prospects. Science **349**(6245), 255–260 (2015)
3. Bostanabad, R., et al.: Review of the state-of-the-art techniques. Prog. Mater Sci. **95**, 1–41 (2018)
4. Holm, E.A., et al.: Overview: computer vision and machine learning for microstructural characterization and analysis. Metall. Mater. Trans. A **51**(12), 5985–5999 (2020). https://doi.org/10.1007/s11661-020-06008-4
5. Ma, W., et al.: Image-driven discriminative and generative machine learning algorithms for establishing microstructure–processing relationships. J. Appl. Phys. **128**, 134901 (2020)
6. Naik, L., Sajid, U., Kiran, R.: Texture-based metallurgical phase identification in structural steels: a supervised machine learning approach. Metals **9**(5), 546 (2019)
7. DeCost, L., Holm, A.: Computer vision approach for automated analysis and classification of microstructural image data. Comput. Mater. Sci. **110**, 126–133 (2015)
8. Gola, J., et al.: Objective microstructure classification by support vector machine (SVM) using a combination of morphological parameters and textural features for low carbon steels. Comput. Mater. Sci. **160**, 186–196 (2019)
9. Abouleta, B.: Classification of copper alloys microstructure using image processing and neural network. J. Am. Sci. **9**, 213–223 (2013)
10. Chowdhury, A., Kautz, E., Yener, B., Lewis, D.: Image driven machine learning methods for microstructure recognition. Comput. Mater. Sci. **123**, 176–187 (2016)
11. Papa, P., Nakamura, M., Albuquerque, H., Falcao, X., Tavares, M.: Computer techniques towards the automatic characterization of graphite particles in metallographic images of industrial materials. Expert Syst. Appl. **40**, 590–597 (2013)
12. Sarkar, S., Ansari, S., Mahanty, A., Mali, K., Sarkar, R.: Microstructure image classification: a classifier combination approach using fuzzy integral measure. Integr. Mater. Manufact. Innov. **10**, 286–298 (2021). https://doi.org/10.1007/s40192-021-00210-x
13. Bhaskaran, A., Kane, G., Biggs, K., Hull, R., Lewis, D.: Adaptive characterization of microstructure dataset using a two stage machine learning approach. Comput. Mater. Sci. **177**, 1–10 (2020)
14. Gupta, S., Sarkar, J., Kundu, M., Bandyopadhyay, R., Ganguly, S.: Automatic recognition of SEM microstructure and phases of steel using LBP and random decision forest operator. Measurement **151**, 107–224 (2020)
15. Ramola, A., Shakya, K., Pham, V.: Study of statistical methods for texture analysis and their modern evolutions. Engineering Reports **2**, e12149 (2020)
16. Aggarwal, N., Agrawal, K.: First and second order statistics features for classification of magnetic resonance brain images. J. Signal Inf. Process. **3**(2), 19553 (2012)
17. Haralick, M., Shanmugam, K., Dinstein, I.: Textural features for image classification. Man Cybern. SMC-3 **6**, 610–621 (1973)
18. Jothi, A., Rajam, A.: Effective segmentation and classification of thyroid histopathology images. Appl. Soft Comput. **46**, 652–664 (2016)
19. Hall-Beyer, M.: GLCM texture: A Tutorial v. 3.0. University of Calgary (2017)
20. Kotsiantis, B., Zaharakis, D., Pintelas, E.: Machine learning: a review of classification and combining techniques. Artif. Intell. Rev. **26**, 159–190 (2006)

21. Kotsiantis, B.: Supervised machine learning: a review of classification techniques. Informatica (Slovenia) **31**, 249–268 (2007)
22. Breiman, L.: Random forests. Mach. Learn. **45**, 5–32 (2001)
23. Breiman, L.: Bagging predictors. Mach. Learn. **24**, 123–140 (1996)
24. Chen, T., Guestrin, C.: XGBoost: a scalable tree boosting system. In: Proceedings of the 22nd ACM SIGKDD International Conference on Knowledge Discovery and Data Mining, pp. 785–794. ACM (2016)

Early Diagnosis of Alzheimer's Disease from MRI Images Using Scattering Wavelet Transforms (SWT)

Deepthi Oommen$^{(\boxtimes)}$ and J. Arunnehru

Department of Computer Science and Engineering, SRM Institute of Science and Technology, Vadapalani, Chennai, Tamilnadu, India
do6523@srmist.edu.in

Abstract. Alzheimer disease (AD) is an incurable, irreversible brain disorder. It impairs thinking capacity and memory loss. Computer-aided diagnosis techniques with image retrieval have developed a new potential in magnetic resonance imaging, which helps to retrieve relevant images and train to detect AD and its stages. Recently, advanced machine learning techniques have successfully exhibited high scale performances in numerous fields. This paper proposed four machine learning techniques such as Support Vector Machine (SVM), K-Nearest Neighbour (K-NN), Naïve Bayes, and Decision Tree using Brain MRI to identify AD stages. The models encompassed scattering wavelet transform for extracting the relevant features from MRI. While most of the existing techniques focus on binary classification, the current work focused on multi-class classification by classifying the stages of Alzheimer disease, namely healthy controls, very mild AD, mild AD and moderate. The SVM classifiers obtained a superior performance with an average accuracy of 98.10% in diagnosing the early stages of AD for the early onset category.

Keywords: Alzheimer's disease · Brain-disorder · SVM · KNN · Decision tree · Naïve Bayes

1 Introduction

Alzheimer's Disease is a catastrophic and fatal neuro disorder that will lead to the death of neurons and slowly tear down the cognitive capabilities. It's a memory decaying disease not because of normal ageing, but because of the deposits of amyloid plaque proteins in the brain cells. The condition can affect people of any age after 40, but it's evident in aged people more vigorously. It's a type of dementia and causes problems with thinking power, memory skills and behavioural pattern. The indications of AD usually develop sedately and aggravate in their later stage. Compared with other neurological diseases, AD has been a considerable cost burden in Asia, North America, and worldwide [1].

According to research studies, it is calculated that by the year 2050, the count of affected people can be .64 billion [2]. The estimations show that by

© Springer Nature Switzerland AG 2022
K. K. Patel et al. (Eds.): icSoftComp 2021, CCIS 1572, pp. 249–263, 2022.
https://doi.org/10.1007/978-3-031-05767-0_20

2050, one out of eightyfive people worldwide can affect by AD [3]. Pinpointing this particular disease at its beginning stage makes a remarkable difference in the patient's life expectancy. For indications of AD at its dawn stage, biomarkers have a crucial role, as they can measure what is happening inside the body through image results or laboratory test results. Image tests are the significant biomarkers for the identification of AD and classifying its current stage. Recently applying image tests are Cerebrospinal fluid (CSF), magnetic resonance imaging (MRI), single-photon emission Tomography (CT-scans), positron emission tomography (PET), and the electroencephalogram (EEG) signal [4].

The temporal part of the brain is the primary focus area for identifying AD at its dawn. The hippocampus and cerebral cortex of the temporal lobe are responsible for memory, learning, emotions and feelings, language processing sensation etc. Progression of Alzheimer's Disease can lead to thinning of the cerebral cortex and loss of volume in the hippocampal area. The biomarkers are the tool to find the changes in these two areas and help the practi 25 tioner diagnose the disease. The empowered, intelligent technologies have brought noticeable changes in the health care sector [5].

The machine learning approach has shown a clear-cut advantage in diagnosing various diseases by analyzing medical data. Researchers have found that machine learning techniques help in diagnosing and classifying Alzheimer's Disease. The most widely used classifiers are SVM, KNN, Decision Tree algorithm, and Naïve Bayes, which is helpful in clinical diagnosis.

The proposed work plan is to analyze brain MRI data using Scattering Wavelet transform with a Machine learning model for detecting and classifying Alzheimer's Disease. Machine learning research with neuroimaging datasets has become an indomitable diagnostic pinpoint for the brain's segmentation and classification of brain MRI. The person affected with AD's hippocampal area will start to shrinks and enlarge ventricles of cortical thickness [6]. These areas are considered as the broadcast between the brain and the human body. The AD makes a cease between neurons and synapses as they can't communicate anymore and leads to severe cognitive inabilities. [7,8] and the MRI could highlight the affected area in low- intensity. Figure 1 shows a set of brain images affected by AD and its different stages.

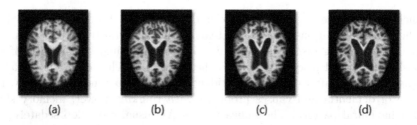

Fig. 1. Different brain MRI to depicts the different phases of AD: (a) Healthy Control (b) very mild AD (c) Mild AD (d) Moderate AD

The variations in MRI helps to distinguish Alzheimer disease from healthy controls. But only an extensive in-depth knowledge with experience can indicate a healthy subjects MRI from an early AD MRI. Furthermore, a productive robust automated computer-aided machine learning model can provide researchers, scientists, and medical practitioners with immense help in diagnosing and classifying the disease [9]. And also, it ultimately could be a timely assist to medical personnel for providing proper treatment to AD patients. In the proposed work, four machine learning models are selected to identify different stages of the disease. The experiments evaluated the model's performance on the Alzheimer Disease Neuro Initiative (ADNI) database [10], and clinical datasets which provides T1-weighted MRI scans.

The paper is composed in the following manner: Sect. 2 describes the summary of related works. Section 3 and 4 explain the dataset and the feature extraction technique used in our work. Section 5 illustrates the workflow model and the experimental methods used in the paper. Section 6 is about the performance measures used to evaluate the model, and Sect. 7 discusses the results acquired. Finally paper ends with a conclusion followed by references.

2 Related Work

In medical imaging, the detection of AD through MRI is relevant in maintaining a person's cognitive capability and for consideration of public health. AD is considered a public health issue, and research indicates that 5.5 million crowd in the age bracket 75–90 years in the United States of America are AD patients [11]. AD causes problems with human memory skills and thinking capabilities, and other daily life activities. It's a slowly progressive disease, which destructs the nerve cells, and eventually, the person loses control over their life as it's gets affected severely on all the parts of the brain [12]. Imaging techniques are powerful tools for identifying AD symptoms in the brain. The collaboration of these images with machine learning techniques can be performed more efficiently in the early diagnosis of AD compared with manual systems. In [13] focused on Alzheimer disease early detection with MRI using image processing techniques. Brain atrophy is analyzed and computed through Wavelet, Watershed, K-Means algorithms, a valuable measure for diagnosing the early stages of AD. The paper [14] introduced AD detection techniques using gradient-echo plural contrast imaging (GEPCI) with MRI, as it can find the brain's damaged tissues due to AD. In the GEPCI method, the affected area gets enhanced resolution in MRI; it's a supporting tool for the practitioner to identify the affected tissues.

The paper [15] developed a method to diagnose AD by using a probabilistic neural network (PNN) with imaging techniques such as MRI, PET etc. In the primary stage, the brain volume and atrophy rate are computed, followed by feature extraction for the classification purpose. PNN surpasses SVM and KNN in terms of performance.

In the paper, [16] introduced a method of AD detection using MRI scans. This paper has applied a PCA called Principal Component Analysis and Singular

value decomposition methods to extract the feature from the images. Classification is achieved by fusing SVM and Decision Tree algorithms to acquire remarkable result. The paper [17] have introduced a framework called multi-textural
(MTL) for extracting features. In this method, MTL computes Structural information from MRI. The researchers have performed various texture grading processes for the extracted data through fusion methods. This novel method has
shown significant results in performance.

In this paper [18] feature extraction was carried out through wavelet entropy
and Hu moment. The classification has been done through SVM with extracted
proximal eigenvalues. The accuracy rate of classification has increased with the
radial base function (RBF) of SVM. In [19] MRI images undergo Voxel preselection and Brain Parcellation before inputting the networks. Voxel preselection
is to erase the voxels with less significance and to diminish computational cost.
The brain parcellation is to split the brain region into patches with the help of
AAL (Automated Anatomical labelling). The paper developed two different Deep
Belief Networks (DBN) alternatives. The first one is DBN-voter, which means
an ensembling of DBN classifiers with a voting strategy. Four voting techniques
have been used for the prediction process: Majority Voting, Weighted Voting,
Classification fusion using SVM, and Classifier using a DBN. The second one
is feature extraction using DBN and SVM called FEDBN-SVM. The training
samples are extracted voxels, which undergoes a fusion process by SVM, and it
will compute the feature's relevance on each stage and cut down the urgency for
the voting phase. Both the networks showcased significant results in prediction
and classification.

3   Dataset

Data applied for our work combines the Alzheimer's Disease Neuroimaging Initiative (ADNI) database and clinical datasets. The ADNI initiative aims to
develop indicators for the timely diagnosis of Alzheimer's disease. It collects clinical, imaging, biochemical, and genetic biomarkers to detect and classify AD. It
has launched by the year 2004 under the guidance of Dr Michael W. Weiner,
MD [20]. This work has considered a whole set of 3209 Healthy Controls (HC),
2240-Very Mild (VM), 896-Mild (M) and 64 Moderate (Mo) participants for
identifying different stages of AD among the early onset category (see Table 1
for demographic details. The dataset contains patient's demographic information
such as age, gender, handedness.

4   Feature Extraction

Wavelets are mathematical functions. It divides the data into frequency components and resolves the components based on their matching scale. Wavelet
Scattering is a set of wavelets capable of achieving linearization and translation
invariance of small diffeomorphisms. It can preserve and achieve stable deformation, to make an effective classification. Wavelet Scattering is an effective

Table 1. Dataset demographics for ADNI.

Parameter	ADNI
MRI	T1 weighted
Subjects	Healthy Controls, Very Mild-,Mild-,Moderate
Sex	Male: 3709, Female: 2700
Age	40 yrs to 65 yrs

tool for feature extraction and accurate data representations. They can be used with most classification algorithms to acquire higher performance. The wavelet scattering transform can extract reliable features of the input data that can be used in conjunction with deep neural networks. Wavelet scattering transform has three primary operations: convolution, nonlinearity and averaging described in Fig. 2, with an input signal $Y = [y1, y2, y3....yn]$. It is 30-dimensional vector data. The main two components of the Wavelet Scattering Transform (WST) are wavelets and scaling function. The first step is to calculate the convolutions and makes the translation covariant. Then the scattering transform computes the non-linear invariants with the help of modulus and low pass filter.

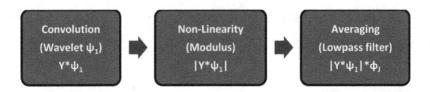

Fig. 2. Wavelet scattering transform processes for the input signal Y

In this paper, scattering wavelet transform is used for feature extraction from the MRI images. The process is achieved by extracting the given input y from a lower frequency to a higher frequency level. Such as $w_y = y * \varphi_i(u)$ is the lower frequency information of the image y with scaling factor i and $w_y = y * \psi_{i,d}(u)$ is the higher frequency information. Where i is the scaling factor and d is the direction. Here $\varphi_i(u)$ is the scaling function, and $\psi_{i,d}(u)$ is the directional wavelet form. The transformation can be expressed as:

$$w_y = \begin{bmatrix} y * \varphi_i(u) \\ y * \psi_{i,d}(u) \end{bmatrix} \tag{1}$$

Even though wavelet transform can extract the features with higher frequency from different directions, it doesn't have the property of translation invariance because of the convolution operation of the wavelet transform. [21–24]. But can achieve it by multiplying non-linear operations called the modulus of the high-frequency coefficients $|y * \psi_{i.d}(u)|$ and with a low pass filter $\varphi_i(u)$ to get the invariant feature such as

$$|y * \psi_{i,d}(u)| * \varphi_i(u) \tag{2}$$

The obtained feature is invariant from multiple directions in multiple scales but with losing its high frequency. To recuperate the lost frequency apply wavelet decomposition and also achieve stability in the feature coefficients continue the low pass filtering and modulus operations, and the result is

$$||y * \psi_{i.d}(u)| * \psi_{i+1,d}(u)| * \varphi_i(u) \tag{3}$$

The increase in iterations produces more invariants. In this work, the given input image normalized into a size of 258×258. By applying a scattering wavelet with scattering level $n = 1, d = 8$ is the scattering directions at each scale and wavelet decomposition $I = 2$. So the wavelet scattering can be expressed as

$$w_y = \begin{bmatrix} T_n \\ S_n \end{bmatrix} \tag{4}$$

The subscript 'n' of the Eq. (4) represents the levels of scattering wavelet, it shows the count of scattering wavelet decomposition to achieve the high frequency. The scattering propagation operator T_n with results

$$T_n = ||y * \psi_{i,d}(u)|......* \psi_{in,d}(u)|When - (n \geq 1); T_n = 0; When - (n \geq 1) \tag{5}$$

and the scattering coefficient

$$S_n = T_n * \varphi_i \tag{6}$$

Scattering wavelet transform can obtain a robust representation of MRI Images by minimizing the noise features. Also, it maintains the discriminability between the affected and normal regions. Even though wavelet scattering and Convolutional Neural Network structure are alike, they can be distinguished through their filter weights and computational complexity. In the wavelet, the filter weights are known and fixed. The coefficient's energy decreases in a scattering network as each layer's level increases; the model used two scattering networks for feature extraction to overcome the scenario. It also reduces the complexity in computation significantly [25–28].

5   Experimental Methods

In this work, the classification of different stages of AD is achieved with four machine learning models, such as SVM, KNN, Decision Tree, Naïve Bayes. In addition, the papers aim to investigate the accuracy of a multiclass classifier for the disease classification. Figure 3 illustrates the model's operational mechanism.

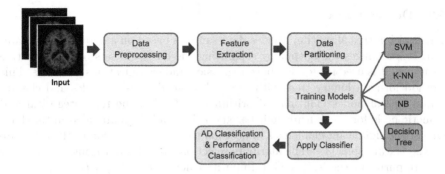

Fig. 3. Workflow diagram for AD categorization

5.1 Support Vector Machine (SVM)

The SVM is a well-known method for classification and regression. Its usage has been incredible and beneficial for many applications, especially in the medical field for disease prediction. Using non-linear or linear kernel functions, SVM maps predictor vectors to high-grade dimensional planes and classifies the data with a hyper-plane. Consider n training datasets for a classification problem can be expressed as $Y = (s1, t1), (s2, t2)....(sn, tn)$, where si $\pounds\ R^d$ feature vectors and ti the class label. [29,30]. For multi-class classification, the problem gets divided into multiple binary classifications called one vs one, such as ti $\pounds(-1, 1)$. One vs one process divides the data points in class 'y' and rest. The no: of classifiers required for this approach defines by the formula $m * (m - 1)/2$, where m is the no of classes. The classification gains accuracy by maximizing the margin and can be expressed as $\min_{w,b}(W^TW/2)$. SVM's resolution function is denoted by the following notation: $f(y) = sign(W^Ty + b)$, where W is the weight vector, y is the input, and b is the bias.

5.2 K-Nearest Neighbour

KNN is a classifier algorithm; its calculation is based on nearest neighbours for prediction instead of developing a model. It's also called as a lazy learner. The KNN Prediction analysis technique locates the data samples' k-nearest neighbours. The distance between neighbours is calculated with the Euclidean function, which helps to find similarities between the two points. A given data $D = (x1, x2, ...xn)$ are the set of labelled data samples. The nearest neighbour classifier assigns test point 'C', the label associated with its closest neighbour in D. The K-nearest neighbour classifier classifies C by assigning it the label most frequently represented among the most immediate samples. Here, X is the algorithm's crucial parameter. The distance between training points xi and testing data points x is computed using Euclidean distance $d_E = (x, x_i) = \sum_{i=0}^{n} |x - x_i|$ [31,32].

5.3   Decision Tree

The decision tree algorithm divides data samples based on a determined parameter. It maximizes the output by separating the data and sets output in a tree structure. It is used for solving both regression and classifications problems. This method forms a binary tree, with the nodes called decision nodes and classification or leaf nodes [33]. The algorithm tries to find the top categorical and numeric node for the feature split based on the conditions provided in the data. One of the significant challenges in decision tree is identification of the attribute for the feature split [34]. The attribute selection can achieve through Entropy or Gini Impurity methods. The Gini Impurity calculated as $\sum_{j=1}^{m} fi(1 - fi)$ where m corresponds to the set of distinctive labels and frequency fi is for the label y at the given node, and for Entropy, it calculates by $\sum_{j=1}^{m} -fi \log(fi)$.

5.4   Naïve Bayes

Naïve Bayes is an active and efficient learning algorithm for machine learning. It is one of the recommended classification algorithms among the other classification methods due to its solid independent assumption. It's a supervised learning algorithm influenced by the Bayes theorem [35]. Based on the set of known probabilities, an attribute finds the occurrence of the particular feature of the attribute. Bayes theorem calculates an event's probability with the probability of an already occurred event. Mathematically it is stated as: $P(M/N) = P(M) * P(N/M)/P(N), where\, P(M|N)$ is a posteriori probability of N, $P(M)$ is the prior probability of M.

6   Performance Measures

The performance measure is used to evaluate the model's performance in classification. The parameter is calculated with a confusion matrix, which will visualize the performance of each model for binary and multiclass classification [36]. True Positive translates the model prediction into true and positive values. False Positive refers to a situation in which the model predicts a positive result but the actual value is false. False Negative refers to a situation in which both the model prediction and the actual value are negative and false. The final one is True negative, which indicates that the model's prediction is incorrect and the actual result is correct. The performance metrics based on positive and negative values can calculate as

$$\text{Accuracy} = \left(\frac{\text{TP+TN}}{\text{TP+TN+FP+FN}}\right) \tag{7}$$

$$\text{F1-Score} = \left(\frac{2 \times \text{TP}}{2 \times \text{TP+FP+FN}}\right) \tag{8}$$

$$\text{Recall} = \left(\frac{\text{TP}}{\text{TP+FN}}\right) \tag{9}$$

$$\text{Precision} = \left(\frac{TP}{TP+FP} \right) \tag{10}$$

7   Results and Discussion

Our experiment was done in MATLAB 2019b using an Intel i7 processor and 16 GB of RAM. The proposed method has used the combination of scattering wavelet transform for extracting features and machine learning techniques used for classifying the disease stages as non-demented, very mild AD, Mild AD and moderate AD. MRI images were normalized, extracted and reduced, then trained on selected Machine learning models as SVM, KNN, Naive Bayes, and Decision tree using 10 fold cross validation for training and testing. The model has used Gabor wavelets for the wavelet decomposition, and its lowpass filter ϕ is a Gaussian function. The input image with a size of 258×258 and an invariance scale of 129. The constructed network is a two-layer network to retain the coefficient energy. The number of rotations per wavelet per filter bank in the scattering network is fixed with 8 for both layers. The scattering coefficients are downsampled on a time basis with a lowpass filter and to acquire an 8- time window for each scattering path. This continuous sequence of convolution, non-linearity, and pooling enables the scattering wavelet network to extract characteristics from brain MRI.

The SVM is a commonly used machine learning approach for classification of binary and multi-class data. The Process of non-linear mapping data to a high dimensional space with a kernel function achieves an optimized classification plane when separating the individuals separately. The research work used the radial basis function (RBF) kernel. Figure 4 represents the confusion matrix of the SVM classifier, and it has given a precise match with its true value and predicted value. SVM with RBF has produced a prediction accuracy rate with an average of 98.10%, and its depicted in Table 2.

		Predicted Label			
		Mild	Moderate	Non	VeryMild
True Label	Mild	856	0	12	28
	Moderate	0	64	0	0
	Non	5	0	3168	36
	VeryMild	9	0	31	2200

Fig. 4. SVM confusion matrix obtained for ADNI dataset

The second machine learning classifier used in the research work is K - Nearest Neighbour (KNN) algorithm. In this process value of k is fixed as 1. Figure 5 illustrates the confusion matrix obtained by the KNN classifier. The classifier matched the true value and predicted value to a higher one in classifying moderate and mild but lack in very mild. KNN acquired an average accuracy of 96.3%.

The third method used for classification is the Decision Tree (DT) technique, which creates the model in the form of the tree structure, the algorithm used in DT is the C4.5 algorithm. It uses the Information Gain mechanism to construct the tree such that it allows computing the degree of information acquired during the classification process. Figure 6 describes the confusion matrix with the classifier Decision Tree. It shows only an average matching with true value and predicted value, as it couldn't classify accurately for mild and very mild stages of AD. The classifier classified the phases of AD with an average accuracy of 89.0% and its depicted in Table 2. The final classifier used under the machine learning technique is Naïve Bayes'. It is a probabilistic method; The Bayes network tries to trace data edges of the given distribution. The details obtained are analysed in the classifier based on the conditions provided on the testing data. The testing data are classified to where the classifier algorithm attains the maximum probability. Figure 7 showcases the confusion matrix with Naïve Bayes. The classifier lacked in predicting accurately for all the stages, and it could achieve only an average accuracy of 88.7%. The classifiers accuracy comparison is highlighted in Table 2.

Predicted Label

		Mild	Moderate	Non	VeryMild
True Label	Mild	830	0	51	15
	Moderate	0	62	2	0
	Non	4	2	3170	33
	VeryMild	9	1	118	2112

Fig. 5. KNN confusion matrix obtained for ADNI dataset

Predicted Label

		Mild	Moderate	Non	VeryMild
True Label	Mild	679	0	99	118
	Moderate	0	48	0	16
	Non	1	0	3053	155
	VeryMild	2	0	297	1941

Fig. 6. Decision tree confusion matrix obtained for ADNI dataset

The performance metrics for the multi-class classification with SVM, KNN, Naïve Bayes, and Decision Tree are depicted in Table 3. The study indicates that the SVM model stands out in the classification of AD stages in comparison with other classifiers. The SVM could achieve an average performance rate for Accuracy of 99%, Precision rate of 98.25%, Recall with 98.5% and F1 score of 98.5%. The SVM revealed its capability again in the image classification with a higher performance rate. In this work, the SVM has opened a pavement towards

Predicted Label

		Mild	Moderate	Non	VeryMild
True Label	Mild	724	0	119	53
	Moderate	1	53	4	6
	Non	13	3	3122	71
	VeryMild	24	1	431	1784

Fig. 7. Naive Bayes confusion matrix obtained for ADNI dataset

Table 2. Average accuracy obtained for the AD classification.

Evaluation models	Accuracy
SVM	98.10%
K-NN	96.30%
Decision Tree	89.00%
Naïve Bayes	88.70%

Table 3. Model performance measures obtained for the AD classification

Models	Class	Accuracy	Precision	Recall	F1 score
SVM	Mild	0.99	0.96	0.98	0.97
	Moderate	1.00	1.00	1.00	1.00
	Non	0.98	0.99	0.99	0.99
	VeryMild	0.97	0.98	0.97	0.98
KNN	Mild	0.99	0.93	0.98	0.95
	Moderate	0.99	0.97	0.95	0.96
	Non	0.94	0.99	0.95	0.97
	VeryMild	0.93	0.94	0.98	0.96
Naive Bayes	Mild	0.91	0.81	0.95	0.87
	Moderate	0.86	0.83	0.93	0.88
	Non	0.89	0.97	0.85	0.91
	VeryMild	0.89	0.8	0.93	0.86
Decision Tree	Mild	0.83	0.76	1.00	0.86
	Moderate	0.95	0.75	1.00	0.86
	Non	0.95	0.95	0.89	0.92
	VeryMild	0.84	0.87	0.87	0.87

the insight of medical health care applications, especially with medical images. The model used ten-fold cross-validations for training the SVM with features extracted by scattering wavelet transform. Usage of scattering transform with SVM developed stability to deformation. It's a powerful feature in image classification and also for obtaining high accuracy.

To evaluate the suggested method's efficiency, the proposed methods' results are quantitatively compared to the most advanced findings and its result, as shown in Table 4.

Table 4. State of the art recognition accuracy for ADNI dataset.

Paper	Method	Accuracy
Proposed	SVM	98.10%
Kruthika et al. [5]	SVM +KNN+BN	92.0%
Ortiz et al. [30]	LVQ-SVM	90.00%
Eldeeb et al. [29]	SVM	89.0%

8   Conclusion

The recent evolution in biomedical engineering shows that the analysis of medical images is a significant area of research work in the current arena. Applying a Machine learning algorithm in analysing the different types of medical images is one of the core reasons for the evolution. In this work, the paper used four machine learning techniques viz SVM, K-NN, Naïve Bayes, Decision Tree for analyzing and classifying the stages of Alzheimer disease on the early onset. As a conclusion of the results, it can be inferred that SVM-based models are suitable for prediction and classification owing to their robustness.

The results of this study demonstrate that scattering wavelet transform with SVM model outperforms other classification models when performance metrics are varied. While the bulk of studies are concerned with binary classification, the suggested work is concerned with multi-class classification. The new study is useful for the early detection of Alzheimer's disease in individuals aged 40 to 65. This category of people who are affected by AD has a significant impact on their life expectancy. Moreover, this research work can classify the disease into four stages as very-mild, mild, moderate and healthy controls. Though the work only concentrated on the AD dataset but can assure that the models can work successfully on the prediction and classification in other medical domains. In the future, the model can work on predicting the time gap required to convert from very mild stage to moderate stage, as it could help the practitioner and the patients follow up the proper treatment.

References

1. Sado, M., et al.: The estimated cost of dementia in japan, the most aged society in the world. PLoS ONE **13**(11), e0206508 (2018)
2. Islam, J., Zhang, Y.: Brain MRI analysis for Alzheimer's disease diagnosis using an ensemble system of deep convolutional neural networks. Brain Inform. **5**(2), 1–14 (2018)

3. Rathore, S., Habes, M., Iftikhar, M.A., Shacklett, A., Davatzikos, C.: A review on neuroimaging-based classification studies and associated feature extraction methods for Alzheimer's disease and its prodromal stages. Neuroimage **155**, 530–548 (2017)
4. Kumar, S.S., Nandhini, M.: A comprehensive survey: early detection of Alzheimer's disease using different techniques and approaches. IJCET **8**(4), 31–44 (2016)
5. Kruthika, K.R., Maheshappa, H.D., Alzheimer's Disease Neuroimaging Initiative, et al.: Multistage classifier-based approach for Alzheimer's disease prediction and retrieval. Inform. Med. Unlocked **14**, 34–42 (2019)
6. Sarraf, S., Tofighi, G., Alzheimer's Disease Neuroimaging Initiative, et al.: DeepAD: Alzheimer's disease classification via deep convolutional neural networks using MRI and fMRI. BioRxiv, p. 070441 (2016)
7. Arunnehru, J., Kalaiselvi Geetha, M.: Automatic human emotion recognition in surveillance video. In: Dey, N., Santhi, V. (eds.) Intelligent Techniques in Signal Processing for Multimedia Security. SCI, vol. 660, pp. 321–342. Springer, Cham (2017). https://doi.org/10.1007/978-3-319-44790-2_15
8. Warsi, M.A.: The fractal nature and functional connectivity of brain function as measured by BOLD MRI in Alzheimer's disease. Ph.D. thesis (2012)
9. Kumar, S., Oh, I., Schindler, S., Lai, A.M., Payne, P.R., Gupta, A.: Machine learning for modeling the progression of Alzheimer disease dementia using clinical data: a systematic literature review. JAMIA Open **4**(3), ooab052 (2021)
10. Arunnehru, J., Kalaiselvi Geetha, M.: Difference intensity distance group pattern for recognizing actions in video using support vector machines. Pattern Recognit Image Anal. **26**(4), 688–696 (2016)
11. Mathew, N.A., Vivek, R.S., Anurenjan, P.R.: Early diagnosis of Alzheimer's disease from MRI images using PNN. In: 2018 International CET Conference on Control, Communication, and Computing (IC4), pp. 161–164. IEEE (2018)
12. Varatharajan, R., Manogaran, G., Priyan, M.K., Sundarasekar, R.: Wearable sensor devices for early detection of Alzheimer disease using dynamic time warping algorithm. Cluster Comput. **21**(1), 681–690 (2018)
13. Patil, C., et al.: Using image processing on MRI scans. In: 2015 IEEE International Conference on Signal Processing, Informatics, Communication and Energy Systems (SPICES), pp. 1–5. IEEE (2015)
14. Gorji, H.T., Haddadnia, J.: A novel method for early diagnosis of Alzheimer's disease based on pseudo Zernike moment from structural MRI. Neuroscience **305**, 361–371 (2015)
15. Sankari, Z., Adeli, H.: Probabilistic neural networks for diagnosis of Alzheimer's disease using conventional and wavelet coherence. J. Neurosci. Methods **197**(1), 165–170 (2011)
16. Zhang, Y., et al.: Multivariate approach for Alzheimer's disease detection using stationary wavelet entropy and predator-prey particle swarm optimization. J. Alzheimer's Disease **65**(3), 855–869 (2018)
17. Nawaz, H., Maqsood, M., Afzal, S., Aadil, F., Mehmood, I., Rho, S.: A deep feature-based real-time system for Alzheimer disease stage detection. Multimedia Tools Appl. **80**, 1–19 (2020)
18. Zhang, Y., Wang, S., Sun, P., Phillips, P.: Pathological brain detection based on wavelet entropy and hu moment invariants. Bio-Med. Mater. Eng. **26**(s1), S1283–S1290 (2015)
19. Giorgio, J., Landau, S.M., Jagust, W.J., Tino, P., Kourtzi, Z., Alzheimer's Disease Neuroimaging Initiative, et al.: Modelling prognostic trajectories of cognitive decline due to Alzheimer's disease. NeuroImage Clin. **26**, 102199 (2020)

20. Veitch, D.P., et al.: Understanding disease progression and improving Alzheimer's disease clinical trials: recent highlights from the Alzheimer's disease neuroimaging initiative. Alzheimer's Dement. **15**(1), 106–152 (2019)
21. Bruna, J., Mallat, S.: Invariant scattering convolution networks. IEEE Trans. Pattern Anal. Mach. Intell. **35**(8), 1872–1886 (2013)
22. Andén, J., Mallat, S.: Deep scattering spectrum. IEEE Trans. Signal Process. **62**(16), 4114–4128 (2014)
23. Sarhan, A.M., et al.: Brain tumor classification in magnetic resonance images using deep learning and wavelet transform. J. Biomed. Sci. Eng. **13**(06), 102 (2020)
24. Andén, J., Lostanlen, V., Mallat, S.: Joint time-frequency scattering for audio classification. In: 2015 IEEE 25th International Workshop on Machine Learning for Signal Processing (MLSP), pp. 1–6. IEEE (2015)
25. Leonarduzzi, R., Liu, H., Wang, Y.: Scattering transform and sparse linear classifiers for art authentication. Signal Process. **150**, 11–19 (2018)
26. Bruna, J., Mallat, S.: Classification with scattering operators. In: CVPR 2011, pp. 1561–1566. IEEE (2011)
27. Arunnehru, J., Geetha, M.K.: Vision-based human action recognition in surveillance videos using motion projection profile features. In: Prasath, R., Vuppala, A.K., Kathirvalavakumar, T. (eds.) MIKE 2015. LNCS (LNAI), vol. 9468, pp. 460–471. Springer, Cham (2015). https://doi.org/10.1007/978-3-319-26832-3_43
28. Sujatha Kumari, B.A., Yadiyala, A.G.V., Aruna, B.J., Radha, C., Shwetha, B.: Early detection of mild cognitive impairment using 3D wavelet transform. In: Jeena Jacob, I., Kolandapalayam Shanmugam, S., Piramuthu, S., Falkowski-Gilski, P. (eds.) Data Intelligence and Cognitive Informatics. AIS, pp. 445–455. Springer, Singapore (2021). https://doi.org/10.1007/978-981-15-8530-2_36
29. Eldeeb, G.W., Zayed, N., Yassine, I.A.: Alzheimer's disease classification using bag-of-words based on visual pattern of diffusion anisotropy for DTI imaging. In: 2018 40th Annual International Conference of the IEEE Engineering in Medicine and Biology Society (EMBC), pp. 57–60. IEEE (2018)
30. Ortiz, A., Górriz, J.M., Ramírez, J., Martínez-Murcia, F.J., Alzheimer's Disease Neuroimaging Initiative, et al.: LVQ-SVM based cad tool applied to structural MRI for the diagnosis of the Alzheimer's disease. Pattern Recognit. Lett. **34**(14), 1725–1733 (2013)
31. Charbuty, B., Abdulazeez, A.: Classification based on decision tree algorithm for machine learning. J. Appl. Sci. Technol. Trends **2**(01), 20–28 (2021)
32. Balamurugan, M., Nancy, A., Vijaykumar, S.: Alzheimer's disease diagnosis by using dimensionality reduction based on KNN classifier. Biomed. Pharmacol. J. **10**(4), 1823–1830 (2017)
33. Saputra, R.A., Agustina, C., Puspitasari, D., Ramanda, R., Pribadi, D., Indriani, K., et al.: Detecting Alzheimer's disease by the decision tree methods based on particle swarm optimization. In: Journal of Physics: Conference Series, vol. 1641, p. 012025. IOP Publishing (2020)
34. Miah, Y., Prima, C.N.E., Seema, S.J., Mahmud, M., Shamim Kaiser, M.: Performance comparison of machine learning techniques in identifying dementia from open access clinical datasets. In: Saeed, F., Al-Hadhrami, T., Mohammed, F., Mohammed, E. (eds.) Advances on Smart and Soft Computing. AISC, vol. 1188, pp. 79–89. Springer, Singapore (2021). https://doi.org/10.1007/978-981-15-6048-4_8

35. Awasthi, S., Kapoor, E., Srivastava, A.P., Sanyal, G.: A new Alzheimer's disease classification technique from brain MRI images. In: 2020 International Conference on Computation, Automation and Knowledge Management (ICCAKM), pp. 515–520. IEEE (2020)
36. Liu, L., Zhao, S., Chen, H., Wang, A.: A new machine learning method for identifying Alzheimer's disease. Simul. Model. Pract. Theory **99**, 102023 (2020)

Constraint Pushing Multi-threshold Framework for High Utility Time Interval Sequential Pattern Mining

Sumalatha Saleti[✉] [ID], N. Naga Sahithya, K. Rasagna, K. Hemalatha,
B. Sai Charan, and P. V. Karthik Upendra

SRM University AP, Amaravati, Andhra Pradesh, India
sumalatha.s@srmap.edu.in

Abstract. This paper aims to detect high utility sequential patterns including time intervals and multiple utility thresholds. There are many algorithms that mine sequential patterns considering utility factor, these can find the order between the items purchased but they exclude the time interval among items. Further, they consider only the same utility threshold for each item present in the dataset, which is not convincing to assign equal importance for all the items. Time interval of items plays a vital role to forecast the most valuable real-world situations like retail sector, market basket data analysis etc. Recently, UIPrefixSpan algorithm has been introduced to mine the sequential patterns including utility and time intervals. Nevertheless, it considers only a single minimum utility threshold assuming the same unit profit for each item. Hence, to solve the aforementioned issues, in the current work, we proposed UIPrefixSpan-MMU algorithm by utilizing a pattern growth approach and four time constraints. The experiments done on real datasets prove that UIPrefixSpan-MMU is more efficient and linearly scalable for generating the time interval sequences with high utility.

Keywords: Sequential patterns · Time intervals · Utility mining · Multiple utility thresholds

1 Introduction

Sequential Pattern Mining (SPM) [1–6] is a salient research issue. The major purpose of mining sequential patterns is to discover the frequent sequences which satisfy the user-defined minimum support threshold. Even the decision makers can derive the patterns with high frequency, they are unable to find the patterns with high profit. So, to find a solution to this issue, high-utility itemset mining (HUIM) [7] has been initiated which can mine the high utility itemsets while both the quantity and profit of items are taken into consideration. High-utility sequential pattern mining (HUSPM) [8] was discovered for mining sequential patterns by considering the ordered sequences in practical circumstances. Finally, considering time periods HUTISP [15] was introduced. The main intention of

© Springer Nature Switzerland AG 2022
K. K. Patel et al. (Eds.): icSoftComp 2021, CCIS 1572, pp. 264–273, 2022.
https://doi.org/10.1007/978-3-031-05767-0_21

HUTISP is to discover the sequences that include the time period between the purchases of each item. For example, let us consider a shop which sells some groceries like Grains, Milk, Yogurt, Bread and Eggs. Consider these items as a set of items in the database. Now, our main aim is to find the time period between the purchases of particular items that are being sold. From this, the shop owner can easily maintain the stock of completed items according to time period. For example, an output sequential pattern including time intervals of the form $\langle item\_x, 3, item\_y, 5, item\_z \rangle$ indicates that a customer who purchased $item\_x$, also bought $item\_y$ after 3 months and visited the store again after 5 months to buy $item\_z$. This issue of HUTISP was introduced in [15]. However, it considers the same utility threshold for each of the item in the database, which shows that each item is assumed to have the same unit profit. This is not convincing as each item is different in real time applications and should not be treated equally. For example, the sales of "Gold bangles" will produce more profit than the sales of "Cotton Jeans". In view of this, in this paper, we propose UIPrefixSpan-MMU.

The contributions of the current paper are detailed below:

- A novel idea of mining sequential patterns considering utility, time intervals and multiple thresholds is presented.
- UIPrefixspan-MMU algorithm is proposed, which follows the efficient pattern growth approach.
- Four constraints are integrated with the proposed algorithm to generate efficient time interval patterns.
- Experiments have been run to assess UIPrefixspan-MMU with and without time constraints including run time, memory utilization and scalability.

The remaining sections of the paper are assembled as follows. A brief description of literature is reviewed in Sect. 2. Section 3 defines the problem. The proposed algorithm is detailed in Sect. 4. The experimental work is given in Sect. 5. The future enhancements and the conclusions are mentioned in Sect. 6.

2   Literature Review

This section presents the existing works on mining sequential patterns, sequential patterns with time intervals and multiple threshold utility mining. Sequential pattern mining [1–6] is used in various research areas of data mining. Usually, there are two classes of algorithms for solving SPM: They include Apriori and Pattern growth approach. Apriori was the first algorithm which was used in pattern mining to generate candidate sequences. It is considered that a frequent itemset of all non-empty subsets should be frequent and this property is known to be anti-monotonic property. Agarwal and Srikant introduced the GSP algorithm [1] which is based on Apriori and Breadth First Search in 1996. GSP has two steps: They are Candidate Generation and frequency calculation. In the year 2001, Zaki introduced the algorithm called SPADE [4] which is based on vertical mining approach. Pattern growth methods like Freespan [2] and Prefixspan [3] were discovered by Han. Constraint based sequential pattern mining has been proposed by Garofalakis, Rastogi and Shim in 1999 and developed SPIRIT algorithm [6]. After researching sequential pattern mining, researchers focused on time interval related patterns. For the time interval mining, Chen and Huang proposed novel approaches, namely, I-Apriori and I-Prefixspan in 2003 [9]. Chen et al. extended the past work and proposed FTI-Apriori and FTI-Prefixspan considers the concept of fuzzy theory to divide the time intervals [10].

High utility itemset mining [7] is an extension of mining frequent itemsets. Utility mining was proposed when non binary transactions of items were taken into thought and then utility mining became a major research issue in the data mining research field. At first, mining high utility itemsets with multiple thresholds was initiated in [14]. Two algorithms called HUI-MMU and HUI-MMUTID are developed based on high utility itemset mining having multiple item thresholds. Considering real world situations, time interval sequential pattern mining including high utility was proposed by Wang et al. [15]. In addition to the centralized approach which was proposed in [15], a MapReduce algorithm was proposed in [16] for discovering sequential patterns with high utility including time intervals, it considers a single threshold for all the items. However, the above mentioned approaches [15,16] considers each item as equally important, thus assuming a single minimum utility threshold. Considering this, Lin et al. [14] introduced HUI-MMU algorithm for discovering the high utility itemsets including more than one utility threshold. However, it is related to HUIM and cannot deal with the problem of HUSP. Recently, a novel framework has been proposed [12,13] to mine high utility sequential patterns that considers individual thresholds for each item. But, they cannot mine the time intervals between each item. Hence, in the current work, we propose an algorithm that mines the sequential patterns including more than one threshold and time constraints.

Table 1. Quantitative sequence dataset.

SID	Sequence
S_1	$\langle 0, b[3]\rangle\langle 1, a[4]b[2]c[1]\rangle\langle 2, f[3]\rangle\langle 3, b[4]\rangle\langle 4, d[1]\rangle$
S_2	$\langle 0, e[2]\rangle\langle 1, a[2]c[4]\rangle\langle 2, d[3]\rangle\langle 3, e[2]\rangle$
S_3	$\langle 0, c[3]f[1]\rangle\langle 1, c[2]\rangle\langle 2, e[4], f[3]\rangle$
S_4	$\langle 0, b[2]\rangle\langle 1, c[6]e[1]\rangle\langle 2, a[2]b[3]\rangle\langle 3, f[5]\rangle$
S_5	$\langle 0, a[4]d[1]\rangle\langle 2, f[2]\rangle$

Table 2. External utility.

Item	a	b	c	d	e	f
External utility	2	3	1	6	2	1

3   Problem Definition

Definition 1. To represent the mining process we need to create a sequential dataset. Now, let us consider a distinct itemset $I \subseteq X$ in a lexicographic order which contains set of itemsets $X = \{i_1, i_2, \ldots, i_m\}$ which occur in a quantitative sequential dataset $D = \{S_1, S_2, S_3, \ldots, S_n\}$, and each sequence has its own identification called SID. Suppose $\mid I \mid = q$ then itemset I is called a q-itemset. The time interval set with interval extended sequence TI is given as $\langle (t_{1,1}, I_1), (t_{1,2}, I_2), \ldots, (t_{1,n}, I_n) \rangle$ (i.e. it occurred with a list of itemsets which is followed by time). Let us consider an itemset $I_i (1 \leq i \leq n)$ and $t_{(\alpha,\beta)}$, the time interval in the middle of the itemsets I_α and I_β , where $t_{(\alpha,\beta)} = I_\beta time - I_\alpha time$. Here, in every itemset, the existence of an item is allocated with a quantity and every item in the dataset is assigned a profit value.

Definition 2. For an item i, its external utility denotes the profit of the item, denoted as $E(i)$ and internal utility represents quantity of an item which is given as $IU(i, S_n)$ where, i is an item in an itemset I that belongs to a sequence S_n. For Example, from Table 1, Item a in S_5 has internal utility $IU(a, S_5) = 4$ and its external utility given in Table 2 is $E(a) = 2$.

Definition 3. Given a sequence S_a with an ordered list of items, each item's i utility in S_a is defined as $u(i) = u(i, S_a) = IU(i, IS_a) \times E(i)$. From Table 1, let us find a's utility in S_2 i.e. $u(i, S_2) = IU(a, S_2) * E(a) = 2 * 2 = 4$.

Definition 4. $P = \langle (t_{1,1}, I_1), (t_{1,2}, I_2), \ldots, (t_{1,n}, I_n) \rangle$ is a utility pattern of length n and $P \subseteq S_a$ and now the sequence utility of P in S_a is known to be $SU(P, S_a)$ and it is defined as

$$SU(P, S_a) = max\{\sum_{i \in P} SU(i, S_a), \forall P \in S_a\} \tag{1}$$

Definition 5. Given a sequence S_a, its utility is identical to the aggregation of the item utilities in S_a and it is defined as

$$SU(S_a) = \sum_{i \in S_a} SU(i, S_a) \tag{2}$$

Definition 6. Given a pattern P, its utility in a dataset D is represented as $U(P, D)$ and it is expressed as

$$U(P, D) = \sum_{s_a \in D} SU(P, S_a) \tag{3}$$

Definition 7. For a dataset D, its utility is expressed as

$$U(D) = \sum_{s_a \in D} SU(S_a) \tag{4}$$

Definition 8. Let us consider time constraints (C_1, C_2, C_3, C_4) in an time interval extended sequence $P = \langle (t_{1,1}, I_1), (t_{1,2}, I_2), \ldots, (t_{1,n}, I_n) \rangle$ and I_i, I_{i+1} be two adjacent itemsets, now the time constraints C_1, C_2, C_3 and C_4 are defined as follows:

C_1 = minimum time interval between I_i and I_{i+1}.
C_2 = maximum time interval between I_i and I_{i+1}.
C_3 = minimum time interval between I_1 and I_n.
C_4 = maximum time interval between I_1 and I_n.

Definition 9. For every item in the dataset, an utility threshold is defined and represented collectively using multiple minimum utility threshold table (MMU-table), i.e. $\{mu(i_1), mu(i_2), \ldots, mu(i_m)\}$. The MMU-table of the sample dataset is given in Table 3.

Definition 10. For a sequence P, its minimum utility threshold is denoted as $MIU(P)$ and is defined as $MIU(P) = minmu(i) \mid i \in P$. For instance, $MIU(\langle c, d \rangle) = min\{mu(c), mu(d)\} = min\{20, 10\} = 10$.

Problem Statement: Given a quantitative sequence dataset with time intervals, external utilities, multiple thresholds, and constraints, the problem of mining constraint based high utility sequential patterns including time intervals and multiple utility thresholds is to generate all the time interval sequential patterns whose utility is greater than or equal to its MIU.

Table 3. MMU table

mu(a)	mu(b)	mu(c)	mu(d)	mu(e)	mu(f)
15	10	20	10	30	10

4   Proposed Methodology

In the proposed algorithm, our principal aim is to impel time constraints and consider multiple utility thresholds maintaining downward closure property. Here, we need to project a database, first we have to arrange the items in order to avoid scanning all the combinations of candidate sequences i.e. it is better to list them alphabetically.

Definition 11. We need to define a prefix and postfix for $P = \langle (t_{1,1}, I_1), (t_{1,2}, I_2), \ldots, (t_{1,n}, I_n) \rangle$, Where I_β is an itemset with integer $j (1 \leq j \leq n)$ which fulfills the condition $I_\beta \subseteq I_j$ and $t_{1,\beta} = t_{1,j}$. Prefix $(P, I_\beta, t_{1,\beta}) = (t_{1,1}, I_1), (t_{1,2}, I_2), (t_{1,3}, I_3), \ldots, (t_{1,j}, I_\beta)$. Postfix $(P, I_\beta, t_{1,\beta}) = (t_{j,j}, I_j), (t_{j,j+1}, I_{j+1}), \ldots, (t_{j,n}, I_n)$.

Definition 12. Given a sequence in projected database $\beta = \langle(t_{1,1}, I_1),$ $(t_{1,2}, I_2), \ldots, (t_{1,m}, I_m)\rangle$ with respect to β which is represented as $D \mid \beta$ is input sequence S_a for all postfixes in quantitative sequential database D. Here, every time downward closure property may not be detected. Thus, a subsequence of a sequence with high utility may not be a high utility sequence.

Definition 13. For a pattern P, its sequence weighted utility is expressed as

$$swu(P) = \sum_{P \in S_a \cap S_a \in D} SU(S_a) \tag{5}$$

Definition 14. A sequential pattern P is said to be a candidate pattern for all multiple minimum thresholds if it satisfies the condition $swu(P) \geq MIU(P)$ and P satisfies the time constraints C_1, C_2, C_3, and C_4 in the proposed algorithm, at first we have to perform the database scan and have to find the length-1 candidate patterns. After generating candidate patterns, it applies a pattern growth approach to repeatedly generate candidate sequences. At last it checks whether all the patterns with utility for every multiple threshold are higher than minimum sequence utility or not.

Algorithm 1. *UIPrefixSpan-MMU*

Input:
    Quantitative sequential database (D)
    External utility table
    MMU-Table (Multiple minimum utility threshold-Table)
    Time constraints
Output:
    Set of sequential patterns with high utility and time intervals.
    Function *UIPrefixSpan-MMU*
1: Let P be a pattern of the form $\langle(time, item)\rangle$.
2: Let the set of candidate patterns is initially empty $C = \phi$ and set of output sequential patterns is initially empty $L = \phi$.
3: Scan the input D and discover all the items whose swu is no less than its MIU.
4: **for** each item i found in step-3 **do**
5:     Include $P \leftarrow \langle(0, i)\rangle$ in C
6:     **if** $su(P, D) \geq MIU(P)$ **then**
7:         $L \leftarrow L \cup P$
8:     **end if**
9:     Call $C \leftarrow subUIPrefixSpan\text{-}MMU(D|P, C, MMU, C_1, C_2, C_3, C_4)$.
10: **end for**
11: Scan D and for each candidate P in C, check the condition $su(P, D) \geq MIU(P)$, If it is true then include P in the set L.
12: Output L.

Algorithm 2. *subUIPrefixSpan-MMU*

Function *UIPrefixSpan-MMU*

1: Scan $D|P$, find $swu(i)$ for each item i and obtain the item and time interval, denoted as a pair (δ, i) such that each item's swu is no less than the $MIU(i)$, and every pair satisfies the constraints C_1 and C_2.

2: Let $P \leftarrow \langle P, (\delta t, i) \rangle$.

3: **if** P satisfies the constraint C_4 **then**

4:     Cal $C \leftarrow subUIPrefixSpan\text{-}MMU(D|P, C, MMU, C_1, C_2, C_3, C_4)$.

5: **end if**

6: **if** P satisfies C_3 **then**

7:     $C \leftarrow C \cup P$

8: **end if**

9: Return C

The proposed algorithm defines a pattern P of the form $\langle (time, item) \rangle$ and two sets, one is candidate set C and the other is the result set L. Step-3 of the Algorithm 1 finds the swu of each item and considers only the items whose MIU satisfies swu for further processing. The candidates whose su satisfies its MIU are included in the result set C (step-7). The sub function is invoked at step-9 to generate the candidate patterns of length greater than 1. In step-11, we check the sequence utility of each candidate with respect to MIU of the candidate and include the high utility pattern in the result set L (step-12).

Algorithm 2 generates the candidates based on the pattern growth approach. In step-1, Algorithm 2 reads the projected database and finds the item pairs along with the time interval whose MIU satisfies the swu and whose time interval satisfies the constraints C_1 and C_2. Thus, a new pattern is found and it is checked for the constraint C_4 (steps 2–3). If the constraint C_4 is satisfied then the sub function is invoked recursively to generate the new candidates (step-4). If pattern P satisfies the constraint C_3, then it is included in the set of candidate patterns and returned (steps 6–9).

5   Experimental Results

To evaluate the run time of UIPrefixSpan-MMU, experiments were conducted on 2 real datasets, namely, Kosarak and BMSWebview2. The further details of the datasets and setting of time intervals can be referred from [16]. The details of setting multiple utility thresholds can be cited from [13]. The experiments are carried out on a computer which has an i7 processor running Windows 10 and 8 GB of RAM. To the best of our insight, there are no algorithms concerned with the issue of mining HUTISP using more than one utility threshold.

5.1   Run Time Performance

The performance is assessed based on the constraints $C_1 = 0$, $C_2 = 5$, $C_3 = 0$ and $C_4 = 20$. So, we executed the proposed algorithm with and without constraints

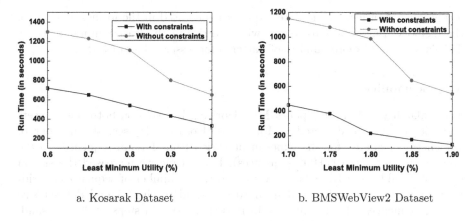

a. Kosarak Dataset   b. BMSWebView2 Dataset

Fig. 1. Performance of UIPrefixSpan-MMU.

for multiple values of least minimum utility (LMU). As presented in Fig. 1(a) and Fig. 1(b), UIPrefixSpan-MMU works faster with time constraints. This is because time constraints lead to the generation of less number of candidates which in turn reduces the search space and improves the performance. Moreover, it is observed that the run time tends to rise for lower values of *LMU*. This is as a consequence of increase in the candidate count for lower values of *LMU*. Further, more time is spent in evaluating the candidates.

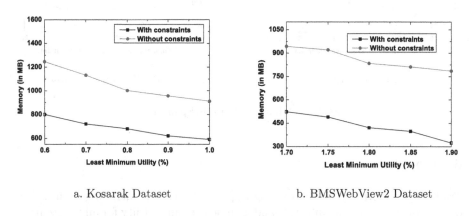

a. Kosarak Dataset   b. BMSWebView2 Dataset

Fig. 2. Memory usage of UIPrefixSpan-MMU.

5.2 Memory Utilization

The memory requirement of UIPrefixSpan-MMU is tested on both the datasets and the results are given in Fig. 2. It is noticed that the algorithm with constraints consume less memory than the algorithm without constraints. This is as a consequence of fewer candidate sequences generated in the former case. Also, the memory usage tends to reduce with the rise in the least minimum utility. This is because of more number of patterns for lesser values of LMU.

5.3   Scalability

The scalability of UIPrefixSpan-MMU algorithm is tested on both the datasets and the results are depicted in Fig. 3. On Kosarak dataset, the number of sequences considered for the experiment varied from 1 Lakh sequences to the original dataset size (990,002 sequences). Every time we increased the size in steps of 2 Lakhs. On BMSWebView2 dataset, the number of sequences considered for the experiment varied from 10 thousand to the original dataset size (77,512 sequences). Every time we increased the size in steps of 20 thousand. The constraints considered for this experiment are $C_1 = 0$, $C_2 = 5$, $C_3 = 0$ and $C_4 = 20$. It is observed that UIPrefixSpan-MMU scales linearly in both the approaches i.e. with constraints and without constraints.

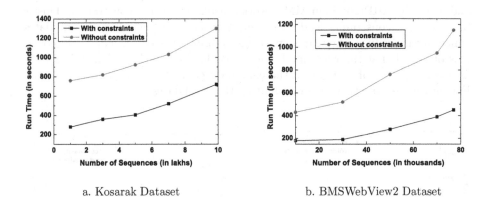

a. Kosarak Dataset   b. BMSWebView2 Dataset

Fig. 3. Scalability of UIPrefixSpan-MMU.

6   Conclusion

Traditional algorithms of discovering sequential patterns including high utility do not consider the time intervals and assume the same utility for all the items in a transaction. In the current work, we have initiated this problem and proposed a constraint pushing multi-threshold framework called UIPrefixSpan-MMU for discovering the sequential patterns which includes high utility and time interval. The proposed algorithm is a candidate pattern growth model based on the prefix projected databases. The performance of UIPrefixSpan-MMU is assessed on two real datasets. From the experimental results, we can derive that the UIPrefixSpan-MMU is more efficient with respect to the constraints. On an average, UIPrefixSpan-MMU with constraints achieve 52% improvement than

UIPrefixSpan-MMU without constraints with respect to the execution time. The algorithm can be further extended to include more efficient pruning strategies. Also, the distributed version of the current framework can be developed in support of big data environment.

References

1. Agrawal, R., Srikant R.: Mining sequential patterns. In: Proceedings of the Eleventh International Conference on Data Engineering, pp. 3–14 (1995)
2. Han, J., Pei, J., Mortazavi-Asl, B., Chen, Q., Dayal, U., Hsu, M.C.: FreeSpan: frequent pattern-projected sequential pattern mining. In: Proceedings of the Sixth ACM SIGKDD International Conference on Knowledge Discovery and Data Mining, pp. 355–359 (2000)
3. Pei, J., et al.: Mining sequential patterns by pattern growth: the PrefixSpan approach. IEEE Trans. Knowl. Data Eng. 16(11), 1424–1440 (2004)
4. Zaki, M.J.: SPADE: an efficient algorithm for mining frequent sequences. Mach. Learn. 42, 31–60 (2001)
5. Fournier-Viger, P., Lin, J.C.W., Kiran, R.U., Koh, Y.S., Thomas, R.: A survey of sequential pattern mining. Data Sci. Patt. Recogn. 1(1), 54–77 (2017)
6. Garofalakis, M.N., Rastogi, R., Shim, K.: Sequential pattern mining with regular expression constraints. In: Proceedings of the 25th VLDB Conference, pp. 223–234. Edinburgh, Scotland (1999)
7. Yao, H., Hamilton, H.J.: Mining itemset utilities from transaction databases. Data Knowl. Eng. 59, 603–626 (2006)
8. Ahmed, C.F., Tanbeer, S.K., Jeong, B.S.: A novel approach for mining high utility sequential patterns in sequence databases. ETRI J. 32, 676–686 (2010)
9. Chen, Y., Chiang, M., Ko, M.: Discovering time-interval sequential patterns in sequence databases. Expert Syst. Appl. 25(3), 343–354 (2003)
10. Chen, Y.-L., Huang, T.C.-K.: Discovering fuzzy time-interval sequential patterns in sequence databases. IEEE Trans. Syst. Man Cybern. Part B Cybern. 35(5), 959–972 (2005)
11. Duong, T.H., Janos, D., Thi, V.D., Thang, N.T., Anh, T.T.: An algorithm for mining high utility sequential patterns with time interval. Cybern. Inf. Technol. 19(4), 3–16 (2019)
12. Gan, W., Lin, J.C.W., Zhang, J., Yu, P.S.: Utility mining across Multi-Sequences with individualized thresholds. ACM 1(2), 1–12 (2021)
13. Lin, J.C.-W., Zhang, J., Fournier-Viger, P.: High-utility sequential pattern mining with multiple minimum utility thresholds. In: Chen, L., Jensen, C.S., Shahabi, C., Yang, X., Lian, X. (eds.) APWeb-WAIM 2017. LNCS, vol. 10366, pp. 215–229. Springer, Cham (2017). https://doi.org/10.1007/978-3-319-63579-8_17
14. Lin, J.C.W., Gan, W., Fournier-Viger, P., Hong, T.P.: Mining high-utility itemsets with multiple minimum utility thresholds. In: Proceedings of the 8th International Conference on Computer Science and Software Engineering, pp. 9–17 (2015)
15. Wang, W.-Y., Huang, A.Y.-Q.: Mining time-interval sequential patterns with high utility from transaction databases. J. Adv. Comput. Intell. Intell. Inf. 20(6), 1018–1026 (2016)
16. Sumalatha, S., Subramanyam, R.B.V.: Distributed mining of high utility time interval sequential patterns using MapReduce approach. Expert Syst. Appl. 141, 1–25 (2019)

Systems and Applications

KTSVidRec: A Knowledge-Based Topic Centric Semantically Compliant Approach for Video Recommendation on the Web

Akhil S. Krishnan[1] and Gerard Deepak[2](✉)

[1] SRM Institute of Science and Technology, Ramapuram, Chennai, India
[2] Department of Computer Science and Engineering, National Institute of Technology, Tiruchirappalli, India
gerard.deepak.christuni@gmail.com

Abstract. In this age of an increasing userbase of the internet on a daily basis, there results in a surplus of video content being available for consumption of the users of the Web. There also exists a problem of the current recommendation systems not being well suited to the semantic structure of the World Wide Web. In this paper, KTSVidRec framework has been proposed which is a semantically compliant approach for video recommendation. The proposed approach is user query centric and incorporates entity population using the Structural Topic Modelling and further entity aggregation from various knowledge bases. The KTSVidRec is a semantically infused machine learning model as it uses algorithms like XGBoost to classify popular video categories and Linked Open Data cloud for the enrichment of categories. Computation of the semantic similarity is carried forward primarily using Kullback-Leibler Divergence and other measures for measuring similarities like Cosine Similarity, Normalized Pointwise Mutual Information and Jaccard similarity. KTSVidRec exhibits higher performance, achieving an accuracy of 93.02% and a high precision percentage for a different number of recommendations, compared to other baseline and benchmark models. The proposed approach was found to be effective as well as semantically compliant for the recommendation of videos on the Web.

Keywords: Entity aggregation · Entity population · LOD cloud · Semantic similarity

1 Introduction

The internet is an ever-growing medium that has seen and continues to see a boom of users daily owing to the decrease in broadband and data prices, also increasing familiarity of usage of the internet has stimulated a wider reach to even the remotest of places on the globe. As a result, the content on the world wide web has skyrocketed. A video is the easiest way to imbibe information as it is an amalgamation of pictures and sound creatively put together to ensure easier comprehension by humans. This makes it difficult for a user to choose a particular video suited to his taste and liking because of the presence

© Springer Nature Switzerland AG 2022
K. K. Patel et al. (Eds.): icSoftComp 2021, CCIS 1572, pp. 277–289, 2022.
https://doi.org/10.1007/978-3-031-05767-0_22

of multiple videos which may be of interest to him. Hence, a video recommender system is essential for providing the user with relevant videos based on the user's interest in the content he consumes.

A recommender system is an algorithm that can be used to suggest relevant content for the users which suit their diverse tastes by taking into account their past interests and habits. The recommender system filters contents to the liking of the user, helping the user to make a decision when met with a variety of choices in cases like the world wide web. A video recommender system makes use of prior data available on the user's interest like favourite channel, the favourite topic of content from a said channel, video length, duration of the video watched by the user, choice of the video after completion of the prior etc. This helps in reducing the time spent by the user to search for videos based on his liking separately by instantly suggesting him related videos based on the content being watched currently.

Motivation: The meteoric rise of users of the world wide web has contributed to an explosion in the amount of data found on the web, hence making it difficult for a user to find relevant content based on their tastes. This directly translates to the user not being able to find videos that suit his interests and is of his liking in the case of websites like YouTube. Thus arises a need for recommendation strategies for the current structure of the web. The existing strategies prevailing don't work particularly well with the current structure, since it is in the midst of Web 2.0 and Semantic Web 3.0. As a result, there is a need for a semantically inclined approach for a recommendation of content from the world wide web. Knowledge representation and reasoning schemes are required to make the recommendation less computationally complex.

Contribution: A semantically compliant video recommendation approach KTSVidRec which uses knowledge representation and reasoning schemes, that help to tackle the highly linked hierarchical data present on the current structure of the world wide web. It uses the Structural Topic Modelling (STM) of entity population, XGBoost and computation of semantic similarity using Jaccard similarity, Kullback-Leibler (KL) Divergence and Normalized Pointwise Mutual Information (NPMI). Which help to tackle the highly linked hierarchical data present on the current structure of the World Wide Web. Firstly, query terms are subject to pre-processing and the query words are obtained. The query words are subject to STM and entity aggregation from knowledge bases like Freebase and other popular video repositories. Categories are obtained from the YouTube trending statistics dataset and enriched using the Linked Open Data (LOD) cloud. Classification based on XGBoost is done on the dataset using the query terms obtained earlier and the top 50% of the classified videos are obtained. Computation of semantic similarity is done between the categories and the entity population of structural modelling; the categories are stored in a HashSet. Further, the semantic similarity of the categories in the HashSet and categories of the classified video categories are compared, ordered in ascending order of semantic similarity and the video is recommended to the user.

Organization: The following portion of this paper's structure is as follows. The works related pertaining to the proposed approach is present in Sect. 2. Section 3 consists of the proposed architecture and further explanation of the KTSVidRec. Section 4 elaborates on the implementation and evaluation of the performance exhibited by the proposed approach. The conclusions drawn from the proposed approach is presented in Sect. 5.

2 Related Works

Davidson et al. [1] put forward a system for recommending the YouTube videos by making use of the user's activity on the site. The recommendations are evaluated using CTR (click-through rate) and recommendation coverage metrics. An easy-to-use user interface is used where thumbnails of videos along with descriptions are present with a link providing an explanation on why the particular video was recommended to the user. Deldjoo et al. [2] have proposed a system to recommend the videos based on the features which are stylistic and also feature sets based on the visual aspects which are extracted from the video. The model involving the extracted visual features are pit against conventional video recommendation systems which make use of features like the genre. The system exhibits better results compared to the conventional model with respect to the metrics of relevance like recall. The addressed system can be used as standalone or in tandem with traditionally prevailing video recommendation systems based on content. Chen et al. [3] have put forward a personalized video recommendation system that makes use of tripartite graph propagation for recommending personalized videos to the user. In addition to the graph approach involving click through data, there are also other graph approaches involving the user queries raised and if the video associated with the query appears on their recommendation lists. The tripartite graph takes into consideration the aforementioned characteristics on a dataset of over 2000 users, 23,000 queries and 55000 videos. Lee et al. [4] propose a large-scale content only video recommendation system which uses a deep learning approach to study the video embeddings and similarity learning, based on the video content is done thereby predicting the relationship between the videos using only visual and audio content on a dataset of a large scale. The proposed system is not completely dependent on the metadata pertaining to the video and can recommend videos based on recently relevant and prior uploaded videos. Tavakoli et al. [5] have suggested a novel system for the recommendation of educational videos for the aid of learners looking for professional jobs. The model makes use of classification and mining of texts on openings for jobs and the skills pertaining to the job and suggests relevant videos based on the quality of the videos by use of an open recommendation system for educational videos. The approach was carried out through interviews, gauging the effectiveness of the videos. Based on more than 250 videos that were recommended, 83% returned to be useful for the users. Soni et al. [6] have put forth a system for recommending videos, which uses facial emotions in comparison to the conventional recommendation systems based on former user statistics. The model carries out the process based on two queries which are, user rating prediction with the use of facial data and the likeliness of the user to predict the video to another. The classifier for deducing the facial information takes into account, the expressions and pulses associated with the skin. Furthermore, it also addresses the influence of the increase in the size of data. Belarbi et al. [7] have come forward with a system for recommending educational videos in accordance with the user's video interests while enrolling in a small-scale online course and looking for other learners with the same interests to recommend videos that may be of relevance to them. The model analyses the user's stream of clicks and designs a user profile to filter based on taste. A similar demographic are clustered together based on clustering in the purview of the K-Means clustering algorithm. Liu et al. [8] have suggested a system for the recommendation of videos based on the tags in the description, which associate a video to a specific category. A framework based on a graph incorporating neural networks is used which combine various parameters like the tags, users, the video and source of media. A loss that depends on the similarity of neighbours is used for encoding the preferences of the users into the diverse representation of nodes. The experimentation based on online and offline assessment is carried out on a Prominent social media network WeChat. Covington et al. [9] propose a system for video recommendation where neural networks are used to recommend videos based on neural rankings. The model follows a two-step approach where the detailing is done based on the generation of the candidates and further designing of a model for ranking of the candidates. Cognizance involving the design and the maintenance of a large-scale recommendation system is addressed. In [10–20] several ontological models and semantic approaches in support of the proposed framework have been discussed.

3   Proposed System Architecture

Figure 1 shows the architecture diagram for entity population, entity aggregation, category enrichment and classification of the proposed knowledge-driven framework for video recommendation, KTSVidRec. The model is semantically compliant as it incorporates semantic similarity strategies, it includes machine learning classification schemes like XGBoost, as a result, it is a form of semantics infused machine learning.

Phase 1 of the proposed architecture consists of query pre-processing. Firstly, the user query, as well as the web usage data, is fed as the input to the framework, subject to pre-processing. The pre-processing of the queries obtained from the user includes tokenization, lemmatization, removal of stop words and named entity recognition. Tokenization is done to obtain individual terms from the queries and the web usage data. Lemmatization involves deriving the base form of the word from its inflectional form. Stop words like of, and, they are eliminated as they tend to increase the noise and don't contribute to aggregating the terminologies. Named entity recognition, signifies which domain and categories the query terms and the web usage data belong to. Web usage data is also normalized in certain cases owing to the metadata available, specifically when the URLs are included the URL normalization is done. Ultimately, the terms form queries and web usage data are obtained.

Phase 2 of the proposed architecture consists of entity population-based on topic modelling, entity aggregation from knowledge bases and video repositories, obtaining and enriching video categories and classifying the categories using XGBoost.

The entity population is done based on topic modelling to increase the mapping of query terms and web usage data terms to the existing categories in the real-life space. The topic modelling used is Structural topic modelling (STM). The STM is similar to the other types of topic modelling but specialises in using metadata on documents to optimize the allocation of latent subjects of the corpus. Furthermore, entity aggregation is done from Freebase. Freebase is a collaborative knowledge base composed of data mainly provided by the users. Freebase is preferred as it's a real-world knowledge source for various terminologies. Entity aggregation is done from popular video repositories like Netflix, Amazon prime, MX player, Vimeo and daily motion. These video repositories are scrapped using Beautiful Soup an HTML parser python package and are crawled

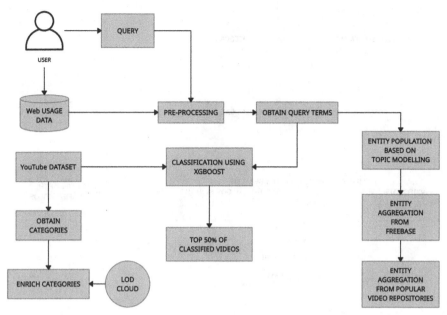

Fig. 1. KTSVidRec architecture for entity population, entity aggregation, video category enrichment and classification

for the extraction of the metadata. The categories are put in a localized repository and enriched.

From the YouTube trending video statistics dataset which is a categorical dataset, the categories for the videos are obtained. These categories obtained are enriched by the LOD cloud API, this LOD cloud stores a graph of real-world entities. Subgraphs relevant to the categories of the dataset is loaded and are enriched. Then, the dataset is classified using the XGBoost classifier and the terms for classification or classification categories are based on the initial query terms and the unique terms obtained from the Web usage data. Under each class, the top 50% of the classified videos are obtained and the categories are loaded. The reason for obtaining only the top 50% data is in this case relevance is of utmost importance. Also of importance is the diversity in results which is taken care of utilizing entity aggregation, topic modelling and enrichment of categories from various heterogeneous sources.

In phase 3 of the proposed approach's architecture, as illustrated in Fig. 2. Computation of the semantic similarity, mapping of categories, arranging the videos in increasing order of similarity and recommendation of the video for the user is carried out.

The computation for semantic similarity is done between the enriched categories obtained from the dataset, entity aggregated from Freebase and popular video repositories, topic modelling of the query terms and the web usage data terms which were obtained in phase 1 and phase 2. The categories are obtained whose semantic similarity is >0.75. Further, the video categories are mapped. The computation of the semantic similarity is done using KL Divergence, Jaccard Similarity, Cosine Similarity and NPMI methodologies.

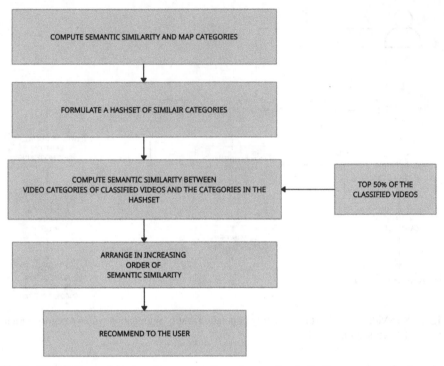

Fig. 2. KTSVidRec architecture for computation of semantic similarity, mapping of categories and recommendation of video to the user.

The KL Divergence is an estimation of how one distribution of the probability is differentiated from another probability distribution which is taken as the reference. The KL Divergence is in simple terms known as the measure of surprise and is employed in diverse fields like fluid mechanics, applied statistics and bioinformatics. Equation (1) illustrates the KL Divergence and P and Q are the probability distributions.

$$D_{\mathrm{KL}}(P\|Q) = \sum\nolimits_{x \in \mathcal{X}} P(x)\log\left(\frac{P(x)}{Q(x)}\right) \tag{1}$$

Jaccard index is used to estimate the measure of similarity in the comparison of two given sets which are of finite size. Jaccard similarity coefficient is illustrated in Eq. (2) where the J is Jaccard distance. A and B are the two given sets

$$J(A, B) = \frac{|A \cap B|}{|A \cup B|} = \frac{|A \cap B|}{|A| + |B| - |A \cap B|} \tag{2}$$

Cosine Similarity is an evaluation of similarity in two vectors of inner product space, it's defined as the cos theta value for the angle which is between the two given vectors and it evaluates if both the vectors point in a similar direction and is essential to gauge the similarity of the documents. The Cosine Similarity is illustrated in Eq. (3).

$$\text{Similarity} = \cos(\theta) = \frac{\mathbf{G} \cdot \mathbf{L}}{\|\mathbf{G}\|\|\mathbf{L}\|} = \frac{\sum_{i=1}^{n} G_i L_i}{\sqrt{\sum_{i=1}^{n} G_i^2}\sqrt{\sum_{i=1}^{n} L_i^2}} \tag{3}$$

Pointwise mutual information is a measure of association between two events namely a and b, instead of focusing on an average of all events like mutual information, point mutual information uses singular events. NPMI can be used to normalize the information between -1 and 1, where -1 is never occurring together, 0 for independent and 1 for occurring together. Equation (4) illustrates NPMI where h (a, b) is the joint self-information.

$$\text{NPMI}(a; b) = \frac{\text{PMI}(a; b)}{h(a, b)} \tag{4}$$

From the mapped categories a HashSet of similar categories is formulated. Further, the semantic similarity of the video categories presents in the HashSet and the top 50% of the classified video categories are computed for a threshold value of >0.75. The videos are arranged in increasing order of semantic similarity and finally recommended to the user.

4 Implementation and Performance Evaluation

The implementation of the proposed semantically compliant video recommendation dataset was done using Google Collab notebook for python, with an i5 processor of 4.0 GHz, 16 GB Ram and 50 GB of hard disk storage. The integration of the R script in Python is achieved using the single subprocess call. This allows the STM library of R to be used with Python. The Freebase API is used to access the Freebase data dump for entity aggregation using the Freebase knowledge base. The entity aggregation which involved popular video repositories like Netflix, Amazon Prime, Vimeo, MX player were called using an HTML parser based on python called Beautiful Soup. It was crawled based on the top trending videos on these platforms. Sklearn machine learning library was used for the implementation of the XGBoost machine learning algorithm. The collection framework HashSet was incorporated from the collection framework of Python. Table 1 presents the algorithm for the proposed KTSVidRec.

The dataset used for the model is the Trending YouTube videos statistic dataset, it incorporates a variety of metrics like the metrics for views, shares, videos liked and comments, it also involves this data collected for several months on these videos and of wide demography which includes countries like USA, Great Britain, India, Germany etc. with up to almost 200 listed trending videos in a day. The different regions are classified into separate files. The dataset's parameters based on which the evaluations are done include the title of the video, name of the channel, time of video upload, various tags involved in the description associating it to a particular domain, the number of views, the number of likes and dislikes for the video, information present in the description and the number of comments. The categories for the videos were clustered based on different regions. The experiment is carried forward for 4047 queries and the ground truth was

Table 1. Algorithm for the proposed KTSVidRec

Input: YouTube Dataset, query and web usage data, Access of API to LOD cloud.
Output: Videos recommended on the increasing semantic similarity order.

Begin

Step 1: The user query data and the web usage data is pre-processed and tokenization, lemmatization and removal of stop words is done and the query terms q_t is obtained.

Step 2: *while* ($q_t$next()! = NULL)

    for each query term qt

        Perform entity population through STM

        Perform entity aggregation from Freebase and popular video repositories

    end for

  end while

Step 3: **for** each query term qt and YouTube dataset ds:

    Perform XGBoosting on the dataset ds and query terms qt

      if (popularity > 50%)

        Obtain video categories

      end if

      for categories in ds:

        Enrich video categories in ds with ds.LODcloud()

      end inner *for*

    end outer *for*

Step 4: Compute the semantic similarity using the user query terms, entity population using STM, entity aggregation using Freebase and enriched categories from the dataset using KL Divergence and other similarity measures.

      if (ssm > 0.75)

      map the video categories

      end if

Step 5: Formulate Hashet H_t of the mapped categories

Step 6: Compute the semantic similarity using the enriched video categories and categories mapped in H_t

      if ($ssm_2 > 0.75$)

      arrange video categories in the increasing order of the semantic similarity

      end if

Step 7: Recommend the videos to the user based on the order of the increase of semantic similarity

End

---

either manually crawled or automatically incorporated from standard recommendation systems. Further, the ground truths were verified by 436 users who knew much about the experimentation domain. To elaborate, each user was given up to 45 queries and were asked to give top 10 categories, as well as recommendations from YouTube and other video sharing websites. The ground truth was obtained of the top categorizations, such that at least five users got a similar set of queries. This was carried out over a period of eight months in two phases. In the first four months, 25 queries were given out and in the next four months 20 queries. Top recommendations for each query from user recommendations were taken and the top 20% were used to verify the ground truth.

The metrics pertaining to performance for the architecture proposed is essential to be evaluated especially in the case of a recommendation system The Precision, Recall, Accuracy, F-Measure, False Discovery Rate (FDR) and Normalized Discounted Cumulative Gain (nDCG) were used in the evaluation metrics. FDR is the ratio of the positive results which were false to the total number of test results that were positive. nDCG is used to estimate the diversity of the results obtained. Standard formulae are used to evaluate the values for Average Precision, Average Recall, Accuracy Percentage, F-Measure, FDR and nDCG. The performance metrics for the proposed KTSVidRec is evaluated and tabulated in Table 2 and is compared with several baseline models and a benchmark approach. It is clear that the KTSVidRec achieves a better performance in comparison to the baseline models and benchmark approaches. The comparison of performance is done between four baseline models namely YVRS [1], CBVRS [2], PVRTGP [3], LSCVR [4] which are implemented in the same environment using the same user query and web usage data and a benchmark approach the Collaborative Filtering approach which is a standard benchmark model, highly famous in recommendation strategies.

Table 2. Performance comparison for the proposed KTSVidRec with other approaches

Search technique	Average precision %	Average recall %	Accuracy %	F-measure %	FDR	nDCG
YVRS [1]	82.14	80.44	81.29	81.28	0.17	0.81
CBVRS [2]	83.48	81.14	82.31	82.29	0.16	0.78
PVRTGP [3]	86.54	83.66	85.1	85.07	0.13	0.79
LSCVR [4]	85.18	82.14	83.66	83.63	0.14	0.81
Collaborative Filtering Approach	81.16	78.14	79.65	79.62	0.18	0.84
Proposed KTSVidRec	**94.87**	**91.18**	**93.02**	**92.98**	**0.05**	**0.96**

It is quite evident from Table 2, that YVRS [1] has a Precision of 82.14%, Recall of 80.44%, Accuracy of 81.29%, F-Measure of 81.28%, FDR of 0.17 and nDCG of 0.81 respectively. The CBVRS [2] approach has a Precision of 83.48%, Recall of 81.14%, Accuracy of 82.31%, F-Measure of 82.29%, FDR value of 0.16 and nDCG of 0.78 respectively. The PVRTG [3] furnishes a Precision of 86.54%, Recall of 83.66%, Accuracy of 85.1%, F-Measure of 85.07%, FDR of 0.13 and nDCG of 0.79 respectively. The LSCVR [4] yields Precision of 85.18%, Recall of 82.14%, Accuracy of 83.66%, F-Measure of 83.63%, FDR of 0.14 and nDCG of 0.81 respectively. The benchmark collaborative filtering approach gives a Precision of 81.16%, Recall of 78.14%, Accuracy of 79.65%, F-Measure of 79.62%, FDR of 0.18 and nDCG of 0.84 respectively. The proposed KTSVidRec yields the highest Precision of 94.87%, Recall of 91.18%, Accuracy of 93.02%, F-Measure of 92.98%, it also yields the lowest FDR of 0.05 and the highest nDCG of 0.96 respectively.

The YVRS [1] uses a co visitation-based traversal of graphs approach where the candidates are generated and a MapReduce is also used for further ranking and recommendation of the videos. User relevance is also an important strategy in this approach. The graph-based approach makes this approach a highly complicated one. As a result, the user relatedness that is co visitation graph based on previously visited videos, the relevance is not computed between the query as well as content in the video, rather main focus is based on previous visits to a specific set of results, which yields a lower relevance score, because it only satisfies previous visitation count. The nDCG values are very low because there is no factor to compute diversity in the result. The CBVRS [2] mainly focuses on stylistic features of the videos and KNN along with similarity is incorporated. This approach is still non-semantic and doesn't work well when the dataset density increases. Moreover, KNN is a quite naïve approach and the nDCG values are quite low and the number of false-positive values becomes pretty high. Stylistic feature extraction requires a proper strategy for the extraction of features, and there is always scope of improvement in such systems based on the above-stated factors. The PVRTGP [3] approach is based on tripartite graph traversal, and a pair of graphs are used to connect users and videos. Most importantly, instead of a query centric approach, the focus is on a user-centric approach where a user is given more importance than the query. While user feedback is important, highly driving something based on only the user is quite a tedious task and tripartite graph traversal is more complicated. There is always a scope for making such computationally expensive approaches efficient. In the LSCVR [4] approach, only the video embeddings are done using a neural network and similarity learning takes place. Since this approach is deep learning, it is very difficult to predict the kind of output that goes to the system. Although, it yields results that are distinct and acceptable it is a blind box approach, where the deep features and deep learning algorithm doesn't contribute to an explainable AI, hence a better approach can be formulated. In the collaborative filter approach, the ratings are required and ratings can be biased. So, content-based approaches can go awry. Hence, there is a need for a semantic approach that is knowledge-driven and query centric but also obtains user feedback. Hence, the proposed approach KTSVidRec is formulated. KTSVidRec having the highest Precision, Recall and F-Measure are because it is semantic and it incorporates topic modelling using STM. Therefore, the density of initial knowledge keeps on increasing based on entity population and entity aggregation from both DBpedia and video repositories crawled, XGBoost for classification makes sure that the approach is inferencing along with learning which is a semantically driven machine learning approach. The KL Divergence to compute Divergence factor and a combination of three similarity measure Jaccard similarity, Cosine Similarity and NPMI are all used to compute semantic similarity thereby contributing to very high relevance, very low Divergence and owing to the density of the knowledge supplied to the approach, the relevance of the approach is also high. As a result, the nDCG is naturally high.

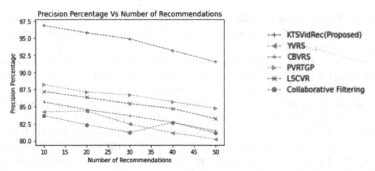

Fig. 3. Precision Percentage vs No of recommendations comparison of the proposed KTSVidRec and other baseline models

Figure 3 presents the percentage of precision for different no of recommendations ranging from 10 to 40 recommendations for KTSVidRec and the other baseline models. It is observable that the precision percentage of KTSVidRec is higher when compared to all the baseline models. This is because this approach uses a semantic approach for the recommendation of videos. Also, the approach uses entity population through STM, and a knowledge base methodology by entity aggregation using Freebase and popular video repositories.

5 Conclusion

A novel semantically inclined video recommendation framework KTSVidRec is proposed, which follows a query-centric approach for video recommendation, based on user queries and web usage data. It also incorporates entity population using the STM and further knowledge representation by entity aggregation from Freebase and different video repositories. YouTube trending video statistic dataset is used and is subject to the classification of the top 50% of video categories and the dataset's categories are further enriched by the LOD cloud. Computation of semantic similarity is done by a variety of methods primarily KL Divergence, Jaccard Similarity, Cosine Similarity and NPMI. It can be concluded that the proposed approach displayed better performance compared to the other baseline models mainly due to its semantic driven approach, entity population through STM and aggregation through a variety of knowledge bases. The KTSVidRec had a very low FDR of 0.05 and the highest nDCG value of 0.96 which makes it an efficient and semantically sound solution for the recommendation of videos on the Web.

References

1. Davidson, J., et al.: The YouTube video recommendation system. In: Proceedings of the Fourth ACM Conference on Recommender systems, pp. 293–296 (2010)
2. Deldjoo, Y., Elahi, M., Cremonesi, P., Garzotto, F., Piazzolla, P., Quadrana, M.: Content-based video recommendation system based on stylistic visual features. J. Data Seman. 5(2), 99–113 (2016)

3. Chen, B., Wang, J., Huang, Q., Mei, T.: Personalized video recommendation through tripartite graph propagation. In: Proceedings of the 20th ACM International Conference on Multimedia, pp. 1133–1136 (2012)

4. Lee, J., Abu-El-Haija, S.: Large-scale content-only video recommendation. In: Proceedings of the IEEE International Conference on Computer Vision Workshops, pp. 987–995 (2017)

5. Tavakoli, M., Hakimov, S., Ewerth, R., Kismihok, G.: A recommender system for open educational videos based on skill requirements. In: 2020 IEEE 20th International Conference on Advanced Learning Technologies (ICALT), pp. 1–5. IEEE (2020

6. Soni, Y., Alm, C.O., Bailey, R.: Affective video recommender system. In: 2019 IEEE Western New York Image and Signal Processing Workshop (WNYISPW), pp. 1–5. IEEE (2019)

7. Belarbi, N., Chafiq, N., Talbi, M., Namir, A., Benlahmar, H.: A recommender system for videos suggestion in a SPOC: A proposed personalized learning method. In: Yousef, F., Laila, M. (eds.) ICBDSDE 2018. SBD, vol. 53, pp. 92–101. Springer, Cham (2019). https://doi.org/10.1007/978-3-030-12048-1_12

8. Liu, Q., et al.: Graph neural network for tag ranking in tag-enhanced video recommendation. In: Proceedings of the 29th ACM International Conference on Information and Knowledge Management, pp. 2613–2620 (2020)

9. Covington, P., Adams, J., Sargin, E.: Deep neural networks for Youtube recommendations. In: Proceedings of the 10th ACM Conference on Recommender Systems, pp. 191–198 (2016)

10. Leena Giri, G., Deepak, G., Manjula, S.H., Venugopal, K.R.: OntoYield: a semantic approach for context-based ontology recommendation based on structure preservation. In: Nabendu, C., Agostino, C., Nagaraju, D. (eds.) Proceedings of International Conference on Computational Intelligence and Data Engineering. LNDECT, vol. 9, pp. 265–275. Springer, Singapore (2018). https://doi.org/10.1007/978-981-10-6319-0_22

11. Deepak, G., Kumar, N., Santhanavijayan, A.: A semantic approach for entity linking by diverse knowledge integration incorporating role-based chunking. Procedia Comput. Sci. **167**, 737–746 (2020)

12. Deepak, G., Rooban, S., Santhanavijayan, A.: A knowledge centric hybridized approach for crime classification incorporating deep bi-LSTM neural network. Multimedia Tools Appl. **80**(18), 28061–28085 (2021). https://doi.org/10.1007/s11042-021-11050-4

13. Vishal, K., Deepak, G., Santhanavijayan, A.: An approach for retrieval of text documents by hybridizing structural topic modeling and pointwise mutual information. In: Mekhilef, S., Favorskaya, M., Pandey, R.K., Shaw, R.N. (eds.) Innovations in Electrical and Electronic Engineering. LNEE, vol. 756, pp. 969–977. Springer, Singapore (2021). https://doi.org/10.1007/978-981-16-0749-3_74

14. Deepak, G., Teja, V., Santhanavijayan, A.: A novel firefly driven scheme for resume parsing and matching based on entity linking paradigm. J. Discr. Math. Sci. Cryptograp. **23**(1), 157–165 (2020)

15. Deepak, G., Kumar, N., Bharadwaj, G.V.S.Y., Santhanavijayan, A.: OntoQuest: an ontological strategy for automatic question generation for e-assessment using static and dynamic knowledge. In: 2019 Fifteenth international conference on information processing (ICINPRO), pp. 1–6. IEEE (2019)

16. Deepak, G., Santhanavijayan, A.: OntoBestFit: a best-fit occurrence estimation strategy for RDF driven faceted semantic search. Comput. Commun. **160**, 284–298 (2020)

17. Arulmozhivarman, M., Deepak, G.: OWLW: ontology focused user centric architecture for web service recommendation based on LSTM and whale optimization. In: Musleh Al-Sartawi, A.M.A., Razzaque, A., Kamal, M.M. (eds.) EAMMIS 2021. LNNS, vol. 239, pp. 334–344. Springer, Cham (2021). https://doi.org/10.1007/978-3-030-77246-8_32

18. Deepak, G., Priyadarshini, J.S.: Personalized and Enhanced Hybridized Semantic Algorithm for web image retrieval incorporating ontology classification, strategic query expansion, and content-based analysis. Comput. Electr. Eng. **72**, 14–25 (2018)

19. Gulzar, Z., Leema, A.A., Deepak, G.: Pcrs: personalized course recommender system based on hybrid approach. Procedia Comput. Sci. **125**, 518–524 (2018)
20. Deepak, G., Sheeba Priyadarshini, J.: A hybrid semantic algorithm for web image retrieval incorporating ontology classification and user-driven query expansion. In: Rajsingh, E.B., Veerasamy, J., Alavi, A.H., Peter, J.D. (eds.) Advances in big data and cloud computing. AISC, vol. 645, pp. 41–49. Springer, Singapore (2018). https://doi.org/10.1007/978-981-10-7200-0_4

Intelligent Facial Expression Evaluation to Assess Mental Health Through Deep Learning

Prajwal Gaikwad$^{(\boxtimes)}$ (ID), Sanskruti Pardeshi (ID), Shreya Sawant (ID),
Shrushti Rudrawar (ID), and Ketaki Upare (ID)

Department of Computer Engineering, AISSMS IOIT, Kennedy Road, Near R.T.O.,
Pune 411 001, Maharashtra, India
{prajwal.gaikwad,principal}@aissmsioit.org
https://aissmsioit.org/

Abstract. Expressions have been one of the most primary forms of non-verbal communication for humans. This is very natural and has formed an irreplaceable part of our lives. The facial expressions have also been highly useful parameters in the understanding of human behavior and analysis of the human psyche. The psycho-therapists have been engaged in this practice for a long time and the rise of awareness about mental health issues have been leading to an increase in the number of patients. Therefore, there is a need for an accurate analysis tool for achieving the facial expression detection. To provide a solution to this problem, this research article illustrates a facial expression technique through the use of Feature Extraction along with Convolutional Neural Networks and Decision tree. The approach has been effectively evaluated for its accuracy which has resulted in highly satisfactory results.

Keywords: YC_BC_R model · Convolution Neural Network · Decision tree · Expression evaluation

1 Introduction

The expressions on one's face and the changes related to the facial patterns gives the information about the emotional state the subject is in and it helps regulate a healthy conversation with the subject. The overall mood of the subject is understood from these expressions, eventually developing a better understanding of the subject. In non-verbal communications and human interactions, an important role is played by facial expressions. The analysis of facial expression deals with analyzing and visually recognizing facial features changes and different facial motions.

The most important aspect of mood detection is to understand the subject. If a subject cannot be understood their problem cannot be solved being able to detect the mode of a person helps one understand the subject in a better way.

© Springer Nature Switzerland AG 2022
K. K. Patel et al. (Eds.): icSoftComp 2021, CCIS 1572, pp. 290–301, 2022.
https://doi.org/10.1007/978-3-031-05767-0_23

Using this method, a doctor can easily be able to understand the patient better. Similarly, a psychologist can easily understand the current mood of the subject and give them the right treatment that would benefit them.

The expressions on one's face plays a vital role in understanding the person, in facilitating communication with humans as well as interacting with them. They also play a major role in medical rehabilitation of a patient and become a base for behavioral studies. Mood detection based on the technique of capturing facial images main provide a very practical approach towards detection of mood non-invasively.

In this system live streaming of the video of the subject is done via a camera placed in front of the subject and processed further. The process of frame grabbing comes next. Here, the frames of the video are extracted and then these frames are processed to extract the region of interest. The region of interest is the area of the image on which further processing is to be done.

In feature extraction and image processing parameters extracted from the images were of two types binary parameters and real value parameters. Depending upon the measured distance the real valued parameters have a definite value which is measured in the number of pixels. Whereas, the binary measures returned either an absent ($= 0$) or present ($= 1$) value. On a whole, 7 binary measures and 8 real valued measures are obtained.

To identify a certain facial expression a number of parameters both, binary and real-valued are analyzed and extracted to decide their effectiveness. Elimination of the features which do not provide any significant information of the expression portrayed is vital. The following binary parameters and real-valued parameters are considered in image processing:

Raised eyebrow distance: It is the distance between the lower and upper eyelid and the lower central point of the eyebrow. Eyebrow to upper eyelid to distance: It is the distance between the eyebrow surface and upper eyelid. Inter-eyebrow distance: It is the distance between the lower central points of the eyebrows. Lower eyelid to upper eyelid distance: It is the distance between the lower and upper eyelid. Top lip thickness: It is the thickness of the upper lip. Lower lip thickness: It is the thickness of the lower lip. Mouth width: It is the distance between the tips of the lips. Mouth opening: It is the distance between the upper surface of the bottom lip and the lower surface of the upper lip.

Upper teeth visible: It is a parameter that indicates whether the upper teeth are visible or not. Lower teeth visible: It is a parameter that indicates whether the lower teeth are visible or not. Forehead lines: It is a parameter that indicates whether wrinkles are present in the forehead's upper side or not. Eyebrow lines: It is a parameter that indicates whether wrinkles are present in the area above the eyebrows or not. Nose lines: It is a parameter that indicates whether wrinkles are present in the area between the eyebrows and over the nose or not.

Chin lines: It is a parameter that indicates whether lines or wrinkles are present in the area below the chin or under the lips or not. Nasolabial lines: It is a parameter that indicates whether thick lines are present on both sides of the nose extending till the upper lip or not. These parameters real-value based and binary, play a major role in creating the base for our calculations to be given to the neural network to compare and analyze the facial expression of the subject whose image has been captured from the video.

The region of interest is analyzed by the Convolutional Neural Network and a Convolution Rate is calculated by the Convolutional Neural Network. The Convolution Rate is analyzed by the Entropy Estimation block where the Entropy List is calculated using the Convolution Rate. This Entropy List is then sent to the Decision Tree where it determines a response about the facial expression of the subject. This expression is then sent further to the Facial Analysis Module where it derives a detailed facial analysis of the subject.

The obtained expression is used by the doctor's application to evaluate the patient's mood in a more technical way. This research mainly focuses on identification of the patient's expression in a different time and helps the doctor to evaluate his/her mental status on each of the visits.

This research article dedicates Sect. 2 for the evaluation of the past works under the name literature survey. And the implemented technique is broadly described under the section Proposed methodology which is numbered as 3. The Sect. 3 discusses the obtained results from the experimental process. And finally, the Sect. 5 concludes this paper along with the scope for the future enhancements.

2   Literature Survey

M. Pantic explains that there has been an increased interest in the human facial characteristics as they are useful for achieving an identification of the inner emotional state of individual [1]. The human beings are highly dependent on such emotional cues which allow for effective communication. These visual cues for the facial expression identification are quite easily identified and extracted by the human beings due to the inherent detection mechanisms in the human brain. This is highly difficult for a computer interface to achieve this identification due to the lack of a useful technique to achieve it. For an improvement in the recognition of the facial expression, an effective approach is determined using temporal, spatial and rule-based reasoning.

Yongqiang Li expresses that there have been multiple techniques that have been instrumental in driving the paradigm of image processing and computer vision forward [2]. This has been crucial in the development and significant improvement in the underlying technique for the detection for the purpose of enabling a useful and effective image processing guideline. The paradigm of achieving a highly useful facial feature extraction approach for the purpose of analyzing the expression of the individual and recognizing it accurately. The approach implements Bayesian networks to achieve the considerable improvements in the facial expression detection mechanism.

Yongmian Zhang elaborates on the concept of facial expression and its use for the purpose of achieving effective interactions with one another [3]. This is highly crucial as human are social beings that need to interact with each other to maintain their mental health in a better position. The effective realization of the automatic detection of the facial expression is needed for utilization in various implementations. This is achieved by the technique presented in this research article through the use of an Image sequences along with dynamic and active fusion of information for the facial expression detection.

Yuta Kihara states that there has been an increased interest in the evaluation of the facial characteristics that have been useful in various implementations that have been usually performed manually [4]. The process of paralysis cannot be determined effectively and discussed by the doctors. The paralysis is highly debilitating and can only be diagnosed by a professional by thorough analysis that takes into account various changes to achieve the accurate evaluation. Therefore, there is a need for an approach that can effectively identify facial parameters in achieving the facial paralysis detection using a database containing dynamic facial expressions using quantitative analysis.

Nazil Perveen discusses that the facial expression evaluation through an image processing approach is useful in achieving various implementations accurately. The facial expression recognition has been realized to allow for greater understanding of the non-verbal communication approaches of human beings [5]. This is highly useful as it can be effective in achieving an enhancement in the various realizations of the detection approach. This publication outlines an effective methodology for the purpose of portraying an improvement in the facial expression recognition through the use of the Gini Index along with the Facial Characteristic points.

Catherine Soladie introduces the concept of assisting the elderly individuals in achieving an effective living alone and achieved other goals. There have been large amounts of interest in achieving an improvement in the detection of the changes in the behavior of the elderly individuals living alone [6]. The most effective approach for the purpose of facial expression detection has been devised taking into account the invariant representation which is useful for achieving an effective and useful implementation. The implementation of such a system is highly useful in achieving an improvement in the recognition of blended expressions.

SuJing Wang narrates that the concept of micro expression is one of the most important and useful indications of concealed or obscured emotions. These emotions have been effective in realizing the emotion of an individual in a high stake's environments. These emotions and the facial expressions, especially micro-expressions are very difficult to detect and identify even for a trained eye [7]. This is due to the fact that these expressions are short lived and can be very fragmented. Therefore, this research article has been utilized to devise an effective micro-expression detection through the use of color Spaces.

Jiannan Yang explains that the process of facial expression recognition has been extremely basic as a large number of individuals can easily do this with a relative ease. This is problematic for the computer to detect expression as it cannot understand the various nuances of the facial expression [8]. The detection of the facial expression has been highly useful for determination of the various implementations that can be highly useful. Therefore, the authors in this approach provide a system that utilizes the facial action unit for the purpose of facial expression detection.

Lei Pang expresses that the human facial expression forms an important aspect of communication between two different individuals [9]. These have been effective in realization of the state and the emotional wellbeing of the individual while performing the communication. Therefore, the use of the effective realization of the automatic recognition of the facial expression needs the implementation of an efficient neural network that have been combined with the use of Gabor Features by the researchers to achieve highly accurate facial expression identification with high accuracy. The neural network used by the authors is the BP neural networks.

Lutfiyatul Fatjriyati Anas elaborates on the concept of non-verbal communication that is one of the major communication mechanisms that are used by humans for the purpose of conveying their message [10]. This consists of emotions that are effectively identified by the target individuals through the highly visible facial expressions. These expressions are difficult to identify by the computer and marks a major problem in computer vision. Therefore, the authors in this research article have proposed the use of Fashion images and landmarks along with the K Nearest Neighbor to achieve the accurate determination of the facial expressions.

Zhang Nan [11] states that there has been an important and highly challenging problem of detection of facial expression in computer vision and image processing. The detection of facial expressions comes naturally by individuals due to the large amount of non-verbal communication that can be extremely challenging for the computer to determine. This problem needs to be solved to implement an effective approach that can be extremely useful in a large selection of implementations. The authors in this research article have proposed the use of Local Facial regions for the purpose of achieving the effective realization of the facial expression.

3   Proposed System for Facial Expression Evaluation

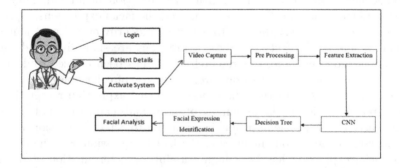

Fig. 1. Facial expression evaluation system overview

The proposed model for facial expression evaluation for the patient's mental health assessment is being depicted in the Fig. 1. The steps that are included to develop the model are detailed in the below mentioned steps.

Step 1: Frame extraction and preprocessing – This is the initial step of the proposed model where the doctor is logged into the system and access the camera to get the picture of the patient. For this purpose, an opencv library is installed for the Java programming language, with the help of this an image of a patient's face is being captured and preprocessed in the image object to store it in jpg format in a destined folder.

Step 2: Feature extraction – This is the important step of the proposed model, where a python program iteratively visits the destined folder and checked for the presence of any image. This image was captured by the doctor through the interactive user interface designed by the java swing framework. If the image is present in the destined folder, then the image is processed to detect the face using the haar cascade properties stored in a .xml file. The HAAR features are used extensively in this xml file which eventually helps to identify the face position in the image. Once the face position is identified in the image, then the coordinates of this facial position are identified to extract only the face in the image object. This image object is subjected to estimate the skin of the face using the YCBCR model.

In the YCBCR model initially each pixel of the image is read to extract the RGB values. These RGB values are extracted to estimate the blue chroma and red chroma components with the below mentioned Eqs. 1 and 2

$$Cb = -0.169 * R - 0.332 * G + 0.500 * B + 12 \qquad (1)$$
$$Cr = 0.500 * R - 0.419 * G - 0.081 * B + 128 \qquad (2)$$

This blue chroma component is evaluated in the range 137 and 177 for the skin pixel. And then again red chroma component is evaluated in the range 77 to 127. Then the combination of these two red chroma and blue chroma components are used to estimate the pixel is skin or not. If the pixel satisfies all the conditions of blue and red chroma components, then it is considered as the skin pixel and marked with the white pixel color value. On the other hand, if it fails to satisfy the condition then it is considered as the non-skin pixel and marked with the black color value. Once this process is iterated for all the pixels of the image the binary image is yielded with skin in white pixel values and non-skin in black pixel values. This process is depicted in the below shown Algorithm 1.

3.1 ALGORITHM 1: Skin Detection Through YCBCR Model

//Input: Input image IIMG
//Output: Skin Detected image SIMG
// function: skindetection(IIMG)
1: Start
2: SIMG = ∅ , count=0
3: for i = 0 to size of breadth of IIMG

```
4: for j=0 to size of Height of IIMG
5: col = IIMG [i,j] ( PIX )
6: R=ol[0]
7: G= col[1]
8: B= col[2]
9: Cb = -0.169 * R - 0.332 * G + 0.500 * B + 128)
10: Cr = 0.500 * R - 0.419 * G - 0.081 * B + 128
11: if (Cr > 137 && Cr < 177), then
12: if (Cb < 127 && Cb > 77), then
13: t = Cb + 0.6 * Cr;
14: if (t > 190 && t < 215), then
16: SIMG [i,j] ( PIX )=[ 255,255,255]
17: end if, else
18: IMG [i,j] ( PIX )=[ 0,0,0]
19: end if, else
20: SIMG [i,j] ( PIX )=[ 0,0,0]
21: end if, else
22: SIMG [i,j] ( PIX )=[ 0,0,0]
23: end if
24: end for
25: end for
26: return SIMG
27: stop
```

Step 3: Convolution Neural Network – The original captured image along with the skin object is subjected to estimate the user expression based on the learned details stored in .h5 files through the CNN model. These trained details are extracted with the help of the dataset downloaded through the internet.

This dataset contains two subfolders like train and test. These two subfolders contain many expression folders like angry, disgusted, fearful, happy, neutral, sad and surprised. Each of these expression folders contains many images belong to that expression. Once these dataset folders are feed to the CNN training model, then they are subjected to set the rescale ratio and then they are rescaled to the width and height of 48 × 48. After this a training batch size of 64 and epochs of 50 is set to train the model. The CNN model is trained by using the keras and tensor flow libraries built for the python programming language. This CNN model is built with the architecture as mentioned below in Table 1.

Once the CNN Model is compiled for the given number of epochs, then the trained data is stored in an .h5 model file. This h5 model file is used to test the input image from the camera which was preprocessed in the step 2.

Step 4: Decision Tree- This is the final step of the proposed system, where trained data is used to predict the emotions for the captured image. For this purpose, another CNN neural network model is created which is used to read the trained data from .h5 file. And then by using the predict method of the model an integer value is obtained which eventually indicates the different emotions with respect to the input directory of the trained data. Based on this integer value a particular an expression is extracted from the parallel array created for the expressions. The obtained expression is stored in the database for the further analysis by the doctor himself.

Table 1. CNN network architecture.

Layer	Activation
CONV 2D 32 × 3 × 3	Relu
CONV 2D 64 × 3 × 3	Relu
MaxPooling2D 2 × 2	
Dropout 0.25	
CONV 2D 128 × 3 × 3	Relu
MaxPooling2D 2 × 2	
CONV 2D 128 × 3 × 3	Relu
MaxPooling2D 2 × 2	
Dropout 0.25	
Flatten	
Dense 1024	Relu
Dropout 0.25	
Dense 7	Softmax
Adam optimizer	

4   Result and Discussion

The proposed methodology helps to detect the expression or emotion of an individual. Detection of facial expression is achieved using deep learning and machine learning algorithms like Decision Tree.

The performance of system is achieved through the realization of intensive experimentation of the methodology. The assessments have been performed to determine accuracy of the facial expression detection approach in thoroughly. This is necessary to attain the performance metrics of the technique to identify if the methodology has been deployed correctly and functions appropriately. The performance evaluation has been elaborated below.

4.1   Performance Evaluation Based on Precision and Recall

The Precision and Recall metrics are being utilized for the purpose of enabling an effective evaluation of the accuracy of the facial expression detection. The precision and recall are essential components to measure the accuracy of the detection mechanism as they are able to determine the real performance of the methodology. The detailed evaluation of the approach has been depicted as follows. The indepth evaluation of the approach determined through the use of precision and recall metric requires the analysis of a number of parameters. The precision is the measure the relative accuracy of the detection whereas the recall metric determines the absolute accuracy of the detection approach.

The precision in this system is the division of the correctly detected facial expressions by the total number of facial expression detections. On the other hand the Recall is measured as the division of total number of facial expression detections by the total number of expected facial expression detections.

The mathematical depiction of the precision and recall metrics has been illustrated below.

A = The number of correctly detected facial expressions
B = The number of incorrectly detected facial expressions
C = The number of facial expressions not detected So, precision can be defined
Precision = (A / (A + B)) * 100
Recall = (A / (A + C)) * 100

The extensive evaluation of the presented approach has been performed through the realization of the equations given above. The evaluation outcomes have been listed in the Table 2 below.

Table 2. Precision and recall measurement table for the performance of facial expression detection.

No. of expected detection	(A)	(B)	(C)	Precision	Recall
38	27	5	6	84.375	81.81818182
67	48	7	10	84.21052632	82.75862069
87	63	10	14	86.30136986	81.81818182
119	89	19	11	82.40740741	89
139	98	20	21	83.05084746	82.35294118

The achieved outcomes for the experimental evaluation through the use of experimentation using the precision and recall have also been converted into a graphical representation given in Fig. 2 above. As it is evident from the line graph, the approach has been effective in realization of the accurate facial expression detection. The precision and recall performance metrics attained in this experimental assessment are 84.06 and 83.54 respectively. These values depict the correct implementation of the facial expression detection approach which achieves satisfactory accuracy measures as expected.

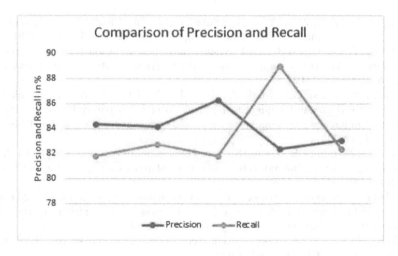

Fig. 2. Comparison of precision and recall for the performance of facial expression detection

5   Conclusion and Future Scope

The presented system for the facial expression detection approach has been achieved through the implementation of Convolutional Neural Networks and Decision Tree. The facial expression detection is a highly useful measure of the actual state of the person's mental status accurately, and can be a valuable tool for the psychoanalysts and psychotherapists. The presented system uses the video frames containing the face of the subject are captured from a camera into the system as an input. These frames are then first preprocessed by resizing and normalization of the frames. These preprocessed images are then provided to the next step for extraction of the features such as, skin detection and Region of interest Evaluation. These features are then provided to the Convolutional Neural Networks for the evaluation of the convolution and the fully connected layers which achieves the convolution rate. These values are then classified using the Decision Tree module. The Decision Tree approach implements if-then rules to achieve accurate determination of the facial expression. The experimental evaluation of the approach yielded satisfactory levels of precision and recall. The future research directions for this facial expression detection approach can be applied to initiate this approach into an API for effective use in a variety of scenarios. And also this can be implemented as mobile application for the fun and effective usage in many real time scenarios.

References

1. Pantic, M., Patras, I.: Dynamics of facial expression: recognition of facial actions and their temporal segments from face profile image sequences. IEEE Trans. Syst. Man Cybern. Part B Cybern. **36**(2), 433–449 (2006)

2. Li, Y., Wang, S., Zhao, Y., Ji, Q.: Simultaneous facial feature tracking and facial expression recognition. IEEE Trans. Image Process. **22**(7), 2559–2573 (2013)

3. Zhang, Y., Ji, Q.: Active and dynamic information fusion for facial expression understanding from image sequences. IEEE Trans. Pattern Anal. Mach. Intell. **27**(5), 699–714 (2005)

4. Kihara, Y., Duan, G., Nishid, T., Matsushir, N., Chen, Y.-W.: A dynamic facial expression database for quantitative analysis of facial paralysis. In: 6th International Conference on Computer Sciences and Convergence Information Technology (ICCIT) (2011)

5. Perveen, N., Gupta, S., Verma, K.: Facial expression recognition using facial characteristic points and Gini index. In: Students Conference on Engineering and Systems (2012)

6. Soladie, C., Stoiber, N., Seguier, R.: A new invariant representation of facial expressions: definition and application to blended expression recognition. In: 19th IEEE International Conference on Image Processing (2012)

7. Wang, S.-J., et al.: Micro-expression recognition using color spaces. IEEE Trans. Image Process. **64**, 6034–6047 (2015)

8. Yang, J., Zhang, F., Chen, B., Khan, S.U.: Facial expression recognition based on facial action unit. In: Tenth International Green and Sustainable Computing Conference (IGSC) (2019)

9. Pang, L., Li, N., Zhao, L., Shi, W., Du, Y.: Facial expression recognition based on Gabor feature and neural network. In: International Conference on Security, Pattern Analysis (2018)

10. Anas, L.F., Ramadijanti, N., Basuk, A.: Implementation of facial expression recognition system for selecting fashion item based on like and dislike expression. In: International Electronics Symposium on Knowledge Creation and Intelligent Computing (IES-KCIC) (2018)

11. Nan, Z., Xue, G.: Facial expression recognition based on local facial regions. In: 4thIET International Conference on Wireless, Mobile and Multimedia Net-works (ICWMMN 2011) (2011)

12. Edwards, J., Jackson, H.J., Pattison, P.E.: Emotion recognition via facial expression and affective prosody in schizophrenia: a methodological review. Clin. Psychol. Rev. **22**(6), 789–832 (2002)

13. Chu, H.-C., William, Tsai, W.-J., Liao, M.-J., Chen, Y.M.: Facial emotion recognition with transition detection for students with high-functioning autism in adaptive e-learning. Soft Comput. **22**, 1–27 (2017)

14. Clavel, C., Vasilescu, I., Devillers, L., Richard, G., Ehrette, T.: Fear-type emotion recognition for future audio-based surveillance systems. Speech Commun. **50**(6), 487–503 (2008)

15. Saste, S.T., Jagdale, S.M.: Emotion recognition from speech using MFCC and DWT for security system. In: 2017 International Conference of Electronics, Communication and Aerospace Technology (ICECA), vol. 1, pp. 701–704. IEEE (2017)

16. Khorrami, P., Paine, T., Huang, T.: Do deep neural networks learn facial action units when doing expression recognition? In: Proceedings of the IEEE International Conference on Computer Vision Workshops (2015)

17. Kahou, S.E., et al.: Emonets: multi-modal deep learning approaches for emotion recognition in video. J. Multi-modal User Interfaces **10**(2), 99–111 (2016)

18. Krizhevsky, A., Sutskever, I., Hinton, G.E.: Imagenet classification with deep convolutional neural networks. In: Advances in Neural Information Processing Systems, pp. 1097–1105 (2012)

19. Zhang, J., Yin, Z., Cheng, P., Nichele, S.: Emotion recognition using multi-modal data and machine learning techniques: a tutorial and review. Inf. Fusion. **59**, 103–126 (2020)
20. Wilson, P.I., Fernandez, J.: Facial feature detection using HAAR classifiers J. Comput. Small Coll. ročník 21 č., **4**, 127–133 (2006). ISSN 1937-4771
21. Zhao, G., Pietikäinen, M.: Dynamic texture recognition using volume local binary patterns. In: Vidal, R., Heyden, A., Ma, Y. (eds.) WDV 2005-2006. LNCS, vol. 4358, pp. 165–177. Springer, Heidelberg (2007). https://doi.org/10.1007/978-3-540-70932-9_13
22. Das, P.K., Behera, H.S., Pradhan, S.K., Tripathy, H.K., Jena, P.K.: A modified real time A* algorithm and its performance analysis for improved path planning of mobile robot. In: Jain, L.C., Behera, H.S., Mandal, J.K., Mohapatra, D.P. (eds.) Computational Intelligence in Data Mining - Volume 2. SIST, vol. 32, pp. 221–234. Springer, New Delhi (2015). https://doi.org/10.1007/978-81-322-2208-8_21
23. Sally, J.D., Paul, S.: Chapter 3: Pythagorean triples. Roots Res. Vertical Dev. Math. Probl. Am. Math. Soc. Bookstore. 63 (2007). ISBN 0821844032
24. Ouyang, W., et al.: Deepid-net: deformable deep convolutional neural networks for object detection. In: Proceedings Conference on Computer Vision and Pattern Recognition, pp. 2403–2412 (2015)
25. Tang, X.: Image super-resolution using deep convolutional networks. IEEE Trans. Pattern Anal. Mach. Intell. **38**(2), 295–307 (2016). https://doi.org/10.1109/tpami.2015.2439281

Intelligent Mobility: A Proposal for Modeling Traffic Lights Using Fuzzy Logic and IoT for Smart Cities

Gabriel Gomes de Oliveira[1]([✉])(iD), Yuzo Iano[1](iD), Gabriel Caumo Vaz[1](iD),
Pablo David Minango Negrete[1](iD), Juan Carlos Minango Negrete[2](iD),
and Euclides Lourenço Chuma[1](iD)

[1] School of Electrical and Computer Engineering, State University of Campinas,
Campinas, Brazil
{oliveiragomesgabriel,euclides.chuma}@ieee.org, yuzo@unicamp.br
[2] Instituto Tecnológico Universitario Rumiñahui, Sangolqui, Ecuador
juancarlos.minango@ister.edu.ec

Abstract. One of the greatest challenges for a traffic control system is
to synchronize the flow of vehicles to prevent traffic jams. This issue gets
worse when there are priority vehicles, such as ambulances, trying to
move through the traffic. Given the current situation, with the COVID-
19 pandemic, and the trends of smart cities, in this work, we propose
and simulate a traffic control system that prioritizes ambulances within
large urban centers, using Fuzzy logic and IoT devices. The simulation
of our proposed model was performed on the software Dojot, which is an
open platform for IoT modeling. It addressed a real situation, in a path
that is usually used by ambulances to get to a reference hospital in the
city of Campinas, Brazil. The proposed traffic control system can also
be used after the COVID-19 pandemic is over in order to improve traffic
flow for other priority vehicles (e.g., firefighters and police) and increase
people's life quality within smart cities.

Keywords: Fuzzy logic · Traffic control · Smart cities · IoT · MQTT ·
Dojot

1 Introduction

In recent years, cities around the world have been expanding, what has led to
an increase in the size of the population living in urban areas. By 2019, around
54.5% of the world's population resided in urban areas [1]. The United Nations
(UN) has reported that probably this number will increase to 66% by 2050 [2].
Thus, governments need to explore and analyze possible solutions to improve
city traffic jams, waste generation, and contamination.

Traffic jams are one of the major problems in big cities, given the daily
increase in the number of vehicles. In this sense, likewise, congestion increases

K. K. Patel et al. (Eds.): icSoftComp 2021, CCIS 1572, pp. 302–311, 2022.
https://doi.org/10.1007/978-3-031-05767-0_24

fuel consumption, delays the mobilization between different places inside the cities, which expend the energy of the population. Methods to manage traffic have been proposed in order to make traffic dynamic as well as vehicle speed and junction's specifics more efficient [3].

Thus, the concept of smart cities has emerged as a solution to maximize Quality of Life (QoL) by offering services that integrate information and communications technologies [4] such as the Internet of Things (IoT) and fifth generation (5G) networks. Intelligence mobility is a communication technology and advanced information used to improve transportation by achieving enhanced mobility and safety for the urban population. Furthermore, smart cities try to reduce the environmental impact produced by CO_2 vehicle emissions [5].

In light of the sanitary emergency triggered by the Coronavirus pandemic (COVID-19), prioritizing emergency vehicles to respond to different kinds of emergencies in the least possible time has increased. During the pandemic scenario in Brazil, there were not rigid lockdowns decreed in the country on the whole. For that reason, the workgroup proposes this article. In our proposed method, we have selected the Pontifical University Campinas (PUC) Hospital located beside John Boyd Dunlop Avenue, in Campinas City - Brazil. The choice of PUC-Hospital is due to the fact that it presents only one street access with approximately thirteen traffic lights during the path from Anhanguera highway to the PUC-Hospital. For this reason, our Modeling Traffic Sings proposes and evaluates a priority method for ambulances, police, and fire brigade based on a Fuzzy logic. To the best of our knowledge, this is the first time that the Dojot platform has been employed in transportation.

The remainder of this paper is structured as follows. Section 2 features a brief explanation on the main concepts about smart cities and intelligent transportation systems. In Sect. 3, our proposal of Modeling Traffic Sings is discussed. The simulation results and discussion are available in Sect. 4. Finally, in Sect. 5 the conclusions are presented.

2   Background

In this section, we will present basic concepts, which guide this research, to explain clearly and sharply its importance in a national and international scope, when we deal with Smart Cities.

2.1   Smart Cities

Although there is not just one accepted definition, the contemporary understanding of a smart city [6,7] takes up a coherent urban development strategy managed by the city's government, which tries to plan and align, in the long-term, the management of the infrastructure assets and services of the city aiming to improve the citizen's life quality [8,9]. For this reason, over the XX century, the idea that a city could be intelligent was a scientific fiction depicted in the popular media but, with the massive proliferation of new technologies and smart devices, the perspective that a city can become smart and even sentient is quickly coming true [10]. The convergence of information and communication technologies is creating urban environments that are different from everything that has already been tried until now. When cities become smart, it is possible to automate functions that serve people, buildings, and traffic systems, and also monitor, understand, analyze, and plan the city to improve the efficiency, equity, and citizen life quality in real-time [10]. It is changing the way people plan on many time scales, that is, there is a gain in perspective that cities can become smarter in the long-term through thinking and actions in the short-term. Smart cities are frequently depicted as constellations of instruments connected through many networks that provide continuous data about the movement of people and materials in function of the flow of decisions about the physical and social shape of the city [10].

2.2 Intelligent Transportation Systems

For the reasons presented above, the Intelligent Transportation System (ITS) concept was created in the beginning of the XXI century and applied worldwide in all kinds of transportation systems (ground, water, and air) [11]. The ITS has been receiving increasing attention in the academic and industrial areas [12].

The ITS integrates information and technologies of data communication and applies them in the transportation field to develop an integrated system of people, roads and vehicles. It can establish a transportation management system that is fully functional, accurate, and efficient [13,14].

Intelligent Transportation Systems work with information and control technologies that are the core of their functions [15].

2.3 Fuzzy Logic

Fuzzy logic is a concept that arose in the 1960s and is applied in many fields, from engineering to artificial intelligence. It consists of approximately modeling a problem, instead of using a precise approach, as it occurs in human reasoning in many situations [16–20].

When the Fuzzy logic is applied, it is interesting to compare it with a Likert scale because, different from binary research, the results may vary according to the degree of agreement when a person answers a given question. In other words, whereas a binary logic answer can assume just two values (0 or 1), in Fuzzy logic, it is possible to achieve an answer that is between them and work with uncertain pieces of information that are usually used in the human daily routine, such as answering a survey with "maybe", "more or less", or "sometimes", for example [16–20].

Figure 1 illustrates how the Fuzzy logic can be applied in an intelligent transportation application.

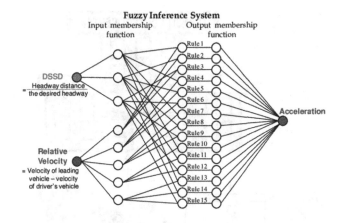

Fig. 1. Application of fuzzy logic in intelligent transport system.

3   Theoretical Proposal

COVID-19 accelerated the need for solutions in large urban centers to use the concepts of smart cities. In this work's proposal, we present a practical solution to prioritize vehicles such as ambulances, police, and fire brigade in heavy traffic using Internet of Things (IoT) devices and Fuzzy logic. Figure 2 graphically presents the proposed solution.

The ambulance, police or fire brigade vehicles (known as priority vehicles) calls support to Transit Control Center (TCC) informing their destination using a GPS IoT device. During the call, the TCC gets the current position of the priority vehicle as well as the traffic information (rush time, accidents, etc.) using IoT traffic cameras.

The Transit Control Center calculates the best route and sends it to the priority vehicle. The TCC calculates the IoT traffic lights time switching to accelerate the transit flow and increase the average velocity of the priority vehicles.

IoT traffic lights are switched by a command received from Transit Control Center, and the current position of the priority vehicle are constantly monitored.

The Transit Control Center uses a Fuzzy logic controller to determine whether to extend or terminate the current green phase after a minimum green time has been elapsed to accelerate the traffic flows and allow the priority vehicle to cross the traffic light.

If the green time is extended, the fuzzy logic controller will determine whether to extend the green phase after the time interval Δt. In this paper, the time interval Δt is taken as 5 s.

The number of approaching vehicles for each approach during a given time interval can be estimated using the detectors. Fuzzy variables considered arriving vehicles (A), queuing vehicles (Q) and extension (EXT) time of the green phase.

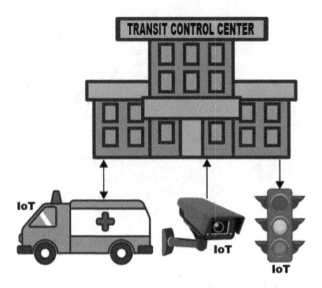

Fig. 2. Proposed solution.

4   Methodology

This section presents the metrics applied. For this purpose, several approaches were used to make an excellent article, of maximum clarity and scientific detail for the readers understanding and future application.

Scientific research was carried out in the main databases, and up-to-date references were selected. This research aims to scientifically demonstrate the importance of the theme worldwide and identify conventional metrics in this field.

The study starts from the principle that a large city, in this case, Campinas, can take advantage of a methodology for intelligent urban traffic focused on the optimization of priority services (e.g., ambulances) to ensure they can move around the city in a safer, faster, and more efficient way.

It is worth mentioning that this theme has gained much more relevance in recent years, due to the COVID-19 pandemic, which unfortunately affects the whole world.

Taking into account the factors presented previously, the research group searched for a hospital in the city of Campinas that best suits this research. The following metrics were applied in the selection:

- The hospital must currently treat cases of COVID-19.
- The hospital must assist both the Unified Health System (the Brazilian public health system) and the private health care plans. This requirement is important to maintain impartiality in the selection of the hospital.

The chosen hospital is located beside John Boyd Dunlop Avenue, which is one of the main avenues in Campinas and has many traffic lights, which makes

---

Algorithm 1. Fuzzy logic to each traffic light in the route

After a minimum green (5s)

```
0: INPUT
0: if (A) is True then
0:    (EXT) is zero
0: else
0:    if (A) is False AND (Q) is (many) then
0:       (EXT) is large.
0:    else
0:       if (A) is False AND (Q) is (medium) then
0:          (EXT) is medium.
0:       else
0:          if (A) is False AND (Q) is (small) then
0:             (EXT) is small.
0:          else
0:             if (A) is False AND (Q) is (few) then
0:                (EXT) is few.
0:             end if
0:                OUTPUT
```

---

it an interesting setting for the research. In Fig. 3, we can observe that there is just one route from point A (green point - Anhanguera Highway) to point B (red point - PUC Campinas Hospital), so both the population and the ambulances are forced to take this path to get to the hospital.

The workgroup used an IoT tool to bring an innovative idea and pave the way for readers and researchers who want to apply and move forward with this study.

The IoT tool applied in this work is Dojot [21], which is a platform that enables the development and demonstration of technologies applied in smart cities. It was designed and distributed by CPqD [22], a Brazilian center for research and development located in Campinas. Being a regional technology, it is widely used at our university.

The Dojot is an open-source platform, and all researchers and professionals who wish to develop works in the field of IoT can apply it in the development of the project. This characteristic opens the possibility for readers and researchers to use this free and easy-to-use platform, as well as this working group.

The Dojot platform works with digital representations of the physical devices, and, in each representation, there are assigned attributes that correspond to real variables. When all the virtual devices are created, it is possible to represent the whole system in the platform with real-time information. The events are handled with flows, which contain entry points (the trigger of the flow), processing blocks, and exit points (where the results should be forwarded to). The communication between the platform and the devices is made with Message Queuing Telemetry Transport (MQTT) protocol [23], which is a standard for IoT messaging.

Fig. 3. Studied route.

Initially, a member of the group went to the designated location and identified all the traffic lights, for both pedestrian crossing and crossroads, in the route from point A to point B described above. Then, with the use of Google Maps [24], the geographical coordinates of all those traffic lights were collected compiled. Table 1 shows all parameters inserted into the simulation environment.

Table 1. Parameters for the simulation environment

Point A	Access loop from Anhanguera highway to John Boyd Dunlop avenue
Point B	PUC ampinas hospital on John Boyd Dunlop avenue
Distance from A to B	2.5 km
Number of traffic lights (pedestrian crossing)	3
Number of traffic lights (crossroads)	5
Simulated traffic lights on The main and adjacent streets	15

In the Dojot platform, two templates were created: one for the traffic lights and one for a vehicle (in this case, an ambulance). Then all traffic lights were inserted into the map, with the collected geographical coordinates, and the studied route was mapped, as shown in Fig. 4, to indicate the only trajectory that the ambulance will use to assist the population in cases of urgency and emergency.

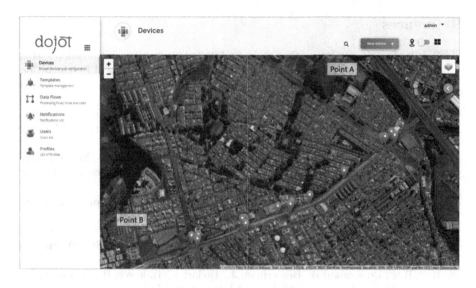

Fig. 4. Dojot map with the traffic lights and the start and end points.

The virtual traffic lights behave as output devices and are monitored in the platform itself. In contrast, the ambulance behaves as an input device and its geographical position is simulated with an application that sends some fixed values of coordinates within the route via MQTT protocol.

5   Conclusion

A new system for smart traffic lights with IoT and Fuzzy Logic concepts was proposed, aiming at a future application within the Dojot Platform (CPqD).

This article presents a literature review of the main concepts behind the proposed system and identifies a real situation in which it can be applied and validated.

One can note that, with this proposal, the traffic lights system of a city becomes more effective and safer, allowing the ambulances to arrive faster at its destination and the patient to be treated earlier, improving his healing expectation. It is worth mentioning that this subject area has been even more important in light of the global pandemic (COVID-19), a situation in which thousands of people are infected and lose their lives daily.

The proposed system is still capable of many applications besides those tested for this article, as the possibility of establishing a priority ranking to enable other essential services (e.g., police and fire brigade) within a smart city.

References

1. Hettikankanama, H., Vasanthapriyan, S.: Integrating smart transportation system for a proposed smart city: a mapping study. In: 2019 International Research Conference on Smart Computing and Systems Engineering (SCSE), pp. 196–203. IEEE (2019)
2. Derawi, M., Dalveren, Y., Cheikh, F.A.: Internet-of-things-based smart transportation systems for safer roads. In: 2020 IEEE 6th World Forum on Internet of Things (WF-IoT), pp 1–4. IEEE (2020)
3. Kamdar, A., Shah, J.: Smart traffic system using traffic ow models. In: 2021 International Conference on Artificial Intelligence and Smart Systems (ICAIS), pp. 465–471. IEEE (2021)
4. Al-Shariff, S.M., Alam, M.S., Ahmad, Z., Ahmad, F.: Smart transportation system: mobility solution for smart cities (2019)
5. Přibyl, O.: Transportation, intelligent or smart? On the usage of entropy as an objective function. In: 2015 Smart Cities Symposium Prague (SCSP), pp. 1–5. IEEE (2015)
6. Hall, R.E., Bowerman, B., Braverman, J., Taylor, J., Todosow, H., Wimmersperg, U.: The vision of a smart city. In: 2nd International Life Extention Technologies Work 7 (2000)
7. Hollands, R.G.: Will the real smart city please stand up? Intelligent, progressive or entrepreneurial? City **12**(3), 303–320 (2008)
8. Galdon-Clavell, G.: (Not so) smart cities?: the drivers, impact and risks of surveillance-enabled smart environments. Sci. Public Policy **40**(6), 717–723 (2013). https://doi.org/10.1093/scipol/sct070
9. Schaffers, H., et al.: Integrating Living Labs with Future Internet experimental platforms for co-creating services within Smart Cities. In: 2011 17th International Conference on Concurrent Enterprising, pp. 1–11 (2011)
10. Batty, M.: Smart cities of the future. Eur. Phys. J. Spec. Top. **214**(1), 481–518 (2012). https://doi.org/10.1140/epjst/e2012-01703-3
11. Stoate, C., et al.: Ecological impacts of early 21st century agricultural change in Europe-a review. J. Environ. Manage. **91**(1), 22–46 (2009)
12. Moisio, L., Hillmansen, S.: Intelligent transport system-RRUKA special issue. IET Intell. Transp. Syst. **10**(1), 1 (2016)
13. De Brucker, K., Macharis, C., Verbeke, A.: Two-stage multi-criteria analysis and the future of intelligent transport systems-based safety innovation projects. IET Intell. Transp. Syst. **9**(9), 842–850 (2015). https://doi.org/doi.org/10.1049/iet-its.2014.0247
14. Hou, Z., Zhou, Y., Du, R.: Special issue on intelligent transportation systems, big data and intelligent technology. Transp. Plan Technol. **39**(8), 747–750 (2016). https://doi.org/10.1080/03081060.2016.1231893
15. Jarašūniene, A.: Research into intelligent transport systems (ITS) technologies and efficiency. Transport **22**(2), 61–67 (2007). https://doi.org/10.1080/16484142.2007.9638100
16. Sato, T., Akamatsu, M.: Understanding driver car-following behavior using a fuzzy logic car-following model. In: Fuzzy Logic-Algorithms Techniques and Implementations, pp. 265–282 (2012)
17. Baker, A.: Simplicity: The Stanford Encyclopedia of Philosophy: Metaphysics Research Lab (2016)

18. Combinando modelos de Machine Learning com Lógica Fuzzy-Parte 1. In: Medium (2021). https://medium.com/creditas-tech/combinando-modelos-de-machine-learning-com-l%C3%B3gica-fuzzy-parte-1-b5a9f0761a5d. Accessed 11 Aug 2021
19. Singh, P.: FQTSFM: A fuzzy-quantum time series forecasting model. Inf. Sci. (Ny) **566**, 57–79 (2021). https://doi.org/10.1016/j.ins.2021.02.024
20. Singh, P., Bose, S.S.: Ambiguous D-means fusion clustering algorithm based on ambiguous set theory: special application in clustering of CT scan images of COVID-19. Knowl. Based Syst. **231**, 107432 (2021). https://doi.org/10.1016/j.knosys.2021.107432
21. dojot Soluções para IOT - Plataforma de Desenvolvimento para IOT. In: Dojot - Desenvolvimento de Sistemas para IOT (2021). https://dojot.com.br/. Accessed 11 Aug 2021
22. Cpqd.com.br (2021). https://www.cpqd.com.br/. Accessed 11 Aug 2021
23. MQTT - The Standard for IoT Messaging. In: Mqtt.org (2021). https://mqtt.org/. Accessed 11 Aug 2021
24. Google.com (2021). https://www.google.com/maps/@-0.0457148,-78.4725815,15z. Accessed 11 Aug 2021

Crop Disease Prediction Using Multiple Linear Regression Modelling

Hudaa Neetoo⬤, Yasser Chuttur$^{(\boxtimes)}$⬤, Azina Nazurally⬤, Sandhya Takooree,
and Nooreen Mamode Ally

University of Mauritius, Reduit 80837, Mauritius
{s.neetoo,y.chuttur}@uom.ac.mu, {azina.nazurally,
sandhya.takooree1,nooreen.mamode1}@umail.uom.ac.mu

Abstract. Agriculture is a key player in the economic growth and sustainability of Small Island Developing States like Mauritius. However, during the past decade, climatic variations in Mauritius have caused economically important crops such as onion, potato, and tomato, to become more vulnerable to diseases, representing a severe threat to food security. Diseases caused by fungal microorganisms are expected to increase since climatic factors such as temperature, humidity, rainfall, and radiation directly affect fungal infection, growth and spread. Thus, mechanisms for efficient crop disease prediction based on climatic data are more than ever needed. Several applications exist for climatic data prediction but none of them have been adapted for Mauritius. In this paper, we have recorded climatic variables and collected crop disease incidence data from regions in different agro-climatic zones of Mauritius over a period of two years. We have subsequently developed an android mobile application implemented in Mauritian Kreol which applies Multiple Linear Regression modelling to the collected data for crop disease prediction. The application also acts as an "information window" which provides planters with a catalogue of diseases to aid diagnosis, cultivation tips, and appropriate treatments. Our findings indicate that the models developed were able to fit the collected data with fair R-squared values ranging from 0.14–0.28. We anticipate that our application can help farmers in Mauritius and in other similar countries to better forecast incidents of crop diseases and take remedial actions accordingly.

Keyword: m-Agriculture · Prediction · Regression · Plant diseases · Potatoes · Tomatoes · Onions

1 Introduction

Mauritius, like most Small Island Developing States (SIDS), is thought to be particularly vulnerable to the effects of global warming due to climate change, given its physical size and geographical isolation [1]. This is further compounded by its proneness to natural disasters and extreme weather events, its reliance on imports and its low adaptive capacity [1]. According to the Mauritius Meteorological Services [2], over the last 100 years (1906–2005), global surface temperature has increased by 0.74 °C. This amount albeit seemingly insignificant represents a globally averaged value. In fact, [3] have reported

© Springer Nature Switzerland AG 2022
K. K. Patel et al. (Eds.): icSoftComp 2021, CCIS 1572, pp. 312–326, 2022.
https://doi.org/10.1007/978-3-031-05767-0_25

that the climate change impacts on the agricultural sector of Mauritius will include amongst others (i) changes in soil moisture status, (ii) shifts in agricultural zones from lower to higher altitudes, (iii) changes in cropping pattern and crop cycle, (iv) heat stresses, which will lead to lower crop productivity, and (v) increased incidence of agricultural pests and diseases. [4] mentioned that diseases, especially those caused by fungal microorganisms, will likely increase thereby having a profound impact on food security [5].

The three most economically important non-sugar food crops for Mauritius are onions, tomatoes, and potatoes. Potato is considered as a strategic commodity by the Mauritian government [6]. It is an important component of the Mauritian diet. The most common potato varieties in Mauritius are Spunta and Delaware. Potatoes are a widely consumed and is a nutritional powerhouse in terms of vitamin C, potassium, fiber, vitamin B6 and iron. It is a complex carbohydrate and contains several antioxidants. Potato cultivation is carried out by small, medium, and large-scale growers and they produce both ware and seed potatoes.

Tomatoes, and onions represent are also widely consumed vegetables in Mauritius after potatoes. Indeed, the value of the tomato industry is estimated to be around Rs 300 M with an annual production of 14 700 t over an area of 935 ha and at a market price ranging from Rs 13.00 to 105.00/kg [8]. The average per capita consumption of fresh tomato is around 12 kg/year. The onion industry is also prominent with an annual value of Rs 126 M and the crop is grown over ca. 300 ha with an average production of about 9000 t. The annual per capita consumption of onion in Mauritius was estimated at 12–13 kg between 2012 and 2013.

Climatic factors can greatly exacerbate fungal diseases of strategic crops of Mauritius [7, 9] resulting in significant yield losses and causing the agricultural sector to incur great economic loss. Since food crop production in Mauritius is dominated by small-scale farming, this will negatively affect food production, food prices and farmers' livelihoods. Hence, research is warranted to ascertain the climatic impacts on local agriculture so that "climate-proofing" measures can be taken in a timely manner.

To the best of our knowledge, no studies have been carried out till now in Mauritius to investigate the relationship between key climatic variables (air temperature, relative humidity, rainfall and windspeed) and fungal disease incidence affecting onion, tomato, and potato cultivation. Moreover, currently local planters do not avail of any tailored disease-forecasting tool to assist them in decision-making and in mounting appropriate preparedness plans in the event of adverse weather conditions. With the increasing use of smartphones among planters, the proposed system will also provide a corresponding mobile application, which can further help in alerting farmers of fungal disease outbreaks to allow farmers and concerned authorities to adopt any preparedness plans. Currently, planters subscribe to a *SMS Alert System* devised which usually issues within less than 48 h of an outbreak.

The objectives of this research were therefore two-fold: (i) to determine the relationship between climatic factors and incidence of key fungal diseases affecting onion, tomato, and potato crops in Mauritius and (ii) to develop a disease-forecasting tool for outbreaks of fungal diseases.

2 Related Work

2.1 Fungal Diseases

Fungal diseases are directly affected by climatological or climatic factors [9]. In the open field, climatological factors are generally understood to mean moisture, light, temperature, and wind. These factors may either independently affect the host and parasite, or they may affect the interrelations of these organisms [9].

Many fungal diseases of crops are directly associated with abundant precipitation, or a humid atmosphere. For instance, brown rot of stone fruits, black rot of grapes and late blight of potatoes are diseases that occur in moist weather. Moisture generally augments the production of spores and enhances their germination. It may also promote the susceptibility of the host to attack [9].

Table 1. Common fungal pathogens for Mauritius

Potato		Onion		Tomato	
Pathogen	Disease	Pathogen	Disease	Pathogen	Disease
Alternaria solani	Early blight, Alternaria rot, Tuber rot	*Alternaria porri*	Purple blotch	*Alternaria alternata Alternaria spp**	Black shoulder
Fusarium spp Fusarium solani	Fusarium wilt Fusarium dry rot	*Botrytis spp*	Neck rot, Botrytis leaf blight	*Penicillium spp*	Blue mold rot
Phytophthora infestans	Late blight	*Macrophomina phaseolina*	Charcoal rot	*Phytophthora infestans*	Late blight
Spongospora subterranea	Powdery scab	*Macrophomina phaseolina*	Charcoal rot	*Verticillium albo-atrum*	Verticillium wilt
Verticillium albo-atrum	Verticillium wilt	*Penicillium spp*	Blue mold rot		
		Phoma terrestris	Pink root disease		
		Stemphylium vesicarium	Stemphylium blight		

Fungi constitute the largest number of plant pathogens and are responsible for a range of serious plant diseases. In fact, diseases affecting onion, tomato, and potato crops, are primarily caused by fungal pathogens. Sources of fungal infections include infected seed, soil, crop debris, nearby crops, and weeds [10]. While some fungal pathogens such as *Alternaria*, *Fusarium* and *Phytophthora* have a broad host spectrum affecting multiple vegetables, others tend to be more host-specific such as *Phoma* and *Botrytis* [11]. Table 1 summarizes the most common fungal pathogens that are the greatest concern for the potato, tomato, and onion industry in Mauritius. These diseases may occur either at the pre-harvest stage or post (i.e., during storage). Of these fungi, only a few are responsible for mycotoxin production on specific crops.

2.1.1 Onions

Major pathogens of onions include *Botrytis* spp. [12], *Penicillium* spp. [13] and *Aspergillus niger* [14, 15]. The primary symptom of infection by A. niger is a black discoloration of tissue. Infected bulbs may show blackening at the neck, black streaks, or spots on or beneath the outer scales, and black discoloration in injured areas. Other species frequently encountered on onions are A. *flavus*, A. *alliaceus* and A. *fumigatus* [13]. Onion mildew (*Macrosporium sarcinula*) is another well- known disease of onions characterized by a "furry violet appearance" of the affected leaves, which subsequently become moldy in character, pale, collapsed and broken.

Another common disease is Stemphylium leaf blight of onion crops (*Stemphylium vesicarium*) which results in small, light yellow to brown, water-soaked lesions on the leaves. These small lesions grow into elongated spots that frequently coalesce resulting in blighted leaves. Pink root diseases of onions caused by *Phoma terrestris* is also common in Mauritius and result in pink discoloration of roots that eventually turn purple. Plants become visibly stunted with occurrence of leaf dieback.

2.1.2 Tomatoes

Tomato plants are susceptible to different plant pathogenic *Alternaria* species [16]. Tomatoes, given their soft skin, are particularly susceptible to invasion with Alternaria alternata [17–19] since the organism enters through injured or weakened tissues for penetration and growth. In the presence of water on the ripening fruit from rain, dew or overhead irrigation, fungal spores can germinate on the surface of the fruit [20]. In addition to *Alternaria* species, *Penicillium tularense*, P. *olsonii* and P. *expansum* have been reported to infect healthy tomatoes [19]. Tomatoes and potatoes are both members of the Solanaceae family and therefore, tomatoes are equally susceptible to Late Blight and Verticillium wilt.

2.1.3 Potatoes

Early blight (*Alternaria solani*) and late blight (*Phytophthora infestans*) are probably the most important fungal foliage diseases of potatoes. Both are dependent on weather conditions for sporulation and spread and the amount of yield loss they cause depends on how early and quickly they destroy foliage. Late blight can attack and destroy foliage of developing plants at any stage of growth given suitable weather. Fusarium wilt (*Fusarium* spp) is a common vascular wilt disease affecting potato crops, exhibiting symptoms similar to Verticillium wilt caused by *Verticillium albo-atrum*. Fusarium dry rot is a very common disease of potato tubers resulting in a variety of colored rots. Powdery scab (*Spongospora subterranea*) is another common disease of potato tubers. Symptoms of powdery scab include small lesions in the early stages of the disease, progressing to raised pustules containing a powdery mass.

2.2 Crop Disease Prediction

Crop disease prediction, unlike disease detection and recognition, has not been widely studied. Most disease predictive models in the past few decades have been based on statistical approaches and more recently, predictive models using Artificial Intelligence were developed. Regression analysis including the simple regression, multi linear regression, and multivariate regression are classified under the statistical approaches adopted in crop disease prediction. Moreover, machine learning algorithms such as the Support Vector Machine (SVM) and the Naïve Bayes, and state of the art deep learning algorithms are widely adopted for disease modelling [21, 22]. With the huge amount of data being made available from sensors among others, artificial intelligence algorithms have achieved great outcomes. In addition, it has been found that supervised machine learning and deep learning-based approaches perform best with time series data, compared with statistical approaches [23].

To predict the presence or severity of diseases, parameters for example visual cues or climate data are considered. Visual cues include pictures of affected areas and are often captured using cameras or even satellite imagery [26, 27, 30]. Climate data on the other hand, can be categorized under time series data. Climatic variables that are the most studied are temperature, relative humidity, precipitation or rainfall, soil topography, leaf wetness, wind speed and solar radiation [24–30]. Coakley [24] developed a multivariate regression analysis model to predict the severity of rust in wheat based on meteorological factors. Diseased wheat samples were collected from 1968 to 1986 at the different growth rates: milk and dough stages. Climate data collected included daily high and low temperature values as well as precipitation. The meteorological data were averaged based on time periods and disease prevalence during these periods were then examined. They used the climate data which has the highest correlation value to build their resulting prediction model. The latter had an adjusted R^2 value greater than 0.75. The authors underlined that the best predictive models are obtained with a minimum of 10 years of data, as this number gives a reasonable range of values for the meteorological factors to be used in model development.

Later, [25] developed a model to predict the presence of Deoxynivalenol (DON) in wheat based on meteorological variables (daily rainfall, temperature, relative humidity, and leaf wetness) using ordinal regression models. They collected data from 18 different locations in Belgium during years 2002 to 2011. The devised model yielded an R^2 of 0.554. Their proposed method made it possible to predict whether certain severity thresholds will be exceeded as time goes by.

In [26], the severity of the Late Blight disease in potatoes was assessed using a neural network. Images with complex background from various farms were included in the dataset, thus there was a need for heavy image pre-processing prior to training and validation of the prediction model. The dataset was divided into three categories: training (238 images), validation and testing phase (51 images each). Their proposed approach obtained an accuracy of 93%. In their work, [21] also experimented with deep learning and came up with a novel approach to predict yellow rust in wheat. Their introduced framework was based on an ensemble and spatio-temporal recurrent neural networks. Data such as bioclimatic, remote sensing, topographic and soil variables were collected and used to predict the presence of yellow rust. The data was trained and validated using a 5-fold cross validation. Their approach obtained a RMSE of 0.1818.

Like [26, 27] used a multilayer perceptron and a CNN to predict the severity of Late Blight in potatoes. Their dataset consisted of images of potato farms in Ventaquemada which were collected using unmanned aerial vehicles (drones) flying at an altitude of 30 m above ground level. The CNN achieved the best results. They deduced that remote sensing approaches to collecting data could predict diseases after planting phase. Fenu and Malloci [28] devised an SVM to predict Late Blight in potatoes based on historical weather data including temperature, humidity, rainfall, wind speed, solar radiation, and risk index. They recorded data over a period of 4 years. The collected data was then trained and validated using stratified cross validation on an SVM-based model. From their work, it could be inferred that rainfall had no impact on the devised predictive model. An accuracy of 78% was obtained when all the climate parameters were used.

Bhatia et al. [29] implemented a hybrid SVM along with logistic regression to predict the presence of powdery mildew in tomatoes. They modeled the disease on a dataset containing 244 records of data. Meteorological variables used to predict the disease include humidity, wind speed, temperature, global radiations, and leaf wetness. An accuracy of 87.2% was achieved. Isip et al. [30] also made use of satellite data, climatic variables, and onion disease records to predict the twister disease in onions. They made use of the Geographically Weighted Regression analysis. The resulting approach was 86% effective.

2.3 m-Agriculture

With the rise of technological advancements in the recent decades, smartphones have been an essential asset in a multitude of fields including agricultural, economic and financial activities. Mobile agriculture (m-Agriculture or m-Agri) is the use of mobile technologies to support agriculture [31]. m-Agriculture is a subclass of e-Agriculture where agricultural services are being provided using technology. Since agriculture is a central economic activity in various African countries, it is of paramount importance to ensure that farmers are provided with the adequate information and resources to assist them in producing quality products to ensure continued economic growth.

m-Agriculture is a game changer in smallholder agriculture [32] as projects and end products can be of varying complexities, ranging from informative applications to diagnosis and predictions. Moreover, m-Agriculture can be powered using numerous and cutting edge technologies, for example, the Global Positioning System (GPS), Radio-frequency Identification (RFID), Internet of Things (IoT), wireless technologies including Bluetooth and infrared, machine and deep learning, among others [33–36]. m-Agri applications are also often used to share knowledge on good farming practices and trends (m-Learning) as well as assisting farmers in taking the right decisions based on information being provided (m-Farming).

3 Methodology

This section introduces the steps in building a tool for prediction of diseases based on MLR models integrated in a mobile application along with a plant and disease knowledge base for the three strategic crops of Mauritius. Being a novel concept in Mauritius, the development of such an application entailed various important processes: (i) collection of meteorological data, (ii) computing of Disease Incidence (DI) data for all three crops in fields located in different agro-climatic zones (iii) assess the level of adherence of planters to *Good Agricultural Practices* (GAP) through a visual assessment and an interview, (iv) devising predictive models, and (iv) setting up of a knowledge base in Mauritian Kreol.

3.1 Study Area

The study sites involved onion. Tomato and potato fields located in different agro-climatic zones (sub-humid, humid, and super-humid) of Mauritius.

3.2 Data Collection

Data collected at each visited site included current climatic data and Disease Incidence (DI) data for key diseases affecting these crops during the period of July 2018 to July 2020 inclusive. The agro climatic data was accessed on the Mauritius Meteorological Services[1] or timeanddate.com[2] websites. In addition, relative humidity and temperature values were also recorded *in situ* using a digital hydrometer and then cross checked with the above-mentioned websites to ensure that the recorded values were accurate. Climatic data recorded include: (i) high and low temperature values in °C, (ii) rainfall/precipitation in mm, and (iii) percentage relative humidity. DI was calculated as the percentage of infected crops in two random plots at each farm. To confirm disease diagnosis, crop specimens were also collected and brought to the laboratory for identification of the causative agent. Adherence of planters to *Good Agricultural Practices* (GAPs) was also determined by both a visual assessment of the fields and an interview with planters and a GAP score was assigned. GAP scores ranged between 0–4, where a score of 0 meant poor adherence while a score of 4 implied an excellent compliance. Field visits were conducted twice or more monthly.

3.3 Data Cleaning

To obtain satisfactory results, raw data collected were cleaned and pre-processed. Data collected was saved in excel files. All null values or missing data were replaced by appropriate data, duplicate entries were eliminated, and all numerical values were set to 3 significant figures for standardization. The resulting files were complete, consistent, relevant, and uniform.

3.4 Disease Modelling Approach

The Multiple Linear Regression (MLR) has been adopted to model the crop disease predictions in past studies [24, 25]. MLR is a statistical technique that is widely used in

---

[1] http://metservice.intnet.mu/.

[2] https://www.timeanddate.com/.

various fields, for example in finance, to predict the outcome of a response variable. It is believed that the response variable is directly related to a linear combination of the exploratory variables.

$$y_i + \beta_0 + \beta_1 x_{i1} + \beta_2 x_{i2} + \ldots + \beta_p x_{ip} + e \qquad (1)$$

where, for $i = n$ number of observations, y_i = dependent variable, x_i = explanatory variables, β_0 = y-intercept, β_p = slope coefficients for each explanatory variable, and e = error term.

In this work, we seek to predict the DI based on climatic data, where the response variable is the DI, and the predictor variables include GAP, Temperature, Rainfall and Humidity values. MLR, being a multivariate model, can easily assess many variables and therefore a decent disease incidence prediction could be obtained even with a limited dataset. On the other hand, machine learning and deep learning models often require huge amount of data and complex network architectures to yield good results. Since the datasets consisted of an average of 60 data entries which is quite small, it was deemed that data was not enough as input to machine learning or deep learning algorithms.

The general formula could be illustrated as:

$$DI = \beta_0 + \beta_1(Temperature) + \beta_2(Humidity) + \beta_3(Rainfall) + \beta_4(GAP) + e \qquad (2)$$

MLR models were implemented for the most severe disease for each crop i.e., late blight for potato, early blight for tomato and pink root rot for onions.

3.5 Knowledge Base

Background information about the three strategic crops as well as disease facts were also gathered. The database contained general information such as varieties of the crop, ideal climate conducive for growth, cultivation tips, as well as a database of diseases identified of relevance in the Mauritian context. Besides, for each disease, pictures of diseased specimens, name of causative agent and appropriate treatment information including chemical and biological were also compiled.

3.5.1 Translation to Mauritian Kreol

The compiled information catalogue was then translated to Mauritian Kreol by consulting the appropriate translation tools: Morisia 1.0[3], a website developed by a team at the University of Mauritius, English Morisyen Dictionary online[4], and Diksioner Morisien, which is a dictionary for the Mauritian creole language.

---

[3] http://www.translatekreol.mu/#source.

[4] https://glosbe.com/en/mfe.

3.6 Proposed Architecture

It is proposed that once the mobile application is launched, the user's current location will be determined using the device's geolocation. The latitude and longitude values will then be used to retrieve current temperature in °C, relative humidity, and precipitation in mm. The disease prediction models will then be used to predict DI based on current meteorological data (Fig. 1).

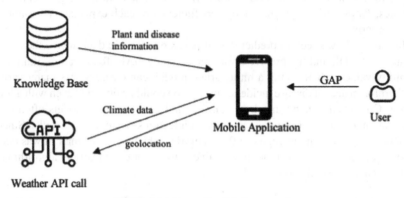

Fig. 1. Mobile application framework

4 Implementation and Discussion

4.1 Prediction Models

The collected data was stored in.csv files. In addition to *current* climatic data collected on the day of the visit, historical climatic data (temperature, rainfall, and relative humidity) for the past 14 days before the actual field visit (D-14 to D0) were retrieved where D0 referred to the day of visit and D-14 referred to the date 14-days back. This information was essential since the onset of diseases as evidenced by apparent symptoms on plants can occur several days after infection or after an episode of adverse weather conducive for the disease. Moreover, the incubation time for different fungal agents is known to vary. The headings in each.csv file are DI, High Temperature, Low Temperature, Average Temperature, Relative Humidity, Rainfall and GAP (Table 2).

For each dataset, MLR was performed for various time spans (e.g., Day of visit (D0), D-7 to D0, D-10 to D0, D14 to D-10) to determine which timeframe could better predict the prevalence of the disease. Weather variables having the highest correlation were chosen to build the resulting model. The MLR models were implemented in R language, one of the most used languages for statistical computing and analysis. The MLR models are obtained using the following R code snippet:

```
diseaseModel = lm(DI~GAP+Humidity+Rainfall+Temperature)
```

Table 2. Dataset headings and their meaning

Heading	Meaning
DI	Estimated disease prevalence on D0 (date on which field visit has been conducted)
GAP	Good Agricultural Practice
Rainfall	Rainfall in mm
Humidity	% Relative humidity
High temperature	Highest temperature recorded
Low temperature	Lowest temperature recorded
Temperature	An average of the high and low temperature values

4.1.1 Pink Root Rot (Onion) Predictive Model

The equation obtained for predicting incidence of Pink Root Rot in onions as a function of climatic variables was as follows (Table 3):

$$DI = 247.547 - 3.550\ (Temperature) - 1.266\ (Humidity) - 4.556\ (Rainfall) - 9.132(GAP)$$
(3)

Table 3. Regression statistics for onion model

Multiple R-Squared	0.1489
Adjusted R-Squared	0.007069
Standard Error	29.72

From the above equation, it can be inferred that the model explained 14.9% of the data although the relationship was not statistically significant ($P > 0.05$).

4.1.2 Early Blight (Tomato) Predictive Model

The equation obtained for predicting incidence of Early Blight in tomatoes as a function of climatic variables was as follows (Table 4):

$$DI = -49.792 - 1.177(Temperature) + 1.808(Humidity) + 7.183(Rainfall) - 13.833(GAP)$$
(4)

From the above equation, it can be inferred that the model explained 27.4% of the data although the relationship was not statistically significant ($P > 0.05$).

Table 4. Regression statistics for tomato model

Multiple R-Squared	0.27364
Adjusted R-Squared	0.19294
Standard error	27.12174

4.1.3   Late Blight (Potato) Predictive Model

The equation obtained for predicting incidence of Late Blight in potatoes as a function of climatic variables was as follows (Table 5):

$$DI = -48.997 - 7.734(Temperature) + 3.028(Humidity) - 17.590(Rainfall) - 9.794(GAP) \tag{5}$$

Table 5. Regression statistics for potato model

Multiple R-Squared	0.2713
Adjusted R-Squared	0.1925
Standard error	26.04

From the above equation, it can be inferred that 27.1% of the variation could be explained by the model although the relationship was not statistically significant ($P > 0.05$). Overall, it can be inferred that the R-squared values were relatively small (15–27%) for all devised regression models as compared to other models generated in the literature [24, 25]. The small R-squared values obtained could be because of the limited number of records in the dataset. Indeed, sample sizes are known to affect identification of important inputs for computer models. It is also important to underline that the data collection phase was performed during a timeframe of 2 years. [28] stressed that for a predictive model, particularly one which involves time series of climatic data and DI to be robust, a longitudinal study spanning over a minimum of 10 years should be conducted [24], with data collection at regular intervals of at least 1 h.

With regards to the tomato MLR model, the F-statistic was greater than the critical value ($F = 3.391$) with a p-value of 0.01877 and was thus considered significant at an α-level of 0.05. This indicates that the overall model was significant and that at least one of the predictor variables was significantly related to the response variable. Similarly, the F-statistic for the potato MLR model was greater than the critical value ($F = 3.444$) with a p-value of 0.01724, thus pointing to a statistically significant model. As for the onion model, the model was not significant ($P. > 0.05$) and this could be attributed to the very limited dataset with only 31 data entries. It is possible that the large variations in the dataset could have masked any possible relationship between the predictor and the response variables.

4.2   Mobile Application Development

The mobile application has been developed using Flutter[5], a flexible cross-platform and open-source framework powered by Google. The translated knowledge base was categorized to enhance data retrieval and was stored in JSON format. Real-time weather data was obtained from a weather application programming interface (API) named the OpenWeather One Call API[6]. Rainfall, humidity, and temperature values were then fed into the various devised prediction models to predict disease incidence. In addition, climatic data for the next five days was also retrieved by invoking the same API call. Screenshots of the mobile application are given in Figs. 2 and 3.

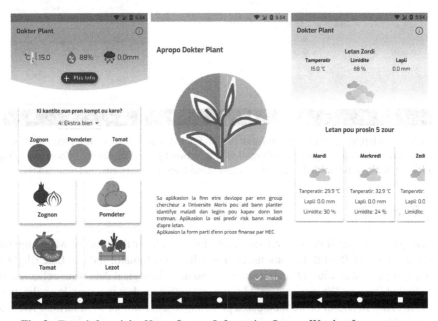

Fig. 2.   From left to right: Home Screen, Information Screen, Weather forecast screen

The application interface has been designed to show the following features: (i) current weather updates, (ii) weather forecast ("Plis info"), (iii) Basic information about the crops("Informasyon"), (iv) cultivation tips ("Bann Bon Pratik") and (v) a database of relevant diseases ("Maladi"). Moreover, the application is equipped with a drop-down button to allow the planter to input a GAP score of 0 - 4, based on his day-to-day practices, to be able to predict risks of these three major diseases in his field. The application also acts as a disease forecaster and depicts disease risks using colored icons, where red indicates an elevated risk (DI $\geq$ 50%), orange indicating a moderate risk (11 < DI < 49%) and a green icon indicating a low risk (DI $\leq$ 10%).

---

[5] https://flutter.dev/.

[6] https://openweathermap.org/api/one-call-api.

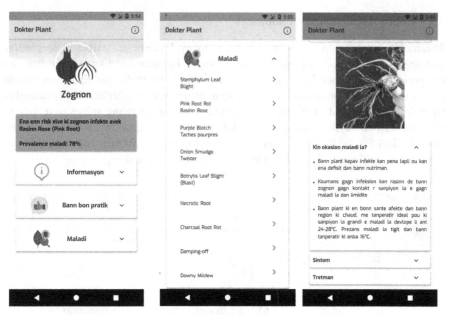

Fig. 3. From left to right: Information about onion (includes disease incidence prediction for the current day and location, general information, cultivation tips and diseases), Information about Pink Root Rot disease (includes causative agent identification, symptoms, and treatments)

5   Conclusion

In this paper, we put forward three different Multiple Linear Regression (MLR) models to predict (i) Pink Root Rot in onions, (ii) Early Blight in tomato, and (iii) Late Blight in potato crops in Mauritius. Diseased samples and meteorological data were collected for a period of 2 years. We also integrated these predictive models in a mobile application, Dokter Plant. In addition, we included a plant and disease library in Mauritian Kreol to educate Mauritian planters and assist them in timely and appropriate decision-making. The results of the MLR showed that meteorological variables do affect disease incidence of the three most strategic crops albeit with a relatively low goodness of fit. As future work, we intend to improve the performance of the predictive models by expanding our dataset through the collection of additional DI and weather data over the coming years, by applying data transformation techniques or by performing a non-linear regression. Moreover, we need to perform a user acceptance test to determine the extent by which the application is useful. Furthermore, additional predictive models will be introduced for other key diseases or other crops.

References

1. United Nations Conference on Trade and Development: World investment report 2014: Investing in the SDGs - an action plan. UN (United Nations Conference on Trade and Development (UNCTAD) World Investment Report (WIR)) (2014). https://doi.org/10.18356/3e74cde5-en

2. Mauritius Meteorological Services (2009). http://metservice.intnet.mu/. Accessed 6 Aug 2021
3. Deenapanray, P.N.K., Ramma, I.: Adaptations to climate change and climate variability in the agriculture sector in Mauritius. Impacts of Climate Change on Food Security in Small Island Developing States, pp. 130–165. IGI Global (2015). https://doi.org/10.4018/978-1-4666-6501-9.ch005
4. Van der Spiegel, M., van der Fels-Klerx, H.J., Marvin, H.J.P.: Effects of climate change on food safety hazards in the dairy production chain. Food Res. Int. **46**(1), 201–208 (2012). https://doi.org/10.1016/j.foodres.2011.12.011
5. Paterson, R.R.M., Sariah, M., Lima, N.: How will climate change affect oil palm fungal diseases? Crop Prot. **46**, 113–120 (2013). https://doi.org/10.1016/j.cropro.2012.12.023
6. "Statistics I Mauritius Chamber of Agriculture. https://chamber-of-agriculture.mu/statistics/. Accessed 13 Aug 2021
7. Sharma, M., Sharma, M.: Influence of environmental factors on the growth and sporulation of Geophilic Keratinophiles from soil samples of public parks. Asian J. Experimen. Sci. **23**(1), 307–312 (2009)
8. Ministry of Agro-Industry and Food Security (2017). http://agriculture.govmu.org/English/AboutUs/Pages/An-overview.aspx. Accessed 26 Nov 2017
9. Duggar, B.M.: Fungous Diseases of Plants, with Chapters on Physiology, Culture Methods and Technique. Ginn and Company, Boston (1909). https://doi.org/10.5962/bhl.title.44582
10. AUSVEG. https://ausveg.com.au/biosecurity-agrichemical/crop-protection/fungal-diseases/. Accessed 25 Nov 2017
11. Turkensteen, L.: An overview of the important fungal pathogens. Report of the Planning Conference: Control of Important Fungal Diseases of Potatoes. Lima Peru (1978)
12. Abdel-Sater, M.A., Eraky, S.A.: Mycopathologia **153**(1), 33–39 (2002). https://doi.org/10.1023/a:1015263932149
13. Overy, D.P., Seifert, K.A., Savard, M.E., Frisvad, J.C.: Spoilage fungi and their mycotoxins in commercially marketed chestnuts. Int. J. Food Microbiol. **88**(1), 69–77 (2003). https://doi.org/10.1016/s0168-1605(03)00086-2
14. Encinas-Basurto, D., Valenzuela-Quintanar, M.I., Sánchez-Estrada, A., Tiznado-Hernández, M.E., Rodríguez-Félix, A., Troncoso-Rojas, R.: Alterations in volatile metabolites profile of fresh tomatoes in response to Alternaria alternata (Fr.) Keissl. 1912 infection. Chilean J. Agric. Res. **77**(3), 194–201 (2017). https://doi.org/10.4067/S0718-58392017000300194
15. Hayden, N.J., Maude, R.B.: The effect of heat on the growth and recovery of Aspergillus spp. from the mycoflora of onion seeds. Plant. Pathol. **43**, 627–630 (1994)
16. Özer, N., Köycü, N.D., Chilosi, G., Magro, P.: Resistance to fusarium basal rot of onion in greenhouse and field and associated expression of antifungal compounds. Phytoparasitica **32**(4), 388–394 (2004). https://doi.org/10.1007/bf02979850
17. Hasan, H.: Alternaria toxins in black rot lesion on tomato fruits: condition and regulation of their production. Mycopathologia **130**, 171–177 (1995)
18. Da Motta, S., Valente Soares, L.M.: Survey of Brazilian tomato products for alternariol, alternariol monomethyl ether, tenuazoic acid and cyclopiazonic acid. Food Add. Contamin. **18**, 630–634 (2001)
19. Andersen, B., Frisvad, J.C.: Natural occurrence of fungi and fungal metabolites in moldy tomatoes. J. Agric. Food Chem. **52**(25), 7507–7513 (2004). https://doi.org/10.1021/jf048727k
20. Barkai-Golan, R.: Alternaria mycotoxins. Mycotoxins Fruits Veg. **185–203** (2008). https://doi.org/10.1016/b978-0-12-374126-4.00008-5
21. Xu, W., Wang, Q., Chen, R.: Spatio-temporal prediction of crop disease severity for agricultural emergency management based on recurrent neural networks. GeoInformatica **22**(2), 363–381 (2017). https://doi.org/10.1007/s10707-017-0314-1

22. Picon, A., Alvarez-Gila, A., Seitz, M., Ortiz-Barredo, A., Echazarra, J., Johannes, A.: Deep convolutional neural networks for mobile capture device-based crop disease classification in the wild. Comput. Electron. Agric. **161**, 280–290 (2019). https://doi.org/10.1016/j.compag. 2018.04.002

23. Khamparia, A., Saini, G., Gupta, D., Khanna, A., Tiwari, S., de Albuquerque, V.H.C.: Seasonal crops disease prediction and classification using deep convolutional encoder network. Circuits Syst. Sig. Process. **39**(2), 818–836 (2019). https://doi.org/10.1007/s00034-019-01041-0

24. Coakley, S.M.: Predicting stripe rust severity on winter wheat using an improved method for analyzing meteorological and rust data. Phytopathology **78**(5), 543 (1988). https://doi.org/10.1094/Phyto-78-543

25. Landschoot, S., Waegeman, W., Audenaert, K., Haesaert, G., Baets, B.D.: Ordinal regression models for predicting deoxynivalenol in winter wheat. Plant. Pathol. **62**(6), 1319–1329 (2013). https://doi.org/10.1111/ppa.12041

26. Biswas, S., Jagyasi, B., Singh, B.P., Lal, M.: Severity identification of Potato Late Blight disease from crop images captured under uncontrolled environment. In: 2014 IEEE Canada International Humanitarian Technology Conference - (IHTC), pp. 1–5 (2014). https://doi.org/10.1109/IHTC.2014.7147519

27. Duarte-Carvajalino, J., Alzate, D., Ramirez, A., Santa, J., Fajardo-Rojas, A., Soto-Suárez, M.: Evaluating late blight severity in potato crops using unmanned aerial vehicles and machine learning algorithms. Remote Sens. **10**, 1513 (2018). https://doi.org/10.3390/rs10101513

28. Fenu, G., Malloci, F.M.: Artificial intelligence technique in crop disease forecasting: a case study on potato late blight prediction. In: Czarnowski, I., Howlett, R.J., Jain, L.C. (eds.) IDT 2020. SIST, vol. 193, pp. 79–89. Springer, Singapore (2020). https://doi.org/10.1007/978-981-15-5925-9_7

29. Bhatia, A., Chug, A., Singh, A.P.: Hybrid SVM-LR classifier for powdery mildew disease prediction in tomato plant. In: 2020 7th International Conference on Signal Processing and Integrated Networks (SPIN), pp. 218–223. Noida, India (2020). https://doi.org/10.1109/SPIN48934.2020.9071202

30. Isip, M., Alberto, R., Biagtan, A.: Forecasting anthracnose-twister disease using weather based parameters: geographically weighted regression focus. Spat. Inf. Res. **29**(5), 727–736 (2021). https://doi.org/10.1007/s41324-021-00386-6

31. Gichamba, A., Lukandu, I.A.: A model for designing m-agriculture applications for dairy farming. Afric. J. Inf. Syst.**4**, 19 (2012)

32. Yusof, M.M., Rosli, N.F., Othman, M., Mohamed, R., Abdullah, M.H.A.: M-DCocoa: m-agriculture expert system for diagnosing cocoa plant diseases. In: Ghazali, R., Deris, M.M., Nawi, N.M., Abawajy, J.H. (eds.) SCDM 2018. AISC, vol. 700, pp. 363–371. Springer, Cham (2018). https://doi.org/10.1007/978-3-319-72550-5_35

33. Ajao, L.A.: A scheduling-based algorithm for low energy consumption in smart agriculture precision monitoring system. Agric. Eng. Int. CIGR J. **22**, 103–117 (2020)

34. Francis, M., Deisy, C.: Mathematical and visual understanding of a deep learning model towards m-agriculture for disease diagnosis. Arch. Comput. Meth. Eng. **28**(3), 1129–1145 (2020). https://doi.org/10.1007/s11831-020-09407-3

35. Prasad, S., Peddoju, S.K., Ghosh, D.: Energy efficient mobile vision system for plant leaf disease identification. In: 2014 IEEE Wireless Communications and Networking Conference (WCNC). Presented at the 2014 IEEE Wireless Communications and Networking Conference (WCNC), pp. 3314–3319. IEEE, Istanbul, Turkey (2014). https://doi.org/10.1109/WCNC.2014.6953083

36. Yuen, J.E., Hughes, G.: Bayesian analysis of plant disease prediction: Bayesian analysis of plant disease prediction. Plant Pathol. **51**, 407–412 (2002). https://doi.org/10.1046/j.0032-0862.2002.00741.x

Radial Basis Function Network Based Intelligent Scheme for Software Quality Prediction

Ritu[✉] and O. P. Sangwan

Department of CSE, Guru Jambheshwar University of Science and Technology,
Hisar 125001, Haryana, India
rituchopra1984@gmail.com

Abstract. With the increasing interest in digital world, various software whether it is in e-commerce or in the entertainment industry, have gained much attention in recent time. It becomes important to access the quality of a software which is usually done with the expert supervision and results into more money and time. Neural network systems have been effectively incorporated by famous companies such as Flip-Kart, Snapdeal, Netflix, and others due to their capacity to expand the amount of quantifiable attributes for prediction. This paper presents a framework of a radial basis function network for software quality prediction and uses Thin-plate spline RBF as its activation function. This is a network with a single layer making it easier to be trained with effective prediction results. The paper presents the theory of its training using gradient descent based back propagation algorithm with the training data-set of 20 samples and MATLAB results are provided to support the theoretical claim. Furthermore, the proposed scheme has been validated for five unknown software samples which quality is required to be predicted and It is demonstrated that the predicted soft quality results by the proposed approach are very close to the actual software quality which shows the effectiveness of the proposed approach.

Keywords: Neural networks · Software quality · Radial basis function network (RBFN) · Back-propagation · Gradient descent algorithm

1 Introduction

Artificial neural networks (ANN) and learning systems has exploded in popularity during the last decade. From academia to industry, various studies are being proposed to model the behaviour of a system where researchers have to deal with data available and thus ANN study becomes crucial when interacting with systems with highly uncertain dynamics and/or several constraint parameters to build a model closely matches with the system's actual behaviour. Various ANN frameworks are developed in literature e.g. Multilayered Neural Network,

© Springer Nature Switzerland AG 2022
K. K. Patel et al. (Eds.): icSoftComp 2021, CCIS 1572, pp. 327–340, 2022.
https://doi.org/10.1007/978-3-031-05767-0_26

Recurrent Neural Network, KSOM, Convolutional Neural Networks etc. on the basis of the targeted application and availability of computation resources. MLNs have been widely used in industry, stock-market [1], weather prediction [2] etc. Recently, various studies has been conducted even for the prediction of COVID-19 cases [3] with the statistical analysis via MLNs.

A simplest form of ANN is a radial-basis function network (RBFN) which has excellent dynamic approximation capability. It is computationally inexpensive also due to having only one hidden layer and linear in output weights [4]. Various studies have been conducted with RBFN in various area, to name a few:

- Authors in [5] use a clustering based RBFN to detect slow-varying signal during data transmission in a digital communication channel. The clustering algorithm along with the RBFNs the provide an automatic compensation for equipment distortion and nonlinear channels.
- Authors in [6] used RBFN to perform fault diagnosis. The authors state that ability of RBFN to produce robust decision surfaces and to also provide an estimate of how close a test case is to the original training data help the RBFNs to present the most plausible classification.
- In [7], authors develop conditions for the approximation of functions in certain general spaces using RBFNs. The authors considerably expand the idea of using RBFNs are universal approximators.
- Authors in [8] investigate the use of RBFNs in short-term system load forecasting. The authors present the accuracy of the RBFNs using the forecasting results of daily peak and total loads, for a period of 1 year, from a large-scale power system.

ANN has been also widely used recently in the field of computer software characteristics modelling and prediction certain features. A Convolutional Neural Network based prediction scheme is proposed in [9] by utilizing deep neural network architecture which is a computationally extensive approach. A cost effective solution has been provided in [10] for software defect prediction by utilizing honey bee swarms algorithm to optimize the ANN weights. [12] describes a reliability model of software by considering the case of software failure. Software reliability prediction framework has been discussed in [11] for real-time data-set. [13] presents object-oriented system based reliability prediction while discussing the limitations on the system for correctly prediction.

[14] developed a methodology to predict which class in a java application will be broken in future releases. The architecture is subsequently confirmed with good accuracy on a future release of the same application. [15] developed a discriminant model and a neural network model of a major telecommunications system, categorising modules as fault-prone or not. The authors also presented a comparison study of their suggested work to the non-parametric discriminant model, finding that the neural network model had superior predictive accuracy. Typically, the literature recommends using historical data to use parametric effort estimation methodologies.

Authors in [17] discuss a new approach for quality assessment of eggs in a poultry farm. A prediction algorithm is developed using RBFN to predict the haugh and air cell height along with PCA for accuracy enhancement.

In [16], the authors recommend using principal component analysis to improve the performance of neural networks. Using software metrics, the authors get raw data from a big commercial software. Two neural nets, one with raw data and the other with PCA data, were trained. When the two nets are compared, it is determined that the data after PCA generates more informative results. The authors of [18] present a study that is comparable to this one. The authors then come to the conclusion that, when compared to statistical methods, neural network models are better at predicting software quality. In [19], neural networks are used to measure service quality effectively. The authors conclude that the perception-only model is more accurate than the perception-plus-expectation model in predicting service quality.

Some studies are there in literature which discusses the software quality prediction as well. [20] proposed a neural-network based software quality prediction scheme based on the four inputs efficiency, maintainability, clarity and extensibility where fuzzy rules for software quality are also taken account for better accuracy. A novel support vector machine based approach has been developed in [21] where quality has been tested for a MRI data software runs for five years. A new clustering scheme has been developed for software quality assessment in [22] which takes in to account various software matrices at various stage of software. However, this scheme [22] only discusses only qualitative assessment for software and doesn't cover the quantitative assessment. Prediction of bug in a software has been investigated in [23] which uses machine learning approaches i.e. ANN, decision tree, naive bias etc. for bug prediction by using previous data of the software. A natural language processing based approach has been proposed in [24] by employing recurrent neural network for software quality assessment with four binary classifiers. The attributes used for quality were 'appropriate, singular, complete, correct'.

Most of the above studies target a particular attribute of a software for example reliability, usability etc. There is still missing a link to unify them in order to evaluate the quality of a software so that the client can choose a particular software by using the proposed model. Thus, motivated from the above discussion, we propose an RBFN based software quality prediction model by presented RBFN architecture. As RBFN based modelling framework for prediction is missing in the literature, this paper investigates the working of the RBFN and examine its implementation over the actual software dataset. This includes theory pertaining to its weight tuning as well as updating its centres. To the best of author's knowledge, this is a first attempt of using RBFN for an early stage software quality prediction.

The organization of the paper is as follows: A detailed description of the architecture and its training algorithm is covered in Sect. 2. This is followed by the proposed methodology, data collection and simulation study in Sect. 3.

Results and analysis is covered in Sect. 4. Finally concluding remarks and future work is given in Sect. 5.

2 Proposed Radial-Basis Function Networks (RBFN) Architecture

A general RBFN architecture is shown in Fig. 1. It consists of a single hidden layer containing neurons that use radial-basis functions as activation functions, as previously stated. The value of radial basis functions is determined by the distance between the input and the fixed point. The euclidean distance is the most common distance computed. The following are some examples of activation functions [25]:

– Gaussian RBF:

$$\Phi(x) = e^{(-\epsilon x^2)} \tag{1}$$

– Multiquadric RBF:

$$\Phi(x) = \sqrt{1 + (\epsilon x)^2} \tag{2}$$

– Inverse quadratic RBF:

$$\Phi(x) = \frac{1}{1 + (\epsilon x)^2} \tag{3}$$

– Inverse Multiquadric RBF:

$$\Phi(x) = \frac{1}{\sqrt{1 + (\epsilon x)^2}} \tag{4}$$

– Thin-plate spline RBF:

$$\Phi(x) = r^2 \log(r) \tag{5}$$

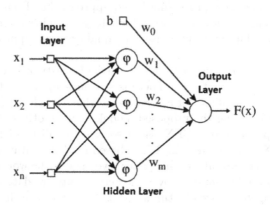

Fig. 1. RBFN architecture

Here, x represents the euclidean distance between the input and the fixed pointed center. To explain the paper further, let us consider the thin-plate spline as the activation function. In the succeeding subsection, we present the weight update formula using gradient descent algorithm [26].

2.1 RBFN Training

As mentioned earlier, the back-propagation algorithm will be used to update the weights and the centers for the RBFN. For a better understanding of the derivation, consider the RBFN shown in Fig. 2. where,

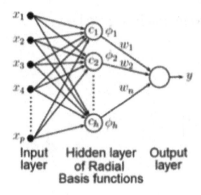

Fig. 2. RBFN under training

- $x_1, x_2, \ldots, x_n$: Represent the inputs such that $x_i \in \mathbb{R}$.
- $c_1, c_2, \ldots, c_h$: Represent the centers of the RBFN. Each center is such that $c_i \in \mathbb{R}^n$.
- $w_1, w_2, \ldots, w_h$: Represent the weights associated with each center such that $w_i \in \mathbb{R}$.
- y: Represents the output.

Feed-Forward Calculations. The feed-forward calculations pertaining to the RBFN shown in Fig. 2 is as mentioned below. Here, the input is defined as $x = [x_1, x_2, x_3, \ldots., x_n]$ and output is calculated as follows:

$$y = w_1(||c_1 - x||)^2 \log(||c_1 - x||) + w_2(||c_2 - x||)^2 \log(||c_2 - x||) + \ldots.. \quad (6)$$
$$+ w_h(||c_h - x||)^2 \log(||c_h - x||)$$

The above equation can be rewritten in compact form as:

$$y = \sum_{i=1}^{h} w_i(||c_i - x||)^2 \log(||c_i - x||) = \sum_{i=1}^{h} w_i f(c_i, x) \quad (7)$$

Back-Propagation Calculations. To perform weight update using gradient-descent, an energy function has to be considered. In the case of a single output RBFN, the energy function is as shown below:

$$J = \frac{1}{2}(y^d - y)^2 \quad (8)$$

The goal is to identify new weights that minimise this energy function. The weight update equation for the gradient descent based back-propagation method is as follows. The gradient-descent approach for weight and centre updating is depicted in Fig. 3.

$$w_{i,\text{new}} = w_{i,\text{old}} - \zeta \frac{\partial J}{\partial w_i} \tag{9}$$

On similar lines, the center can be updated as:

$$c_{i,\text{new}} = c_{i,\text{old}} - \zeta \frac{\partial J}{\partial c_i} \tag{10}$$

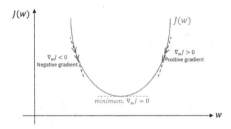

Fig. 3. Gradient descent algorithm

Using the above formula, we present the weight update technique for the RBFN architecture shown in Fig. 2. Consider the weight update for w_1 from Eq. 9:

$$w_{1,\text{new}} = w_{1,\text{old}} - \zeta \frac{\partial J}{\partial w_1} \tag{11}$$

Now using chain rule, the above expression can be rewritten as:

$$w_{1,\text{new}} = w_{1,\text{old}} - \zeta \frac{\partial J}{\partial y} \frac{\partial y}{\partial w_1} \tag{12}$$

Simplifying, we get:

$$w_{1,\text{new}} = w_{1,\text{old}} + \zeta(y^d - y)(||c_1 - x||)^2 \log(||c_1 - x||) = w_{1,\text{old}} + \zeta(y^d - y)f(c_1, x) \tag{13}$$

To generalise,

$$w_{i,\text{new}} = iw_{i,\text{old}} + \zeta(y^d - y)f(c_i, x) \tag{14}$$

The next stage is to update the centers. This can be done using the Eq. (10). Hence, let us consider the update policy for center c_1.

$$c_{1,\text{new}} = c_{1,\text{old}} - \zeta \frac{\partial J}{\partial c_1} \tag{15}$$

As done previously, we will be using chain rule to expand the partial derivative:

$$c_{1,\text{new}} = c_{1,\text{old}} - \zeta \frac{\partial J}{\partial y} \frac{\partial y}{\partial c_1} \tag{16}$$

Simplifying the above equation, we get:

$$c_{1,\text{new}} = c_{1,\text{old}} + \zeta(y^d - y)(||c_1 - x||)(\log(||c_1 - x||^2) + 1)w_1 \tag{17}$$

Generalizing the above formula, we get:

$$c_{i,\text{new}} = c_{i,\text{old}} + \zeta(y^d - y)(||c_i - x||)(\log(||c_i - x||^2) + 1)w_i \tag{18}$$

In certain cases, as in Fig. 2, a bias term would be added at the output layer. In such cases, the Eq. (7), gets modified with an extra term. This can be written as:

$$y = \sum_{i=1}^{h} w_i(||c_i - x||)^2 \log(||c_i - x||) + w_b b \tag{19}$$

To update the weight of the bias, the Eq. 9 is modified as:

$$w_{b,\text{new}} = w_{b,\text{old}} - \zeta \frac{\partial J}{\partial w_b} \tag{20}$$

Hence, in this case, the updated bias weight is given as:

$$w_{b,\text{new}} = w_{b,\text{old}} - \zeta \frac{\partial J}{\partial y} \frac{\partial y}{\partial w_b} \tag{21}$$

Simplifying, we get:

$$w_{b,\text{new}} = w_{b,\text{old}} + \zeta(y^d - y)b \tag{22}$$

Using Eqs. (14), (18) and (22), a RBFN is trained to learn a mapping between input and output. Followed by this mapping, quality of a new software can be predicted.

3   Proposed Approach

Proposed approach comprises of the following sub-tasks e.g. Data preparation, Training of RBFN, Validation of trained model. Each of these are discussed in this section in detail.

3.1   Data Preparation

Data preparation is a crucial factor to success a learning based model e.g. for ANN, Fuzzy logic, Deep learning etc. There is tremendous amount of effort has been made to search for the parameters that can affect the quality of a software. Out of several parameters, following five parameters are sufficient to access the quality of a software:

Table 1. Training data for RBFN

S. No.	Input to RBFN					Output
	Reliability	Usability	Efficiency	Maintainability	Portability	Quality
1	0.3195	0.3421	0.3097	0.3172	0.3545	0.495
2	0.3195	0.7406	0.8396	0.1306	0.4813	0.7
3	0.312	0.2895	0.2425	0.2649	0.1754	0.276
4	0.312	0.2895	0.2425	0.3172	0.6679	0.412
5	0.6353	0.7932	0.7948	0.6007	0.9739	0.505
6	0.7556	0.7256	0.8097	0.2948	0.6007	0.703
7	0.718	0.7105	0.7948	0.6455	0.7127	0.646
8	0.9511	0.9436	0.959	0.3097	0.8321	0.759
9	0.06391	0.07143	0.0709	0.0709	0.08582	0.1
10	0.3195	0.3421	0.3097	0.3172	0.3545	0.495
11	0.718	0.6729	0.6455	0.3097	0.5709	0.638
12	0.1692	0.7256	0.8097	0.3246	0.8022	0.57
13	0.9887	0.9887	0.9664	0.3097	0.9142	0.691
14	0.8985	0.9361	0.7873	0.1679	0.7425	0.89
15	0.8534	0.6504	0.9142	0.653	0.9291	0.664
16	0.5752	0.5677	0.8246	0.3246	0.6903	0.537
17	0.312	0.3195	0.2799	0.3172	0.6679	0.512
18	0.8534	0.6955	0.8545	0.6306	0.9739	0.6
19	0.7556	0.7256	0.8097	0.2948	0.6007	0.703
20	0.8534	0.7932	0.8993	0.04104	0.8619	0.89

1. **Reliability**: It is the ability of a software to maintain its performance at prescribed threshold for certain amount of time under specified conditions. For better quality, reliability of a software should be high.
2. **Usability**: It is the ability of a software that software is made user-friendly, easily operable and controllable by the end users. Usability should be high for a better quality software.
3. **Efficiency**: Under specified conditions, it is the software's ability to provide adequate performance in relation to the quantity of resources used. It should be high for a high quality software.
4. **Maintainability**: The degree to which a software product can be altered which includes corrections, upgrades, or adaptations of the program in a different environment. Further requirements and functional specifications, were also included where required effort has to be modified. Maintainability is kept low for a better quality software.
5. **Portability**: The software product's capacity to be moved from one environment to another where the environment could comprise additional hardware or software as well. A better quality software should have high portability.

Hence, five inputs are utilised to forecast software quality: Reliability, Efficiency, Usability, Maintainability, and Portability. The output of RBFN is software quality, which is calculated using these five inputs. A total of 25 input-output pairs data sets are obtained from various software, with 20 data sets used for RBFN training and 5 data sets utilised for testing. Various studies contributed to the prepared data set and was double-checked by experts as well. The data-set used is shown below as in Table 1.

It is accomplished in this paper via MATLAB environment on a PC with intel i5 processor. We have written whole training algorithm in MATLAB from scratch and then trained the RBFN. There are several parameters for training the RBFN that can have a significant impact on the accuracy of the trained RBFN. For example, a very small value of learning rate may result in slower convergence of parameters, i.e. slow learning, whereas a larger value may result in RBFN parameter instability, i.e. RBFN instability, so these should be carefully selected. We experimented with the following settings, starting with a small value as a starting point:

Learning Rate for weight parameters update: 0.3
Learning Rate for centers update: 0.2
No. of Centers in hidden layer: 50
No. of weight parameters: 50

3.2 Training of RBFN

All of the 50 weight parameters are initialized with small value randomly selected between −0.1 to 0.1. A total of 50000 iterations are used to train the RBFN where one iteration corresponds to the one passage of whole data-set (20 data-set are used here) to RBFN. All the weights are updated at each instant of data-set. Stopping criteria for RBFN training is either root mean square error

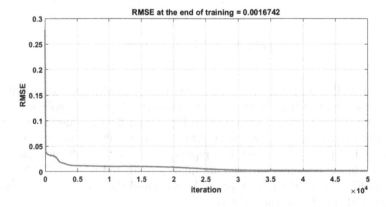

Fig. 4. Error plot of RBFN training

$(RMSE) < 0.001$ or 50000 iterations i.e. the training algorithm is stopped wither RMSE reaches under 0.001 or 50000 is reached.

The RMSE per iteration has been shown in Fig. 4 where we see that RBFN has been trained successfully as RMSE is reached to 0.0016742.

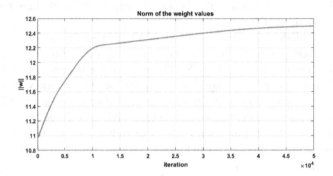

Fig. 5. Evolution of weights

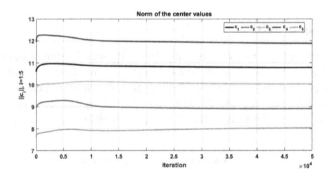

Fig. 6. Evolution of centers

Because all of the weights and centres are updated in real time, all of the weights and centre values converge to their ideal values. The evolution of norm of the weights and centers are shown in Fig. 5 and 6 which justifies the former statement for parameter convergence. The training results for all 20 examples are presented in Fig. 7, where 'red' line represents the actual software quality, and the 'blue' line represents the software quality predicted by the RBFN. We can see from Fig. 7 that the RBFN properly evaluated the target quality, resulting in successful training of the RBFN.

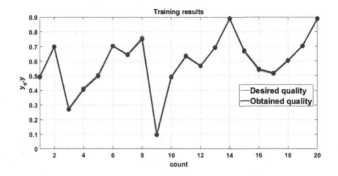

Fig. 7. RBFN training results

4   Results and Analysis

The ideal weights, i.e. centers of hidden layer and hidden to output layer weights, are stored once the RBFN has been trained. These weights are then used to determine software quality based on five criteria: reliability, efficiency, usability, maintainability, and portability. Five sample for different software are taken which quality needs to be estimated using the proposed RBFN.

Table 2. Testing results of trained RBFN

S. No.	Input to RBFN					Quality	
	Reliability	Usability	Efficiency	Maintainability	Portability	Obtained	Actual
1	0.9887	0.9887	0.9888	0.3246	0.9515	**0.6374**	0.633
2	0.3346	0.3271	0.3172	0.6381	0.3321	**0.4382**	0.412
3	0.9511	0.9436	0.8769	0.3022	0.6828	**0.8717**	0.8590
4	0.8459	0.891	0.9515	0.153	0.8097	**0.8829**	0.89
5	0.8534	0.6955	0.7948	0.6157	0.9739	**0.5617**	0.575

The estimated quality using RBFN and actual quality has been shown in Table 2 above.

Testing results for these five input samples are also shown in Fig. 8. From Table 2 and Fig. 8, we see that the estimated quality by RBFN closely matches with the actual quality of software (desired quality). As a result, we may conclude that the suggested RBFN architecture accurately forecasts the quality of an unknown data set that was not included in the RBFN's training. As a result, the proposed technique has the potential to be used in real-time to save a customer's money and time.

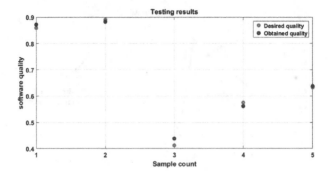

Fig. 8. Testing results

5   Conclusion and Future Scope

This research proposes the RBFN framework, to forecast software quality based on five parameters: reliability, usability, efficiency, maintainability, and portability. Because of RBFNs universal approximation capability and linear parameter weights, RBFNs has been used to create a simpler form of an ANN. A total of 25 softwares were collected, with six factors (reliability, usability, efficiency, maintainability, portability, and quality) stored for each software, with 20 data sets being used for RBFN training. To test the efficacy and correctness of the trained RBFN, five data sets of unknown software quality were used. The simulation results show that the proposed RBFN structure accurately predicts software quality. In a nutshell, the proposed technique can be extremely useful for the clients (who wish to choose a specific software) to anticipate software quality in real-time based on only five parameters.

In future, our aim is to test and improve the accuracy of the proposed approach on a larger data-set of various software. Moreover, we plan to include linguistic rules with the proposed approach to improve accuracy and compare the approach with some existing methods in terms of computation and accuracy on a larger data-set.

References

1. Zakaryazad, A., Duman, E.: A profit-driven Artificial Neural Network (ANN) with applications to fraud detection and direct marketing. Neurocomputing **175**, 121–131 (2016)
2. Mignan, A., Broccardo, M.: Neural network applications in earthquake prediction (1994–2019): meta-analytic and statistical insights on their limitations. Seismol. Res. Lett. **91**(4), 2330–2342 (2020)
3. Liu, X., Zhu, Z., Yu, Z.: Applications of ANN in COVID-19. Int. Core J. Eng. **7**(2), 133–140 (2021)
4. Bodyanskiy, Y., Pirus, A., Deineko, A.: Multilayer radial-basis function network and its learning. In: 2020 IEEE 15th International Conference on Computer Sciences and Information Technologies (CSIT), vol. 1. IEEE (2020)

5. Chen, S., Mulgrew, B., Grant, P.M.: A clustering technique for digital communications channel equalization using radial basis function networks. IEEE Trans. Neural Netw. 4(4), 570–590 (1993)
6. Leonard, J.A., Kramer, M.A.: Radial basis function networks for classifying process faults. IEEE Control Syst. Mag. 11(3), 31–38 (1991)
7. Zemouri, R., Racoceanu, D., Zerhouni, N.: Recurrent radial basis function network for time-series prediction. Eng. Appl. Artif. Intell. 16(5–6), 453–463 (2003)
8. Ranaweera, D.K., Hubele, N.F., Papalexopoulos, A.D.: Application of radial basis function neural network model for short-term load forecasting. IEE Proc. Gener. Transm. Distrib. 142(1), 45–50 (1995)
9. Li, J., et al.: Software defect prediction via convolutional neural network. In: 2017 IEEE International Conference on Software Quality, Reliability and Security (QRS). IEEE (2017)
10. Arar, Ö.F., Ayan, K.: Software defect prediction using cost-sensitive neural network. Appli. Soft Comput. 33, 263–277 (2015)
11. Zheng, J.: Predicting software reliability with neural network ensembles. Expert Syst. Appl. 36(2), 2116–2122 (2009)
12. Cai, K.-Y., et al.: On the neural network approach in software reliability modeling. J. Syst. Softw. 58(1), 47–62 (2001)
13. Budur, A., şerban, C., Vescan, A.: Predicting Reliability of Object-Oriented Systems Using a Neural Network. Studia Universitatis Babes-Bolyai Informatica, vol. 64, issue number 2 (2019)
14. El Emam, K., Melo, W., Machado, J.C.: The prediction of faulty classes using object-oriented design metrics. J. Syst. Softw. 56(1), 63–75 (2001)
15. Khoshgoftaar, T.M., et al.: Application of neural networks to software quality modeling of a very large telecommunications system. IEEE Trans. Neural Netw. 8(4), 902–909 (1997)
16. Khoshgoftaar, T.M., Szabo, R.M.: Improving neural network predictions of software quality using principal components analysis. In: Proceedings of 1994 IEEE International Conference on Neural Networks (ICNN 1994). Vol. 5. IEEE (1994)
17. Aboonajmi, M., et al.: Quality assessment of poultry egg based on visible-near infrared spectroscopy and radial basis function networks. Int. J. Food Prop. 19(5), 1163–1172 (2016)
18. Kumar, R., Rai, S., Trahan, J.L.: Neural-network techniques for software-quality evaluation. In: Annual Reliability and Maintainability Symposium. 1998 Proceedings. International Symposium on Product Quality and Integrity. IEEE (1998)
19. Huang, J., et al.: Cross-validation based K nearest neighbor imputation for software quality datasets: an empirical study. J. Syst. Softw. 132, 226–252 (2017)
20. Pizzi, N.J., Summers, A.R., Pedrycz, W.: Software quality prediction using median-adjusted class labels. In: Proceedings of the 2002 International Joint Conference on Neural Networks. IJCNN 2002 (Cat. No. 02CH37290), vol. 3. IEEE (2002)
21. Xing, F., Guo, P., Lyu, M.R.: A novel method for early software quality prediction based on support vector machine. In: 16th IEEE International Symposium on Software Reliability Engineering (ISSRE 2005). IEEE (2005)
22. Yang, B., et al.: Software quality prediction using affinity propagation algorithm. In: 2008 IEEE International Joint Conference on Neural Networks (IEEE World Congress on Computational Intelligence). IEEE (2008)
23. Hammouri, A., et al.: Software bug prediction using machine learning approach. Int. J. Adv. Comput. Sci. Appl. 9(2), 78–83 (2018)

24. Gramajo, M.G., Ballejos, L., Ale, M.: Recurrent Neural Networks to automate Quality assessment of Software Requirements. arXiv preprint arXiv:2105.04757 (2021)
25. Broomhead, D.S., Lowe, D.: Radial basis functions, multi-variable functional interpolation and adaptive networks. Royal Signals and Radar Establishment Malvern (United Kingdom) (1988)
26. Heaton, J.: AIFH, volume 3: deep learning and neural networks. J. Chem. Inf. Model. 3 (2015)

Generating the Base Map of Regions Using an Efficient Object Segmentation Technique in Satellite Images

Kavitha Srinivasan$^{(\boxtimes)}$ ⓘ, Sudhamsu Gurijala$^{(\boxtimes)}$ ⓘ,
V. Sai Chitti Subrahmanyam ⓘ, and B. Swetha ⓘ

Department of Computer Science and Engineering, Sri Sivasubramaniya Nadar
College of Engineering, Kalavakkam, Tamil Nadu, India
kavithas@ssn.edu.in, {sudhamsu17170,saichittisubrahmanyam17135,
swetha17174}@cse.ssn.edu.in
https://www.ssn.edu.in

Abstract. Satellite images find various applications today, of which segmenting the objects in a satellite image for map generation is widely used in disaster mitigation planning and recovery. Existing works mostly segment one class of object from satellite images, whereas actual images contain multiple classes of objects. To overcome this shortcoming, this paper describes a system which generates the base map from a satellite image for four classes of objects - Buildings, Roads, Greenery and Water Bodies. Binary and multi-class U-Net architectures are designed and trained with suitable datasets to segment the four classes of objects individually. The trained models can take an input satellite image (RGB, JPG format) of pixel resolution 30–50 cm, individually segment the four classes and generate the base map by combining segmented regions in different colours. The performance is analysed by comparing the Intersection Over Union (IOU) score and cross entropy loss with validation set and existing models. A real time image acquired from Google Earth Pro is tested and the results are subjectively evaluated to infer that 90% of the regions are segmented correctly.
Source Code - https://github.com/sudhamsugurijala/Sat_Image_Seg.

Keywords: Satellite image · Google Earth Pro · Image segmentation · U-Net · ResNet-101 · VGG16

1 Introduction

The rapid developments in technology have caused significant advancements in image acquisition capabilities. There are different ways to obtain data about a region, like data obtained from remote sensing images through Unmanned Aerial Vehicles (UAVs), Light Detection and Ranging (LiDAR), images collected from

Supported by Sri Sivasubramaniya Nadar College of Engineering.

K. K. Patel et al. (Eds.): icSoftComp 2021, CCIS 1572, pp. 341–355, 2022.
https://doi.org/10.1007/978-3-031-05767-0_27

aircraft and data obtained from satellite images. Out of these modalities, satellite images provide data of the largest scale, and the growing number of satellites produce enormous volumes of data in the form of visible imagery (RGB images), water vapour imagery, infrared imagery etc., giving rise to a huge number of satellite image databases with varied contrast and resolution in spatial, spectral and temporal domains based on the applications. Satellite images find applications in areas like landscape planning, agriculture, geology, conservation, forestry and regional planning. One such application is segmenting the objects in a satellite image for base map generation, which can be widely used in disaster mitigation planning and recovery from remotely sensed satellite images by rescue teams.

Base maps are the primary layers of any map, which usually provide location references for features that do not change often in time, for example boundaries, rivers, lakes, roads, buildings and highways. We can use image segmentation to segment such features in satellite images and automate the process of creating base maps, which can aid in creating maps or analysing a region. Image segmentation can be defined as the process of assigning a label to every pixel in an image such that pixels with the same label are similar, for example all pixels of a building in a satellite image are similar. Such a technique can be used for finding road networks in a satellite image by segmenting roads alone, study the changing forest area in a region by segmenting the greenery of a region alone, or combine multiple classes of segmentation (buildings, roads, greenery etc.) to obtain the base map of a region itself, which can aid in simultaneously analysing multiple aspects (forest area, road network, area covered by buildings etc.) of the region. The objectives of the proposed work in this paper are as follows:

- To segment four classes of objects - Roads, Buildings, Greenery and Water Bodies from an input satellite image
- To Colour the segmentation outputs of the four classes using suitable colours for each class to make interpretation easy
- To Combine the coloured segmentation outputs of the four classes to generate a base map for the input satellite image.

The next section reviews some related works in the field of satellite image segmentation and their limitations. Section 3 describes the proposed system and the algorithms used in detail. Section 4 lists the experiments performed with different datasets and the results obtained. Section 5 concludes this paper with some discussion about possible future work.

2   Related Works

There has been extensive research in exploring different methods for segmenting objects in satellite images, but most methods had limitations, like the Split and Merge technique [1] used for extracting buildings from high-resolution Ikonos satellite images performed poorly (low accuracy) on low resolution images, a technique which used Guided Filters with Deep Learning Models [2] for efficient building detection failed to segment buildings partially covered by trees, the Shadow Detection Technique [3] used for segmenting buildings failed in regions where shadows were not detected or when buildings cast shadows on other buildings, and the Two-Stage Model technique [4] used for segmenting buildings did not extract small buildings properly or extracted closely distributed buildings as a single building.

Other studies tried to combine different methods for segmenting objects in satellite images, yet had some drawbacks. A study combined Otsu's segmentation method, connected component analysis and morphological operations in the same order [5] for road detection from high resolution satellite images but this method performed poorly on low resolution images. A three module technique called the Coord-Dense-Global (CDG) model [6] was used for extraction of roads from Massachusetts Roads Dataset but it gave discontinuities in the segmentation results for regions having complex backgrounds (like buildings and forests). Another technique applied a combination of K-Means and Cellular Automata algorithms [7] for segmenting greenery and water bodies but consumed a lot of time in learning phase since K-Means is an Unsupervised Learning Algorithm.

While most of the above methods used complex steps for segmenting satellite images, each method had its own limitations and segmented only one class of object (like roads or buildings or greenery), which is a drawback when we consider that actual satellite images have multiple objects of different types. Although some studies have worked with multi-class segmentation in satellite images, they have either chosen a few classes for segmentation (like the study which used Multi-level Context Gating UNet (MCG-UNet) and Bi-directional ConvLSTM UNet model (BCL-UNet) [8] for segmenting roads and buildings or the study which used segmentation for disaster impact assessment [9] by segmenting roads and buildings in satellite images), or have segmented classes which cannot be used for immediate applications (like the Context and Semantically Enhanced UNet (CSE-UNet) [10] which segmented cars but did not segment roads from satellite images).

To overcome this shortcoming, this paper proposes a system which will segment 4 classes of objects (Roads, Buildings, Greenery and Water Bodies) in a satellite image for generating the base map of that region. For the purpose of segmenting objects in satellite images, U-Net architecture based models are used in the proposed system since they are faster compared to other segmentation models [11].

3   Proposed System

This section explains about the overall system architecture, module-split up and algorithms used in the proposed system for generating base maps by segmenting four classes of objects (Roads, Buildings, Greenery and Water Bodies) from an input satellite image.

Figure 1 shows the overall system architecture of proposed system where all the steps for obtaining the base map of a region from its input satellite image are illustrated.

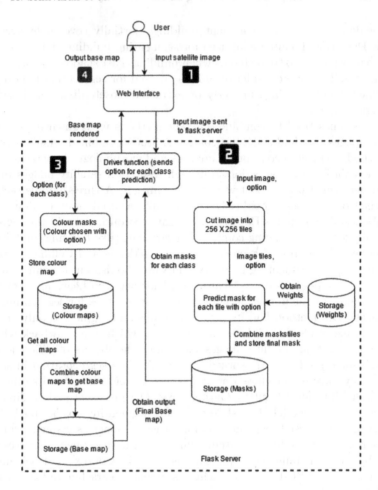

Fig. 1. Overall system architecture of proposed system

3.1 Dataset Collection

The Buildings Dataset was obtained from the mapping challenge competition on AICrowd [17]. The training dataset consists of 2,80,741 satellite images in RGB along with their annotations in MS-COCO JSON format. Each Satellite image is of size 300 × 300 and has pixel resolution of 30–50 cm. The Roads Dataset was obtained from DEEPGLOBE-CVPR18 Road Extraction Challenge [18]. The images were captured by DigitalGlobe's satellites, which provides datasets for academic and research purposes. The training dataset contains 6226 satellite images along with their binary masks. Each satellite image and binary mask is of size 1024 × 1024 and has pixel resolution of 50 cm. The Greenery and Water Bodies Dataset was obtained from Landcover.ai [19]. The dataset consists of 41 images of size between 4200 × 4700 and 9000 × 9500 pixels and the pixel resolution of images vary from 25 to 50 cm. The dataset consists of four classes of images along with their masks. The four classes of objects are Buildings (1), Woodlands (2), Water (3) and Background (0).

3.2 Dataset Pre-processing

The datasets considered in this study are pre-processed with one or more algorithms. The Buildings dataset uses image resizing and PY-COCO tools API (python) for converting MS-COCO annotations to binary masks, the Roads dataset uses only image resizing, the dataset used for segmenting Greenery and Water Bodies uses image cutting, Algorithm 1 for one-hot encoding and Algorithm 2 for one-hot decoding.

Algorithm 1. One-Hot Encoding

Input: Original masks with integer labels
Output: Masks with one-hot encoded labels
1. **function** ENCODE_LABELS(original_masks)
2.   Determine size of one-hot vectors with maximum value of integer label
3.   **for** mask in original_masks **do**
4.   encode integer label to one-hot vector
5.   Repeat step 4 for all integer labels in mask
6.   **end for**
7.   **return** One-Hot encoded masks
8. **end function**

Algorithm 2. One-Hot Decoding

Input: Masks with one-hot encoded labels
Output: Masks with integer labels
1. **function** DECODE_MASKS(one_hot_encoded_masks)
2.   **for** encoded_mask in one_hot_encoded_masks **do**
3.   **for** one_hot vector in encoded_mask
4.   Take argmax of one_hot vector to get integer label
5.   **end for**
6.   **end for**
7.   **return** masks with integer labels
8. **end function**

3.3 Training

The pre-processed data is trained using fully convolutional U-Net Architecture (Vanilla U-Net) and its variants such as U-Net with ResNet-101 or VGG16 as encoder layers to increase the efficiency and accuracy of the model to yield better results. A binary class U-Net model is developed for Buildings and Roads datasets whereas a multi-class U-Net model is developed for Greenery and Water Bodies dataset. All model architectures developed for this study are available in the source code (provided with the abstract). The models are trained with metrics and losses discussed below.

Intersection Over Union (IOU) is defined as the area of intersection between the predicted segmentation and the ground truth divided by the area of union between the predicted segmentation and the ground truth [12]. IOU is a basic metric for any segmentation problem and is defined as:

$$IOU = \frac{Intersection}{Union} = \frac{TP}{TP + FP + FN} \tag{1}$$

where 'TP' is True Positive, 'FP' is False Positive and 'FN' is False Negative.

Pixel Accuracy (or) Accuracy is a metric that shows the percentage of pixels in the image that were correctly classified [13]. It is used as a secondary metric, since it can be affected by imbalance in class labels and is defined as:

$$Pixel\ Accuracy = \frac{TP + TN}{TP + TN + FP + FN} \tag{2}$$

where 'TP' is True Positive, 'TN' is True Negative, 'FP' is False Positive and 'FN' is False Negative.

Binary Cross Entropy loss is used for Road and Building segmentation [14] (binary segmentation problems) and is defined as:

$$H_p(q) = -\frac{1}{N} \sum_{i=1}^{N} y_i \cdot \log\left(p(y_i)\right) + (1 - y_i) \cdot \log\left(1 - p(y_i)\right) \tag{3}$$

where 'N' is the number of samples, 'y' is the predicted probability.

Categorical Cross Entropy (CCE) loss is used for multi-class segmentation of Greenery and Water Bodies classes [15] and is defined as:

$$CCE = -\frac{1}{N} \sum_{i=1}^{N} \sum_{j=1}^{k} t_{i,j} \cdot \log\left(p_{i,j}\right) \tag{4}$$

where 'N' is the number of samples and 'k' is the number of classes. 't' represents truth value and 'p' is the predicted probability.

3.4 Validation and Testing

The Vanilla U-Net, U-Net with ResNet-101 encoder and U-Net with VGG16 encoder which were trained on the four classes of objects (Roads, Buildings, Greenery and Water Bodies) are validated on about 20% of images present in the training dataset and are evaluated based on IOU, pixel-accuracy metrics and cross entropy loss functions. Also, the trained models are tested on a real-time satellite image acquired from Google Earth Pro to segment four classes of objects (Roads, Buildings, Greenery and Water Bodies) and then generate the base map of that region by colouring and combining the segmentation outputs. The base map thus generated is validated using subjective evaluation.

3.5 User Interface Design

A user interface is a web application developed using Flask (in Python), Asynchronous JavaScript And XML (AJAX), HyperText Markup Language (HTML), Cascading Style Sheets (CSS) and JavaScript. Flask is used for back-end server scripting, HTML, CSS and JavaScript are used for designing the front-end web pages and AJAX is used for updating the web pages asynchronously. The user can upload a real-time satellite image as input, for which a base map consisting of objects from the four classes of segmentation (Roads, Buildings, Greenery and Water Bodies) would be displayed as output, where the output is based on the trained U-Net models (binary class U-Net model for Buildings and Roads and multi-class model for Greenery and Water Bodies) encapsulated in the back-end of the Flask server. It is a prerequisite that the height and width of the input satellite image be multiples of 256 pixels.

4 Experiments and Results

This section describes the implementation of various U-Net architectures used for segmentation of four classes of objects i.e., Roads, Buildings, Greenery and Water Bodies in detail. It also analyses the results obtained.

4.1 Dataset Description

In this study, three different types of datasets are used in the training process for segmenting four classes of objects (Buildings segmentation dataset, Roads segmentation dataset and LandCover dataset used for segmenting Greenery and Water Bodies). The outcome is used to segment and generate a base map of any real time region.

The Mapping Challenge AICrowd Buildings dataset consists a total of 2,80,741 images and is divided into two sets - Training and Validation set. The training set consists of 2,52,667 images and the validation set consists of 28,074 images. The sample input and its respective mask is shown in Fig. 2.

The DeepGlobe Roads dataset consists of 6226 images and is divided into two sets for training and validation. The training set consists of 4704 images and

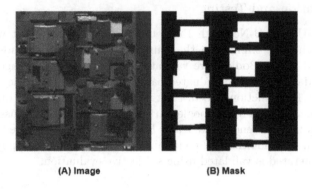

(A) Image   (B) Mask

Fig. 2. Sample from building dataset

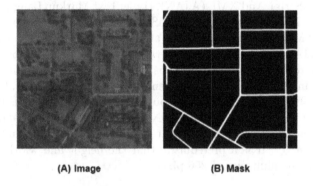

(A) Image   (B) Mask

Fig. 3. Sample from road dataset

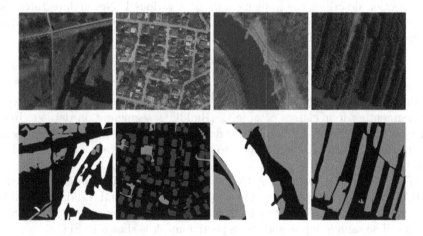

Fig. 4. Samples from Greenery and Water Bodies dataset

the validation set consists of 1522 images. The sample input and its respective mask is shown in Fig. 3.

The Landcover Greenery and Water Bodies dataset consists of 43,292 images and is divided into two sets - training and validation set. The training set consists of 34,634 images and the validation set consists of 8658 images. Four sample images and their respective masks are shown in Fig. 4.

4.2 Results of Pre-processing

The Buildings dataset consists of 2,80,741 input images and its building annotations in MS-COCO JSON format. Using PYCOCO Tools, a python API which is used to load, parse and visualize annotations in MS-COCO format, is used to generate the mask images for 2,80,741 input satellite images from their annotation values in JSON format. Then, the 2,80,741 images and masks are resized to 256×256 pixels. Each image of the Roads dataset of size 1024×1024 pixels is resized to 256×256 pixels. The input images and masks of Greenery and Water Bodies dataset are split into 256×256 pixels tiles with an average entropy of around 0.04. The masks have integer values which are One-Hot encoded (example for four classes here, 1 is encoded to [0 1 0 0]). The output masks are also in One-Hot encoded form. Argmax function is used to find the class label (0 to 3). Each label is replaced by its colour pixel in the mask. For example, label '2' denoting greenery is replaced by a pixel [0 255 0], which denotes GREEN in [R G B] format. The number of training and validation images used from the dataset for each class of object is summarized in Table 1.

Table 1. Dataset statistics for different classes of objects

Class	Total number of images	Number of training images	Number of validation images
Buildings	280741	252667	28074
Roads	6226	4704	1522
Greenery and Water bodies	43292	34634	8658

4.3 Results of Training and Validation

Three binary class neural network models are trained on Buildings dataset after pre-processing based on the U-Net architecture, namely Vanilla U-Net model, U-Net with ResNet-101 encoder and U-Net with VGG16 encoder. The Vanilla U-Net and U-Net with ResNet-101 encoder are trained for 10 epochs in batches of 64 using "adam" optimizer, "binary_crossentropy" as loss function, "Accuracy and IOU (Intersection over Union)" as evaluation metrics. The U-Net with

VGG16 encoder model is trained for 10 epochs in batches of 32 using "adam" optimizer, "binary_crossentropy" loss function, "Accuracy and IOU (Intersection over Union)" evaluation metrics. The trained models are validated on 10% of buildings dataset (28074 images) and its results are recorded and summarized in Table 2.

Table 2. Comparison of models trained with buildings (AICrowd) dataset

Model	Optimizer	Epochs	Batch size	Training			Validation		
				Loss	Accuracy	IOU	Loss	Accuracy	IOU
Vanilla U-Net	adam	10	64	0.1329	0.9446	0.738	0.1281	0.9472	0.7279
U-Net with ResNet-101 encoder	adam	10	64	0.0925	0.9604	0.8121	0.0948	0.9598	0.8085
U-Net with VGG16 encoder	adam	10	32	0.0877	0.9626	0.8231	0.0946	0.9604	0.8163

In the same way, three binary class neural network models (Vanilla U-Net, U-Net with ResNet-101 and U-Net with VGG16 encoder) are trained on Deep-Globe Roads dataset after pre-processing where each model is trained for 50 epochs using "adam" optimizer, "binary_crossentropy" loss function, "Accuracy and IOU (Intersection over Union)" evaluation metrics. The trained models are validated on about 20% of roads dataset (1522 images) and its results are analysed and summarized in Table 3.

Table 3. Comparison of models trained with roads (DeepGlobe) dataset

Model	Optimizer	Epochs	Batch size	Training			Validation		
				Loss	Accuracy	IOU	Loss	Accuracy	IOU
Vanilla U-Net	adam	50	32	0.0761	0.9692	0.5388	0.126	0.9579	0.4609
U-Net with ResNet-101 encoder	adam	50	64	0.0612	0.9903	0.4902	0.1962	0.9574	0.3462
U-Net with VGG16 encoder	adam	50	64	0.0175	0.9926	0.8665	0.2693	0.9523	0.4955

Three multi-class U-Net models (Vanilla U-Net, U-Net with ResNet-101 and U-Net with VGG16 encoder) are trained on the Landcover (Greenery and Water Bodies) dataset after pre-processing where each model is trained for 10 epochs in batches of 32 using "categorical_crossentropy" loss function, "Accuracy and

IOU (Intersection over Union)" evaluation metrics. The models are trained with two optimizers ("adam" and "sgd") separately. The trained models are validated on 20% of Greenery and Water Bodies dataset (8658 images) and the results are recorded and summarized in Table 4.

Table 4. Comparison of models trained with Greenery and Water Bodies (LandCover) dataset

Model	Optimizer	Epochs	Batch size	Training			Validation		
				Loss	Accuracy	IOU	Loss	Accuracy	IOU
Vanilla U-Net	adam	10	32	0.3243	0.8965	0.6623	0.5207	0.7888	0.6623
Vanilla U-Net	sgd	10	32	0.4919	0.8380	0.6616	0.6624	0.7236	0.6624
U-Net with ResNet-101 encoder	adam	10	32	0.1735	0.9467	0.6624	0.1942	0.9352	0.6624
U-Net with ResNet-101 encoder	sgd	10	32	0.2620	0.9275	0.6616	0.2434	0.9274	0.6623
U-Net with VGG16 encoder	adam	10	32	0.2175	0.9359	0.6622	0.4143	0.8375	0.6623
U-Net with VGG16 encoder	sgd	10	32	0.6086	0.7572	0.6616	0.6163	0.7379	0.6623

4.4 Performance Analysis

In this section, the results obtained from the trained models are analysed and compared with other existing models. It is evident that the use of pre-trained models such as ResNet-101 and VGG16 as encoder layers in the U-Net architecture have resulted in significant increase in Pixel Accuracy and IOU (Intersection over Union) values as shown in Fig. 5.

The U-Net with VGG16 encoder architecture has highest pixel accuracy and IOU for Buildings (96.26% accuracy, 0.8231 IOU) and Roads (99.29% accuracy, 0.8665 IOU). For Greenery and Water Bodies, U-Net with ResNet-101 architecture has highest pixel accuracy of 94.67% and IOU of 0.6624 when compared to other models.

The U-Net with VGG16 encoder architecture in this study used for segmenting Buildings achieved better IOU (Intersection over Union) of 0.8231 when compared to VGG11 used in TernausNet [14] which had an IOU of 0.686. The U-Net with ResNet-101 encoder architecture in this study used for segmenting Greenery and Water Bodies achieved better pixel accuracy of 94.67% compared to 91.4% reported in ResNet-50 [16] and mean IOU (mIOU, for multiple classes, or simply IOU) equal to 0.6624 compared to 0.649 reported in ResNet-34 [12].

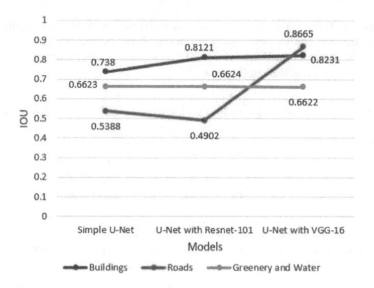

Fig. 5. Comparison between different U-Net models based on IOU

Fig. 6. Image loaded in user interface

4.5   Results of User Interface

The input satellite image provided by the user in the web interface is shown in
Fig. 6 (The satellite image of a region was captured using Google Earth Pro).

The base map generated for the input satellite image is displayed to the
user along with the colour scheme for each label (depicting a class) as shown

YOUR BASE MAP IS READY!

Fig. 7. Base map displayed in user interface

in Fig. 7. The "Validate Map" button can be clicked by the user to impose the base map on the input image for validating the result as shown in Fig. 8, which also shows that this study was successful in segmenting almost 90% of the region correctly (subjective evaluation). Objects belonging to the Buildings, Greenery classes were segmented properly and prominent Roads were correctly segmented. However, there is scope for improvement in segmenting Water Bodies since there have been some false positive segmentation results.

BASE MAP ON INPUT

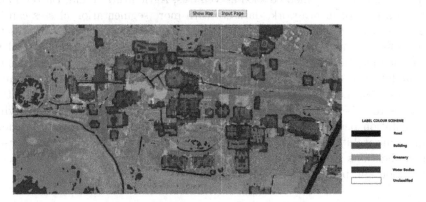

Fig. 8. Validation image shows overlap of base map over input image

5   Conclusion and Discussion

In this paper, the proposed system generated the base map of a region by segmenting four classes of objects separately (Buildings, Roads, Greenery and Water
Bodies) from the input satellite image, colouring each segmentation output with
suitable colours and combining the coloured outputs to get the final base map
(prominent features of a region) which can be used in disaster mitigation and
recovery operations. First, three datasets - Buildings dataset, Roads dataset
and Landcover dataset (Greenery and Water Bodies classes) are collected and
pre-processed using techniques like cutting, resizing and converting MS-COCO
annotations to binary masks. For training, three types of U-Net models are
designed, namely a Vanilla U-Net, a U-Net with ResNet-101 encoder and a
U-Net with VGG16 encoder. Multi-class U-Nets are trained for Greenery and
Water Bodies dataset and binary class U-Nets are trained for Buildings and
Roads classes. For validation, 10–20% of images of each dataset are evaluated
on cross entropy losses, accuracy and IOU metrics. It is found that U-Net with
VGG-16 encoder has highest Pixel Accuracy and IOU for Buildings and Roads,
while U-Net with Resnet-101 encoder has highest Pixel Accuracy and IOU for
Greenery and Water bodies compared to other models. A web application (User
Interface) was developed using Flask, HTML, CSS, JavaScript and AJAX for
providing a satellite image as input and obtaining its base map. For real-time
testing, a satellite image from Google Earth Pro was used as input in the web
application. The base map thus generated was evaluated subjectively and it was
inferred that about 90% of the regions were segmented correctly.

The limitation of this study is that it only classifies four objects from satellite
images (Roads, Buildings, Greenery and Water Bodies). Many other classes of
objects can also be segmented such as Vehicles, Agricultural Land, Barren Land
etc. As a part of future work, datasets with more segmentation classes can be
found or created, alternative pre-processing steps can be applied to datasets,
other loss functions, metrics, optimizers, activation functions can be used to
train the models, different pre-trained models can be used as encoders in the U-
Net architectures or other techniques like Image Fusion (for combining outputs)
and Ensemble Learning (for combining models) can be studied as alternative
methods to this work. While the aforementioned techniques are for improving
segmentation performance, this work can be extended in the future to segment
objects in satellite video feed to produce large scale base maps which can be used
for applications like forest area analysis, study of road networks and disaster
mitigation and planning for large regions.

References

1. Dahiya, S., Garg, P.K., Jat, M.K.: Building extraction from high resolution satellite
   images using Matlab software. In: 14th International Multidisciplinary Scientific
   GeoConference SGEM, Albena, Bulgaria, pp. 71–78 (2014)
2. Xu, Y., Wu, L., Xie, Z., Chen, Z.: Building extraction in very high resolution remote
   sensing imagery using deep learning and guided filters. Remote Sens. **10**(1), 144
   (2018)

3. Yüksel, B.: Automated building detection from satellite images by using shadow information as an object invariant. Master's Thesis, The Graduate School of Natural and Applied Sciences of METU, Turkey (2012)
4. Aamir, M., Pu, Y.F., Rahman, Z., Tahir, M., Naeem, H., Dai, Q.: A framework for automatic building detection from low-contrast satellite images. Symmetry 11(1), 3 (2019)
5. Yadav, P., Agrawal, S.: Road network identification and extraction in satellite imagery using Otsu's method and connected component analysis. In: International Archives of the Photogrammetry, Remote Sensing and Spatial Information Sciences, vol. XLII-5, pp. 91–98 (2018)
6. Wang, S., Yang, H., Wu, Q., Zheng, Z., Wu, Y., Li, J.: An improved method for road extraction from high-resolution remote-sensing images that enhances boundary information. Sensors 20(7), 2064 (2020)
7. Kalyan, M., Rajib, D., Subhasish, D., Anasua, S.: Land use land cover map segmentation using remote sensing: a case study of Ajoy river watershed, India. J. Intell. Syst. 30(1), 273–286 (2021)
8. Gupta, A., Watson, S., Jin, H.: Deep learning-based aerial image segmentation with open data for disaster impact assessment. Neurocomputing 439(7), 22–33 (2021)
9. Abdollahi, A., Pradhan, B., Shukla, N., Chakraborty, S., Alamri, A.: Multi-object segmentation in complex urban scenes from high-resolution remote sensing data. Remote Sens. 13(18), 3710, 1–22 (2021). https://doi.org/10.3390/rs13183710
10. Wang, F., Xie, J.: A context and semantic enhanced UNet for semantic segmentation of high-resolution aerial imagery. J. Phys. Conf. Ser. 1607, 182475–182489 (2020)
11. Ronneberger, O., Fischer, P., Brox, T.: U-Net: convolutional networks for biomedical image segmentation. In: Navab, N., Hornegger, J., Wells, W.M., Frangi, A.F. (eds.) MICCAI 2015, Part III. LNCS, vol. 9351, pp. 234–241. Springer, Cham (2015). https://doi.org/10.1007/978-3-319-24574-4_28
12. Rakhlin, A., Davydow, A., Nikolenko, S.: Land cover classification from satellite imagery with U-Net and lovász-softmax loss. In: Proceedings of the IEEE Conference on Computer Vision and Pattern Recognition Workshops, Salt Lake City, UT, USA, pp. 262–266 (2018)
13. Pan, X., et al.: Building extraction from high-resolution aerial imagery using a generative adversarial network with spatial and channel attention mechanisms. Remote Sens. 11(8), 917 (2019)
14. Iglovikov, V., Shvets, A.: Ternausnet: U-Net with VGG11 encoder pre-trained on ImageNet for image segmentation, pp. 1–5. arXiv preprint arXiv:1801.05746 (2018)
15. Rusiecki, A.: Trimmed categorical cross-entropy for deep learning with label noise. Electron. Lett. 55(6), 319–320 (2019)
16. Ulmas, P., Liiv, I.: Segmentation of satellite imagery using U-Net models for land cover classification, pp. 1–11. arXiv preprint arXiv:2003.02899 (2020)
17. Buildings Dataset Link. https://www.aicrowd.com/challenges/mapping-challenge. Accessed November 2020
18. Roads Dataset Link. https://competitions.codalab.org/competitions/18467. Accessed November 2020
19. Greenery and Water Bodies Dataset Link. http://landcover.ai. Accessed November 2020

KCEPS: Knowledge Centric Entity Population Scheme for Research Document Recommendation

N. Krishnan[1] and Gerard Deepak[2(✉)]

[1] Department of Computer Science and Engineering, SRM Institute of Science and Technology, Ramapuram, Chennai, India
[2] Department of Computer Science and Engineering, National Institute of Technology, Tiruchirappalli, India
gerard.deepak.christuni@gmail.com

Abstract. The data in the Web has increased exponentially, and retrieval from the Web is quite demanding and challenging. There is always a need to improve results in yielding and recommending scholarly articles. An enhanced recommendation system to recommend scholarly articles is best suited to ease the work of a researcher. It is also used to display relevant articles with respect to the user queries. There is a huge probability that the research article recommended might be of no use to the user if the recommendation algorithm does not perform well. This paper proposes a knowledge centric approach for research paper recommendation using semantic similarity and Gated Recurrent Unit along with Cuttlefish optimization algorithm. The Related-Article Recommendation Dataset is used for experimentation. The recommendation is based on the user query and user clicks. The performance is evaluated and compared with the baseline approaches and it is clearly observed that the proposed Knowledge centric recommendation system is superior in terms of performance and attained an average Accuracy, and F-measure of 96.63%, and 96.69% respectively.

Keywords: Cuttlefish Algorithm · Gated recurrent unit · Knowledge centric · Recommendation system

1 Introduction

A research paper is a type of academic work that combines analysis, interpretation, and an argument based on independent research. Before being approved for publishing in an academic conference, most scientific works must go through a peer review procedure. Therefore, scholars must gather knowledge about a topic, express an opinion on it, and support their theory with evidence in a well-organized report while writing research papers. In every discipline, research papers are required for performing scientific investigations. Cross-checking prior research on specific subjects is the only way to enhance studies. Researchers prefer online research libraries over traditional libraries because they allow them to find more relevant material in less time.

© Springer Nature Switzerland AG 2022
K. K. Patel et al. (Eds.): icSoftComp 2021, CCIS 1572, pp. 356–366, 2022.
https://doi.org/10.1007/978-3-031-05767-0_28

To discover relevant research publications, most researchers use keyword-based searches or track references in other papers. And, in most cases, they invest a significant amount of time without achieving satisfying outcomes. Research paper recommendation studies in the past have usually compared articles based on their content and suggest articles that are related to each other. These prior studies had the limitation of recommending the same publications to all researchers.

The fitting of a user query with a set of arbitrary data such that the data in the document is relevant to the user query this process is characterized as document retrieval. Documents maintained in an online document management system have high availability as they are retrievable at any time. These documents can also be accessed from anywhere across the world. As the recent trend is moving towards work from home, the document retrieval system is one of the key components which facilitates companies to share data securely and easily with its employees. For security various user management and access for each of them according to their work hierarchy is provided. For most businesses, this is a significant productivity increase that does not jeopardize document security.

Motivation: As technology advances at a breakneck pace, so does the amount of data generated. Therefore, we need strong recommendation algorithms since massive amount of papers are released every year in various conferences and publications. One issue that has arisen as a result of these good advances for science is that researchers are having difficulty finding the papers they are looking for among a huge number of publications. Researchers expend a lot of time and effort in order to locate the most relevant article to their field of research. Recommendation systems seek to improve search rate efficiency depending on the choices of the individual. According to the key phrases supplied by the individual, the choices are assessed that might show the articles relevant to them. However, these approaches need users to devote time to searching for articles, which is time-consuming, and they cannot ensure that they will locate the exact articles relevant to their subject of study.

Contribution: Knowledge Centric Entity Population Scheme for Research Document Recommendation is proposed for research groups. Proposed system will get input query and user clicks on the particular session of interest and recommend related papers based on the similarities of the works. The proposed KCEPS approach attained an average accuracy of 96.23% with an average recall of 97.89%.

Organization: The following is the publication's structure: Session 2 is comprised of relevant study that has already been made on the subject. Session 3 includes proposed architecture. Session 4 provides advice on implementation. Session 5 is composed of a performance review and outcomes that have been seen. Session 6 focuses on the conclusion.

2 Related Works

Sung-Shun et al. [1] described a method to recommend research-related articles considering the interests of the user. By using ontology to create user profiles, the approach uses the user profile ontology as the foundation for reasoning about users' interests. the approach uses the spreading activation model to find additional prominent users in a community network environment and do research on their interests in order to provide recommendations on the relevant material. Marco et al. [2] delivered an approach based on the Citation Graph and random-walker characteristics to recommend related scholarly articles for research. The system sets a preference score for the documents linked with each other by bibliographic quotations and are contained by the digital library applying the Paper Rank algorithm. Joonseok et al. [3] devised a method to recommend related articles for an academic paper. Here, a web crawler is used to retrieve research papers. Then, the papers are compared to find text similarity, define the same, and recommend using collaborative filtering methods. Betül et al. [4] developed a technique for recommending research articles by taking the user's data into consideration. The system takes into consideration the researcher's work field and previous articles and makes recommendations using TF-IDF and Cosinus.

Kwanghee et al. [5] proposed a system that recommends personalized research papers based on user profiles. A renewal of the user profile greatly increased the number of times every domain, topic, and keyword are used whenever gathered research articles by topic are picked. Each occurrence ratio is computed and displayed on User Profile. Cosine similarity is used to determine how similar the topic and articles are and the initial paper for each topic is recommended by the system. Michael et al. [6] devised a unique technique enabling recommendion of data depending on the profile of the user of interest and description based on the query. The profile is often built and modified automatically in response to user input on the usefulness of items provided to them. Joonseok et al. [7] delivered an approach for research article recommendations based on social relationships. The system demonstrates a realistic application, a customized academic paper recommendation system, in which social relationships are tightly linked to taste. The recommendations are given in a manner such that it is personalized for each researcher using a collaborative-filtering based approach.

Hebatallah et al. [8] described a proposed a customized research paper recommendation system that generates suggestions based on the publications that users are interested in as well as explicit and implicit feedback. In order to increase suggestion quality, this study employs RNNs to identify the publications' continuous and latent semantic characteristics. The proposed method has relied on PubMed, which is widely utilized by physicians and scientists. Buket et al. [9] introduced an approach to recommend scholarly articles for research. The approach takes into consideration the past publications of the author and co-authors. The technique evaluates time-awareness based on meta-data from the researcher's previous papers and recommends using TF-IDF frequency-based similarity analysis method.

3   Proposed Architecture

Figure 1 illustrates the proposed research paper recommendation system's design where both user query and current user clicks are taken into consideration for recommending papers to the user. Only user clicks are considered and not the entire user profile because the technical spectrum of the paper and depth of the research topics varies every time. The entire system consists of 2 phases. Figure 1 depicts phase one. When the user submits an input query, the phase 1 begins. The query is pre-processed by appending tokens to each data element; this is called tokenization. The various words from the pool of data

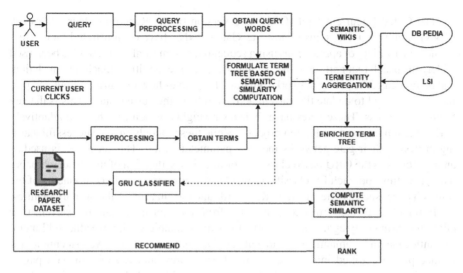

Fig. 1. Diagram of the planned KCEPS system's architecture

are Lemmatized where we are combining different words as a single term. From this irrelevant and stop words are removed and the data is cleaned. The derived query words are then obtained. Using the derived query words, the term tree is formulated based on initial semantic similarity computation using NPMI, Jaccard and cosine similarity. In our unit hood-based method, we utilize Normalized Pointwise Mutual Information (NPMI) to identify the weakest places inside phrases to recommend the optimum position for separating a phrase into two sections. The Jaccard Similarity Coefficient (JSE) approach is employed, to compare the proximity of the data in the process data. JSE depicts the properties of the used objects in a binary manner which enables the user to determine the similarity between the two with ease. Cosine Similarity represents a text as a vector of terms, with the cosine value between two texts' term vectors determining the similarity between them. The similarity value between the user query and the documents is ordered from the highest to the lowest while obtaining documents.

The threshold for each of the similarities is 0.7. To solve an optimal control problem, a divergence is used. A divergence solves a clustering problem, in which a collection of points is separated into numerous clusters containing points that are close to one another in terms of the divergence. The divergence also determines the cluster's center. Bregman proposed the Bregman divergence to handle convex optimization issues. Bregman divergences analysis has been credited with recent breakthroughs in statistical learning, clustering, inverse issues, maximum entropy estimation, and the application of the data processing theorem. The mean, which is an essential characteristic of Bregman divergences, is the minimizer of the expected Bregman divergence given a set of d-dimensional points. The Bregman divergence should be less than 0.5 and the other similarities should be greater than 0.7. The terms obtained from the intersection of all three similarities are used to formulate the term tree.

The Gated Recurrent Unit (GRU) is a version of the Recurrent Neural Network (RNN) paradigm. GRU's, also known as Gated RNNs, have demonstrated their effectiveness in a variety of applications using sequential or temporal data. They've been used widely in speech recognition, music synthesis, speech recognition, machine translation, and other areas. The gating network signaling determines how the current input and past memory are used to update the current activation from the current state. This is the key to GRU's success. These gates have their own weights, which are changed adaptively during the training and evaluation process. While these models enable successful learning in RNNs, their gate networks increase parameterization. This term tree is used as an input class to the GRU based classifier for initial classification of the Research paper dataset. In this approach for classification, we are considering main components of the research paper such as title, abstract, Key words and the citations as the primary features.

In phase 2, To the term tree formulated from the user query and the current user clicks the term entity aggregation is done by using semantic wiki, DBpedia and Lateral semantic indexing. Semantic wikis are wiki engines that use semantic Web3 technologies to incorporate codified information, content, structures, and linkages in wiki pages. Semantic web frameworks are used to depict structured knowledge, making it available and reusable through online applications. The wiki is used to improve characteristics such as better document searching, new link recommendations, recognizing acquaintance networks, dynamic content maintenance, checking and notification, and so on. The basic concept is to make a wiki's intrinsic structure defined by strong page-to-page links available to machines in ways that go far ahead of simple navigation. DBpedia is a multilingual, large-scale knowledge repository derived from Wikipedia. It uses semantic web and Linked Data technologies to extract structured data from Wikipedia. SPARQL querying allows retrieval of datasets from various editions of Wikipedia. It enables deriving Wikipedia and web-linked Wikipedia datasets. It contains millions of RDF connections to various external data sources, allowing users to combine data from these sources with DBpedia data. Querying DBpedia allows the system to retrieve datasets of related research articles. Latent Semantic Indexing (LSI) is a method for topic modeling that uses some of the implicit higher-order relationships between words and text objects to organize data into a semantic network. This allows retrieval of texts based on their concealed data rather than keyword matching. In LSI, topic modeling is based on the deeper level semantic structure rather than merely the surface level word choice, which solves some of the issues of keyword matching. To see if a new article is appropriate, it's folded into the semantic space based on the terms it contains. When it came to making predictions, latent semantic indexing performed better than keyword matching.

Once the entity aggregation is performed. The enriched term tree is modeled. From GRU Index words and Keywords from the documents. Cuttlefish Technique (CFA) which is a nature influenced, meta-heuristic method for optimization. It's color-changing behavior inspired the technique, which is used to tackle numerical global optimization issues.

Cuttlefish are cephalopods that have the ability to change color to blend in with their surroundings or to put on spectacular displays. Distinct layers of cells, including chromatophores, leucophores, and iridophores, produce the patterns and colors seen in cephalopods, and it is the combo of certain cells' processes of reflecting light and matching patterns simultaneously that allows cephalopods to have such a diverse array of patterns and colors. Red, orange, yellow, black, and brown colors can be found in chromatophore cells. Cuttlefish skin, on the other hand, is able to take on all of the rich and varied colors of its surroundings thanks to a collection of mirror-like cells called iridophores and leucophores. Light can be reflected by chromatophores, reflecting cells, or a mix of both, and the cuttlefish's capacity to produce such a diverse range of optical effects is due to the biological variation of the cells.

The CFA searches for the best cluster centers that minimize clustering metrics, the approach proves more effective as the initial values of cluster centers have a big impact on their performance. The research papers from the datasets which is classified by the GRU and Enriched Term Tree are User click-based research paper indexing, and User query-based respectively. The Population used in CFA is a sum of the two. The population is initialized and the best solution is obtained by evaluating the population by a fitness function. The population is divided into four equal groups, which are assigned to six cases. Each group is computed based on their respective cases until the final criteria are returned. The result is returning a set of clusters representing the best possible dataset of research papers. Reranking is performed according to the increasing order of the semantic similarity and initial recommendation is made. This process continues until there are no further user queries.

4   Implementation

The proposed Knowledge based recommendation system architecture was created and executed on Windows 10 operating system. Intel Core i7 10th Gen CPU and 16 GB RAM were used for the implementation. NLTK's Wordnet is used to look up the definitions of terms as well as synonyms. WordNetLemmatizer is used to locate the root word. The python NLP tool packages, Re and Sklearn, were used to pre-process data. A Recurrent neural network model was implemented and trained using Keras, a Python deep learning API package built on top of TensorFlow. RARD: The Related-Article Recommendation Dataset [10] is used for implementation. 4128 Queries were used during the implementation of KCPS whose ground truth was collected both manually and crawled through customized web crawlers.

5   Performance Evaluation and Result

In order to evaluate the proposed performance, Knowledge Centric Entity Population Scheme for Research Document Recommendation that is KCEPS. Precision, Recall, Accuracy, F-Measure, FDR, and nDCG respectively are measured, and utilized as performance review metrics. The Precision, Recall, Accuracy, F-Measure, FDR, and nDCG respectively are computed and plotted as graph 1 and graph 2. Using the Eqs. (1), (2), (3), (4) and (5) respectively.

$$\text{Precision\%} = \frac{Retrieved \cap Relevant}{Retrieved} \tag{1}$$

$$\text{Recall\%} = \frac{Retrieved \bigcap Relevant}{Relevant} \tag{2}$$

$$\text{Accuracy\%} = \frac{\text{Precision} + \text{Recall}}{2} \tag{3}$$

$$\text{F–Measure\%} = \frac{2(\text{Precision} \times \text{Recall})}{(\text{Precision} + \text{Recall})} \tag{4}$$

$$\text{FDR} = 1 - \text{Recall} \tag{5}$$

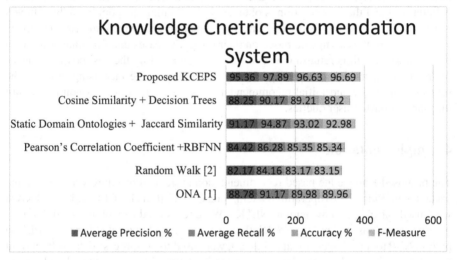

Fig. 2. The suggested architecture's performance metrics versus. percentage of performance measurements

Table 1. The suggested Architecture's performance metrics versus. percentage of performance measurements

Various models	FDR	nDCG
ONA [1]	0.11	0.94
Random Walk [2]	0.18	0.78
Pearson's Correlation Coefficient + RBFNN	0.16	0.8
Static Domain Ontologies + Jaccard Similarity	0.09	0.93
Cosine Similarity + Decision Trees	0.12	0.85
Proposed KCEPS	0.05	0.97

The proposed Knowledge Centric Entity Population Scheme for Research Document Recommendation's performance was obtained and compared with the baseline approaches along with benchmark models. The graph has been plotted accordingly as shown in Fig. 2. On the x-axis are several performance indicators like Average Precision%, Average Recall, Accuracy, F-Measure and on the Y-axis, the models are Proposed KCEPS, ONA [1], Random Walk [2], Pearson's Correlation Coefficient + RBFNN, Static Domain Ontologies + Jaccard Similarity and Cosine Similarity + Decision Trees. FDR, and nDCG are shown in Table 1.

Average Precision%, Recall%, Accuracy%, F-Measure, FDR, and nDCG of the proposed KCEPS approach was formulated to be 95.36, 97.89, 96.63, 96.69, 0.05 and 0.97 respectively.

For ONA [1] Average Precision%, Average Recall%, Accuracy%, F-Measure, FDR, and nDCG was found to be 88.78, 91.17, 89.98, 89.96, 0.11 and 0.94 respectively. For Random Walk [2] the Average Precision%, Average Recall%, Accuracy%, F-Measure, FDR, and nDCG was found to be 82.17, 84.16, 83.17, 83.15, 0.18 and 0.78.

By eliminating a few of the key components, the proposed KCEPS methodology was compared to baseline approaches, and their performance was examined. In terms of performance, the proposed technique outperforms the existing approaches and models. This is because the proposed architecture term tree based on semantic similarity is formulated and GRU classifier is used for classification and for final optimization Cuttlefish Algorithm is used.

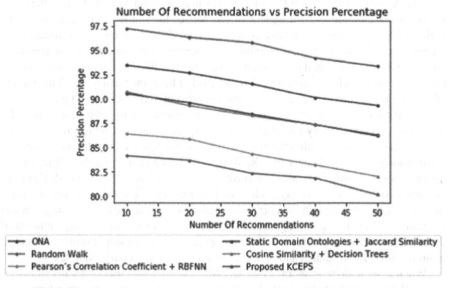

Fig. 3. Number of research paper recommended vs Precision for various models.

Figure 3 shows the amount of research article suggestions vs. the accuracy of those recommendations. Here proposed. KCEPS is baselined with Proposed KCEPS, ONA [1], Random Walk [2], Pearson's Correlation Coefficient + RBFNN, Static Domain

Ontologies + Jaccard Similarity and Cosine Similarity + Decision Trees. Analyzing the graph's trend, the accuracy diminishes as the number of suggested online services grows. Peak accuracy is attained by the proposed KCEPS model when the number of recommendations is 10. The other models ONA [1], Random Walk [2], Pearson's Correlation Coefficient + RBFNN, Static Domain Ontologies + Jaccard Similarity and Cosine Similarity + Decision Trees have an accuracy of 90.55%, 84.15%, 86.39%, 93.47% and 90.74% respectively when the number of recommendations is 10. When the number of suggestions is raised to 50, the suggested KCEPS system has an accuracy of 93.39%, compared to 86.21%, 80.17%, 82.04%, 89.36%, and 86.33% for the other models. When compared to previous models, the suggested method is more accurate even when the number of recommendations is greater.

For categorization, the suggested method employs gated recurrent units. It is an entity population scheme which includes entities from array of data knowledge sources namely the semantic wikis preferably the media wiki has been used here, the DBpedia which holds a lot of relations and topic modelling is done based on Latent Semantic Indexing. In order to achieve the schematic similarity, the NPMI and Jaccard have been used as standard similarity algorithm and to measure the divergence the Bregman divergence is used. In order to maintain the relevance of results much higher, it is optimized under the Cuttlefish optimization algorithm which ensures a combination of the learning scheme and the schematic similarity scheme along with the knowledge basis and a standard optimization scheme make sure the proposed approach is much better than the benchmark algorithms.

The Pearson's Correlation Coefficient was used in order to find the correlation of terms among the entities or instances or the pair of keywords in the approach along with RBFNN. The neural network strategically used the learning of the instances in the approach. The Pearson's Correlation Coefficient was not a true schematic similarity strategy while it contributed to an extent in order to yield the correlation between a set of terms in the approach and the use of RBFNN had to be trained and tested. The training of only relations made it quite insufficient thereby not yielding a very high Precision, Recall, Accuracy, F-Measure and the nDCG was quite average in this case mainly for the reason that no auxiliary knowledge was incorporated into the approach.

The Random Walk technique requires the modelling of a correlation matrix and a correlation graph, and the Paper Rank algorithm, which is a random walk on the correlation graph and traversal in the correlation matrix, has been presented. The most important complexity in this approach is to maintain for the larger the amount of dataset the difficult it is to maintain the correlation matrix as well as the correlation graph and traversal of correlation graph by means of correlation matrix requires a high amount of data structures. There is no learning in this approach. So as a result, it is quite tedious to minimize the data structures and hence it becomes computationally complex.

When a static domain ontology in Jaccard similarity was used, Jaccard similarity is a well-defined schematic similarity scheme so as a result the similarity between a pair of terms or a set of terms was computed and the usage of static domain ontology made sure that high amount of domain knowledge was incorporated into the approach as the background knowledge but the problem with this approach is that the static domain ontology cannot be always modelled for a large set of terms. Going through these things it was definitely a good approach but definitely there is a scope of improving the relevance of the results which is the Precision, Recall, Accuracy, F-Measure percentages and the nDCG is quite good because of the presence of ontology.

6 Conclusion

Research paper recommendation plays a key role in providing relevant data to the academic and research scholars. As the existing systems do not consider the depth of the research conducted by the user it is harder for the system to provide accurate recommendations. Using the GRU classifier and Term entity aggregation, this article presents a knowledge-centric method for Research Document Recommendation The proposed KCEPS approach attained an average accuracy of 96.63% with an average recall of 97.89%. Since the suggested technique uses the GRU classifier in conjunction with the Cuttlefish Technique for optimization, its competency is greater when compared to baseline approaches. From the foregoing conclusion, it is obvious that in the future, a knowledge-based approach to research paper recommendation will be more accurate and provide practical and logical answers.

References

1. Weng, S.S., Chang, H.L.: Using ontology network analysis for research document recommendation. Expert Syst. Appl. **34**(3), 1857–1869 (2008)
2. Gori, M., Pucci, A.: Research paper recommender systems: a random-walk based approach. In: 2006 IEEE/WIC/ACM International Conference on Web Intelligence (WI 2006 Main Conference Proceedings) (WI 2006), pp. 778–781. IEEE (2006)
3. Lee, J., Lee, K., Kim, J.G.: Personalized academic research paper recommendation system. arXiv:1304.5457 (2013)
4. Bulut, B., Kaya, B., Alhajj, R., Kaya, M.: A paper recommendation system based on user's research interests. In: 2018 IEEE/ACM International Conference on Advances in Social Networks Analysis and Mining (ASONAM), pp. 911–915. IEEE (2018)
5. Hong, K., Jeon, H., Jeon, C.: UserProfile-based personalized research paper recommendation system. In 2012 8th International Conference on Computing and Networking Technology (INC, ICCIS and ICMIC), pp. 134–138. IEEE (2012)
6. Pazzani, M.J., Billsus, D.: Content-based recommendation systems. In: Brusilovsky, P., Kobsa, A., Nejdl, W. (eds.) The adaptive web. LNCS, vol. 4321, pp. 325–341. Springer, Heidelberg (2007). https://doi.org/10.1007/978-3-540-72079-9_10
7. Lee, J., Lee, K., Kim, J.G., Kim, S.: Personalized academic paper recommendation system. SRS 2015 (2015)
8. Hassan, H.A.M.: Personalized research paper recommendation using deep learning. In: Proceedings of the 25th Conference on User Modeling, Adaptation and Personalization, pp. 327–330 (2017)
9. Kaya, B.: User profile-based paper recommendation system. Int. J. Intell. Syst. Appl. Eng. **6**(2), 151–157 (2018)
10. Beel, J., Carevic, Z., Schaible, J., Neusch, G.: RARD: the related-article recommendation dataset. arXiv:1706.03428 (2017)

11. Leena Giri, G., Deepak, G., Manjula, S.H., Venugopal, K.R.: OntoYield: a semantic approach for context-based ontology recommendation based on structure preservation. In: Chaki, N., Cortesi, A., Devarakonda, N. (eds.) Proceedings of International Conference on Computational Intelligence and Data Engineering. LNDECT, vol. 9, pp. 265–275. Springer, Singapore (2018). https://doi.org/10.1007/978-981-10-6319-0_22
12. Deepak, G., Kumar, N., Santhanavijayan, A.: A semantic approach for entity linking by diverse knowledge integration incorporating role-based chunking. Procedia Comput. Sci. **167**, 737–746 (2020)
13. Deepak, G., Rooban, S., Santhanavijayan, A.: A knowledge centric hybridized approach for crime classification incorporating deep bi-LSTM neural network. Multimedia Tools Appl. **80**(18), 28061–28085 (2021). https://doi.org/10.1007/s11042-021-11050-4
14. Vishal, K., Deepak, G., Santhanavijayan, A.: An approach for retrieval of text documents by hybridizing structural topic modeling and pointwise mutual information. In: Mekhilef, S., Favorskaya, M., Pandey, R.K., Shaw, R.N. (eds.) Innovations in Electrical and Electronic Engineering. LNEE, vol. 756, pp. 969–977. Springer, Singapore (2021). https://doi.org/10.1007/978-981-16-0749-3_74
15. Deepak, G., Teja, V., Santhanavijayan, A.: A novel firefly driven scheme for resume parsing and matching based on entity linking paradigm. J. Discr. Math. Sci. Cryptograp. **23**(1), 157–165 (2020)
16. Deepak, G., Kumar, N., Bharadwaj, G.V.S.Y., Santhanavijayan, A.: OntoQuest: an ontological strategy for automatic question generation for e-assessment using static and dynamic knowledge. In: 2019 Fifteenth International Conference on Information Processing (ICINPRO), pp. 1–6. IEEE (2019)
17. Krishnan N, Deepak, G.: Towards a novel framework for trust driven web URL recommendation incorporating semantic alignment and recurrent neural network. In: 2021 7th International Conference on Web Research (ICWR), pp. 232–237 (2021) https://doi.org/10.1109/ICWR51868.2021.9443136
18. Krishnan N, Deepak, G.: KnowSum: knowledge inclusive approach for text summarization using semantic allignment. In: 2021 7th International Conference on Web Research (ICWR), pp. 227–231 (2021). https://doi.org/10.1109/ICWR51868.2021.9443149
19. Krishnan, N., Deepak, G.: KnowCrawler: AI classification cloud-driven framework for web crawling using collective knowledge. In: Musleh Al-Sartawi, A.M.A., Razzaque, A., Kamal, M.M. (eds.) EAMMIS 2021. LNNS, vol. 239, pp. 371–382. Springer, Cham (2021). https://doi.org/10.1007/978-3-030-77246-8_35
20. Deepak, G., Santhanavijayan, A.: OntoBestFit: a best-fit occurrence estimation strategy for RDF driven faceted semantic search. Comput. Commun. **160**, 284–298 (2020)

Weighted Hybrid Recommendation System Using Singular Value Decomposition and Cosine Similarity

Sanket Shah[1]([envelope]), Yogesh Raisinghani[2]([envelope]), and Nilay Gandhi[1]([envelope])

[1] Ahmedabad University, Ahmedabad 380009, GJ, India
{sanket.s2,nilay.g}@ahduni.edu.in
[2] Charotar University of Science and Technology, Changa 388421, GJ, India
18ce095@charusat.edu.in

Abstract. A recommendation system (RS) can be defined as a tool that can filter an abundant amount of information and provide suggestions based on the preferences and the behavior of a user. Two of the most popular and widely used RSs are Collaborative Filtering (CF) and Content-based Filtering (CBF). Both of these RSs have their highlights and challenges. CF makes recommendations based on the past ratings given by a user to different items. This makes the recommendations more personalized, but on the downside, if the user or the item is new, CF becomes unreliable. CBF only takes the features of an item into consideration while making recommendations since it assumes that each user is independent. This makes the recommendations common for all the users. But, on the plus side, CBF does not suffer from a new item cold-start problem. To alleviate the individual weaknesses of these systems, a hybrid model that combines the two RSs can be employed. This paper proposes a Weighted Hybrid RS that employs model-based CF using Singular Value Decomposition (SVD), and CBF using cosine similarity. To evaluate the proposed system and compare it with other systems, personalization and diversity have been adopted as the primary evaluation metrics. The experiments conducted in this paper indicate that our proposed system generates a significantly better result than the individual CF and CBF techniques, as well as a cascaded hybrid system in terms of personalization and diversity. Our system also alleviates the new item cold-start problem associated with CF.

Keywords: Recommendation systems · Hybrid recommendation system · Collaborative filtering · Weighted hybrid · Cosine similarity · Singular value decomposition

1 Introduction

Earlier, people used to select an item according to their needs by manually going through the specifications. However, with the rapid growth of ubiquitous

© Springer Nature Switzerland AG 2022
K. K. Patel et al. (Eds.): icSoftComp 2021, CCIS 1572, pp. 367–381, 2022.
https://doi.org/10.1007/978-3-031-05767-0_29

computing and Internet technology in the past years, a colossal amount of digital data is freely available. Such a huge amount of data leads to a problem of "information overload". This creates interference in users' choice by presenting redundant items irrelevant to the user [1,2]. This has made manual scouring through all the available items nearly impossible for an individual and hence hinders the choosing ability and selecting process of the users.

This problem cannot be simply solved by using search engines as they return all the items "matching" the query which in itself is a big list. Also, such systems are not able to incorporate the factor of "personalization" and "users' past behavior" while suggesting items to the user.

RS solves this issue by curating items according to the users' choice and their past behavior hence providing some sense of personalization to the user and satisfying their needs. Such systems have gained huge popularity and have become an integral part of many industrial applications for recommending numerous types of services and products. Examples include e-commerce websites like Amazon [3], movies recommending websites like IMDb [4], etc.

1.1 Contributions of This Paper

In this paper,

- We have proposed a CBF/model-based CF Weighted hybrid RS that uses cosine similarity along with TF-IDF vectorization for the CBF technique, and Singular Value Decomposition for the model-based CF technique.
- We compare our weighted hybrid model with the CBF and model-based CF to highlight the individual drawbacks of the two systems that are alleviated by our proposed model. We have also implemented a simple cascaded hybrid system using the same CBF and CF techniques to compare and evaluate our weighted system and better understand its performance.
- We have evaluated our weighted system based on two evaluation metrics, namely "Personalization" and "Diversity". These metrics have helped us to portray how well our weighted system performs and help us derive inferences regarding the weight allocation of the CBF and CF techniques.

1.2 Structure of This Paper

The remainder of the paper has been divided into 4 main sections, followed by the references. Section 2 briefs the existing study conducted in this domain and also provides a basic knowledge regarding the different recommendation systems that this paper focuses on. Section 3 provides a concise description of the different methods used in this paper, along with the proposed methodology. Section 4 provides a description of the dataset and evaluation metrics used to conduct the experiments and then provides the results of those experiments along with the inferences derived from them. Section 5 contains the final conclusions derived from all the experiments conducted and talks about the pros that the proposed methodology provides.

2   Background Study

Tapestry [5], a manual CF-based mail system was one of the first RS. [6,7] were the first computerized RS prototypes which applied CF. Amazon.com's RS prototypes became the most popular in the late 90s. Hybrid RSs were born when various researchers put together different RSs. The goal was to improve the performance of RSs by eliminating the limitations of the individual RS. This was achieved by combining them and reinforcing their advantages. Fab [8], a meta-level RS, was one of the first hybrid RS. It combined CF and CBF to recommend websites.

[9] presented one of the first detailed analyses of hybrid RS. The author described and compared various RS, hybrid RS, and proposed EntreeC, a new CF and Knowledge-based cascade hybrid system. [10] focuses on the CF technique and presents a systematic survey of various CF-based and hybrid CF-based RS. [11] presents a quantitative review focusing solely on hybrid RS. The authors discusses various problems in different systems and mentions techniques to overcome them.

PTango [12] presents a dynamic weighted CBF and memory-based CF hybrid using a Bayesian network to find the relationship between users, items, and features. [13] presents dynamic weighted CBF and item-based CF hybrid RS. A weighted CBF and item-based CF hybrid is discussed in [14].

In all of the above-mentioned papers, they have created a weighted hybrid system with memory based CF technique. In the era of big data, the number of users and items has become significantly large. This leads to some of the most persisting problems of RS, namely Data sparsity and High dimensionality. According to [10], a model-based CF would be a much better choice to alleviate these problems rather than a memory-based CF. The reason behind this is that a memory-based CF needs to perform some computations over the entire dataset before each prediction. A model-based CF first undergoes a learning phase, where it learns the most optimal parameters needed to make the most accurate predictions. Once this learning phase is over, it can make predictions with great accuracy and speed. One of the most powerful and efficient tools for building a model-based CF is Singular Value Decomposition. According to the documentation of the surprise library [15], SVD has proven to be one of the best models when it comes to accuracy and speed. Hence, SVD-based CF technique can overcome these problems.

2.1   Recommendation Systems

Content-Based Filtering: To make recommendations, a CBF utilizes descriptive keywords associated with each item. It does not necessitate the involvement of other users. Thus, the algorithms employed in a CBF RS are such that it recommends users similar items that the user has liked in the past. Websites like IMDb, Rotten Tomatoes, and Pandora are famous examples [16].

Table 1. Pros and cons of CBF and CF techniques

Technique	Pros	Cons
Content-based	1. Domain knowledge not required 2. Evolutionary 3. Tacit feedback	1. New user cold-start problem 2. Dependent on dataset quality 3. Overspecialization 4. Lack of Personalization
Collaborative	1. Spot cross-genre item 2. Domain Knowledge not required 3. Evolutionary 4. Tacit feedback	1. New user and item cold-start problem 2. Gray sheep problem 3. Dependent on dataset quality

Collaborative Filtering: One of the most widely used, well-established, and mature technologies currently available. Based on inter-user comparisons, CF RS find commonalities among users based on their rating history and provide new recommendations. The primary advantage of CF is that it works well for complex ideas where taste differences account for most variation in preferences. The theory behind CF is that people who have enjoyed similar products in the past would also like similar products in the future. Companies that employ this model include Amazon, Facebook, Twitter, Linked In, Spotify, Google News, and Last.fm [16].

2.2 Hybrid Recommendation Systems

To eliminate these individual flaws of a RS, a hybrid system is created that combines multiple RS. These systems are combined in a way where one system can alleviate the drawbacks of another. There are different ways of creating a hybrid system, this paper will discuss two of them below.

Cascade: The Cascade RS follows a staged process. In this technique, one RS is employed first to produce a list of items and the second technique refines the recommendation from among the sets. This technique falls short as successor's recommendations are restricted by predecessor and the subsequent RS may not introduce additional items. This type of system can be used to customize or refine a list from one significant RS in terms of order. For example if CBF's suggestions are passed into a SVD system, then we do not overcome the limitation of overspecialization in CBF although we refine the list in order as to a user's preference.

Weighted: In a weighted system, a score is computed for each item based on the multiple RS which have been employed. Each system has a certain weight assigned to it based on the types of results they generate. These weights are then used to compute the score for an item. The system then recommends the items with the best score. The simplest example would be a linear combination of the different systems to calculate the score. The P-Tango system [12] uses such a weighted hybrid system. P-Tango gives an initial weight to the different systems and based on the feedback received from a user, it changes the weights to fit the preferences of the user.

3   Methodology

3.1   Content-Based Filtering Approach

The technique chosen for the CBF is the good old vanilla cosine similarity. The cosine of the angle between two vectors projected into multidimensional space is used to calculate cosine similarity between them. It can be used to compute similarity between items in a dataset based on their contents. As indicated in the equation below, the similarity between two vectors (A and B) is obtained by taking the dot product of the two vectors and dividing it by the magnitude value.

$$similarity(A, B) = \cos(\theta) = \frac{A \cdot B}{\|A\|\|B\|} = \frac{\sum_{i=1}^{N} A_i B_i}{\sqrt{\sum_{i=1}^{N} A_i^2}\sqrt{\sum_{i=1}^{N} B_i^2}} \tag{1}$$

As we already know, CBF relies on the features associated with the items, which in our case are movies. The dataset includes many features of the movies such as overview, cast, director, genre, etc. To begin with, we merged the keywords, the main cast, the director, and the genres of the movie into a single string and called it a "soup" of metadata. To reduce noise and get better results, we removed "stop words" and applied stemming on the soups using NLTK [21]. Once the soups were processed, the next steps were to use count vectorizer or TF-IDF vectorizer on every soup generated. A count vectorizer transforms a string to a token count matrix, whereas a TF-IDF vectorizer translates a string to TF-IDF features. Both of these methods returns the vectors required in the formula above.

3.2   Model-Based Collaborative Filtering Approach

A model-based approach requires a learning phase in advance for finding out the optimal parameters before making a recommendation. The algorithm used here, is the Singular Vector Decomposition (SVD), a type of matrix decomposition technique that helps with dimensionality reduction problems that arise due to high sparsity. SVD is a powerful technique to discover hidden or latent correlations and reduce noise. An original rating or utility matrix A can be decomposed into three smaller matrices U, Σ, and V. Multiplying these three matrices gives the closest approximation of the original matrix A.

$$A \approx U\Sigma V^T \tag{2}$$

where A is an $m \times n$ matrix with each row representing a user and each column representing an item. This matrix's elements are the ratings given to items by users. U is an $m \times r$ orthogonal left singular matrix that represents the correlations between users and latent features. Σ, is a $r \times r$ diagonal matrix, which describes the strength of each latent feature in descending order, and V is a $r \times n$ diagonal right singular matrix, which represents the correlation between items and latent features.

The latent features here are the characteristics of an item, for example, the genre of a movie. Most latent features are so subtle or concealed that they cannot be identified by an individual but can only be detected using specific techniques. It is also difficult to even name most latent features, but they do exist.

To understand how SVD is applied in a movie RS, we must first understand that a utility matrix A is split into three submatrices, which yields the closest approximation of A when multiplied together. Since A is an user-item rating matrix, it will contain null values because not all users rate all movies. However, the sub-matrices will not have sparsity when SVD decomposes the utility matrix. Therefore, we can intuit that when multiplied, it will give an approximate ratings for the movies initially unrated by the user. Hence, we can recommend movies based on these predicted ratings.

To implement this, a library called surprise [15] which already incorporates all of the above-explained was used.

Table 2. RMSE and MAE

Metric	Fold1	Fold2	Fold3	Fold4	Fold5	Mean	Standard deviation
RMSE (testset)	0.8970	0.8922	0.8955	0.9057	0.8983	0.8977	0.0045
MAE (testset)	0.6885	0.6877	0.6928	0.6959	0.6939	0.6917	0.0032
First time	4.04	4.07	4.11	4.38	4.38	4.19	0.15
Test time	0.19	0.24	0.15	0.15	0.27	0.20	0.05

SVD is a model-based approach that needs some sort of prior learning. First, we performed k-fold cross-validation as we needed some assurance that our model is going to perform well. As shown in the Table 2, choosing the value of k as 5 results in a mean Root Mean Square Error of 0.8977 which is good enough for our application and hence we move further to train SVD on the whole trainset.

3.3   CBF/CF Cascaded Hybrid Approach

A cascaded hybrid has been implemented by combining the CBF and the model-based CF that was developed above. This hybrid was implemented in order to compare and evaluate the proposed weighted hybrid system and get a better understanding of its performance. The cascade hybrid is basically a series connection of the CBF and the model-based CF. Since a cascade hybrid is order sensitive, the CBF technique is employed first, later accompanied by the model-based CF technique. The CBF returns a list of the top N movies with the highest similarity score. The model-based CF then takes this list of N movies and refines them by sorting them in decreasing order of the rating predicted by the SVD model for the given user. This list of rearranged N most similar movies is presented to the user as the final recommendation list.

3.4 Proposed CBF/CF Weighted Hybrid Approach

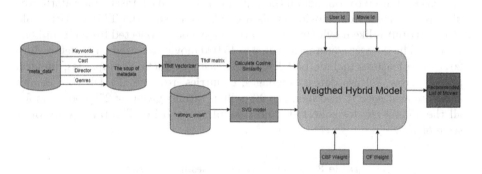

Fig. 1. Proposed system model.

This paper proposes a novel method of a weighted hybrid RS that uses cosine similarity and TF-IDF vectorization approach to develop a CBF RS, and Singular Value Decomposition to develop a model-based CF RS. This algorithm takes two inputs, one is a movie i and the other is an id of a user u. These inputs denote that the user u has watched a movie i and the algorithm is expected to return a list of movie recommendations for this particular user and this particular movie.

First, the cosine similarity matrix generated by the CBF is used to fetch the pair-wise similarity scores of the movie i with the rest of the movies j. Next, the ratings for all the movies j are predicted using the ".predict()" function of SVD. Once we have the similarity scores of movie i with each movie j, i.e., S_{ij}, and a predicted rating P_j for each movie j, which is scaled down from the range of 1–5 to 0–1, a linear combination with the weights assigned to CBF and CF is created. This will serve as the model for calculating the score for each movie j.

$$W(i, j, u) = w_1 S_{ij} + w_2 P_{uj} \tag{3}$$

where, relation between the two weights is

$$w_2 = 1 - w_1 \tag{4}$$

This weighted score, W is then used to sort the movies in a decreasing order and the top N movies are chosen to make the final recommendation list.

The weights used in the proposed system are 0.6 and 0.4 for CBF and CF respectively, the justification for choosing these weights has been given in the next section with the help of an experiment.

4   Experimental Results

4.1   Dataset

The Movies Dataset [18] which is a combination of TMDb [19] and GroupLens [20] has been used to conduct all the experiments. In this dataset, some attributes like credits, keywords, movie details are collected using the TMDb API while other attributes like movie links and ratings have been collected from GroupLens website. The dataset contains features of 45,000 movies released on or before July 2017. The features include plot keywords, cast, production companies, budget, posters, crew, release dates, languages, countries, revenue, TMDb vote counts and vote averages. It also includes 26 million ratings given by 270,000 users for all the 45,000 movies collected from GroupLens website. The ratings are on a scale of 1–5.

Table 3. Sample data of movies_metadata.csv

Id	Title	Genres	Overview	Tagline
8844	Jumanji	['id': 12, 'name': 'Adventure', 'id': 14, 'name': 'Fantasy', 'id': 10751, 'name': 'Family']	When siblings Judy and Peter discover an enchanted board game that...	Roll the dice and unleash the excitement!
15602	Grumpier Old Men	['id': 10749, 'name': 'Romance', 'id': 35, 'name': 'Comedy']	A family wedding reignites the ancient feud between...	Still Yelling. Still Fighting. Still Ready for Love.

Table 4. Sample data of ratings_small.csv

UserId	Movie Id	Rating	Timestamp
1	31	2.5	1260759144
1	1029	3.0	1260759179
1	1061	3.0	1260759182
2	10	4.0	835355493
2	62	3.0	835355749

Table 5. Information about dataset

Name	Description	Dimensions	Size
movies_metadata.csv	The main Movies Metadata file. Contains information on 45,000 movies featured in the Full MovieLens dataset. Features include posters, backdrops, budget, revenue, release dates, languages, production countries and companies	(45466, 2)	34.45 MB
keywords.csv	Contains the movie plot keywords for our MovieLens movies. Available in the form of a stringified JSON Object	(46419, 2)	6.23 MB
credits.csv	Consists of Cast and Crew Information for all our movies. Available in the form of a stringified JSON Object	(45476, 3)	189.82 MB
links_small.csv	Contains the TMDB and IMDB IDs of a small subset of 9,000 movies of the Full Dataset	(9125, 3)	183.37 KB
ratings_small.csv	The subset of 100,000 ratings from 700 users on 9,000 movies	(100004, 4)	2.44 MB

We chose The Movies Dataset because it contains enough data for both our CBF and CF methodologies. It provides metadata or features about movies, such as an overview, cast, director, and so on, that a CBF requires (as shown in Table 3). It also includes information about the user's rating of a particular movie that will be used for CF (as shown in Table 4).

4.2 Evaluation Metrics

To properly evaluate and compare our experiments, we have used different evaluation metrics appropriate for the type of RS. For the model-based CF, we have used RMSE and MAE as the evaluation metrics. While for the CBF, cascaded hybrid, and weighted hybrid systems, we have used "Personalization" and the "Diversity" of a recommendation list as the evaluation metrics. These metrics are explained below and then in later sections, the implemented systems have been tested on these metrics and the results have been discussed in detail.

RMSE and MAE: The SVD predicts the rating that a user might give to the movies they haven't watched yet, the accuracy of these predictions can be evaluated using metrics like RMSE (Root Mean Square Error) and MAE (Mean Absolute Error). Lower the RMSE and MAE, better the predictions.

$$MAE = \frac{1}{N} \sum_{i=1}^{N} |r_i - \hat{r}_i| \tag{5}$$

$$MSE = \frac{1}{N} \sum_{i=1}^{N} (r_i - \hat{r}_i)^2 \tag{6}$$

$$RMSE = \sqrt{MSE} = \sqrt{\frac{1}{N} \sum_{i=1}^{N} (r_i - \hat{r}_i)^2} \tag{7}$$

where N is total number of recommended movies, r_i and $\hat{r}_i$ are actual and predicted ratings of movie i, respectively.

Diversity: For better user satisfaction, the recommendation list generated for a user needs to be diverse. Diversity means the dissimilarity in the movies' content in a user's recommendation list. The higher diversity of a system indicates that it allows the user to explore movies belonging to various genres. The formula used to calculate the diversity is [10].

$$Diversity = 1 - \frac{\sum_{i,j \in R(u)} s_{i,j}}{\frac{1}{2}(N(N-1))} \tag{8}$$

Here, s_{ij} represents the similarity score calculated for the movie i with the movie j. N represents the total number of movies in the recommendation list (R(u)).

Personalization: Personalization is the dissimilarity between the recommended lists generated for two different users by the RS. A lower personalization score means that the recommendations for different users are very similar. The higher personalization of a system ensures that the recommendation list generated for a user is as close as possible to their taste and preferences. The "recmetrics" [17] library was used to calculate the personalization.

4.3   Content-Based Filtering

Table 6 shows the list of top 10 movies that were recommended for the movie "The Dark Knight" by CBF. The recommendations not only includes the movies which are similar to the given movie in terms of the genre or the keywords but also presents the movies which share similar cast members or directors. This is the reason why the diversity calculated for the CBF is approximately 0.85.

Table 6. Recommended list for content-based filtering technique for the input movie: "The Dark Knight"

Rank	Title	Similarity score
1	The Dark Knight Rises	0.407
2	Batman Begins	0.377
3	The Prestige	0.314
4	Inception	0.253
5	Following	0.252
6	Insomnia	0.246
7	Memento	0.230
8	Interstellar	0.185
9	Batman Returns	0.160
10	Batman: Under the Red Hood	0.148

In terms of personalization, the list generated will be exactly the same for a given input movie ("The Dark Knight" in our case) for any user because CBF does not incorporate user's behaviour factor. Hence, the personalization score calculated for this system is approximately 0.

4.4 Model-Based Collaborative Filtering

The mean RMSE of 0.89 and the mean MAE of 0.69 indicate that the Singular Value Decomposition performs very well with accurate predictions of a user's ratings. This means that the CF technique will be able to understand the behavior and preferences of a user more accurately and in turn, it will make the system more personalized for the user.

The average diversity for 30 users selected at random in CF is 0.98. This shows CF gives very diverse recommendations which sometimes might not be a good choice as there's a huge disparity in the similarity of the movies recommended. This makes the recommendations seem random and irrelevant. The personalization score calculated is around 0.7 which means that 70% of of the movies present in the recommended list of a user are not repeated in the recommended lists of other users.

Though, the new item cold start problem and the stability/plasticity dilemma persist in this implementation, which makes the CF less reliable.

4.5 CBF/CF Cascaded Hybrid

Figure 2 displays the values of diversity calculated for the recommended lists generated for 30 users by the cascaded hybrid system. The X-axis depicts the user id and the Y-axis depicts the value of the diversity.

The diversity is constant at 0.85 for all the users because the list generated by a cascade system is just a rearrangement of the list generated by CBF. This

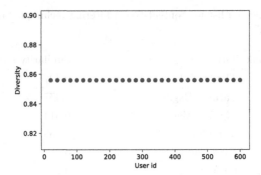

Fig. 2. Diversity of recommended list of different users for Cascaded system

is the reason why the personalization score of the cascade system is 1.1×10^{-16}, i.e., approximately 0. Therefore, a cascaded system suffers from the same issues as the CBF.

The cascaded system, however, alleviates the new item cold-start problem which the CF suffers from, because of the involvement of the CBF which does not require any data except the content of the movie to make recommendations. Hence, this gives the new items in the dataset a much better chance of getting recommended.

4.6   Proposed CBF/CF Weighted Hybrid

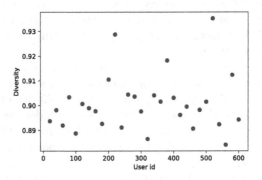

Fig. 3. Diversity of recommended list of different users for weighted system

Figure 3 displays the values of diversity calculated for the recommended lists generated for 30 users by the proposed weighted hybrid system when the weights are 0.6 and 0.4 for CBF and CF techniques respectively. The X-axis depicts the user id and the Y-axis depicts the value of the diversity.

The diversity calculated for each user is different and varies approximately from 0.88 to 0.94. This indicates that the issue of overspecialization has been reduced by a great deal as compared to the CF and cascaded systems. The personalization score for this combination of weights (0.6 for CBF and 0.4 for CF) is approximately 0.38 which means that around 38% of the movies present in the recommended list of a user are not repeated in the recommended lists of other users. Hence, making the lists significantly more personalized compared to the CBF and Cascaded systems but less personalized than CF technique implemented alone. This trade-off is necessary to incorporate the best of both techniques: CBF and CF, as it gives the user more room to explore as well as takes into account their past behaviour. This also helps in solving the cold start problem persisting in the CF technique.

An experiment was done to determine the optimal value of the weights that can result in a satisfying personalization score as well as diversity. A total of 20 users were selected randomly and the personalization score and the value of the diversity were calculated for different sets of weights assigned to CF and CBF techniques. For each sets of weights, the diversity was taken as an average of the diversities calculated for all the 20 users. The sets of weights were taken in the range of 0.1–0.9 with an interval of 0.1. The weights were assigned according to the Eq. 4. The personalization scores and diversity values were plotted against the weights taken, to generate Fig. 4 and Fig. 5.

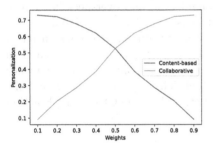

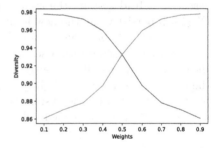

Fig. 4. Variation of personalization with weights

Fig. 5. Variation of diversity with weights

Figure 4 shows that as the weight assigned to the CF technique increases, the personalization increases. This is expected because that would give more priority to the predicted ratings generated by the SVD, which makes the result more personalized for the user.

Figure 5 shows that as the weight assigned to the CF technique increases, the diversity increases. This too is expected because that would give less priority to the similarity scores generated by the CBF technique which in turn increases the dissimilarity between the movies being recommended and diversifies the result.

If the same weight is assigned to both the CBF and CF systems, the extremely popular movies (with a very high number of user ratings) will make their way into the recommendations even if their similarity score might be close to 0. To prevent this from happening, weights of 0.6 and 0.4 are assigned to CBF and CF respectively, which gives the similarity score of a movie a slightly higher priority than its predicted rating. This also alleviates the new item cold-start problem.

Hence, the weights 0.6 and 0.4 assigned to CBF and CF respectively, have proven to generate the best results in terms of alleviating the new item cold-start problem and obtaining good personalization scores and diversity.

5   Conclusion

Both CBF and CF systems have their individual flaws and drawbacks, these can be solved by building a hybrid model that combines the two systems. In a hybrid model, one RS alleviates the drawbacks of the other RS and vice versa. CBF suffers from overspecialization and the lack of personalization, these flaws are alleviated by the model-based CF system. The model-based CF suffers from the new item cold-start problem, but when CBF is employed in combination with CF, this problem is reduced and the chances of a new item being recommended are ameliorated.

The proposed Weighted hybrid RS combines CBF and model-based CF. CBF is implemented using cosine similarity and TF-IDF vectorization, and the model-based CF uses SVD. The experiments that we conducted, indicate that in a cascaded hybrid system, the issues of personalization and overspecialization of CBF still persist while the new item cold-start issue of CF is alleviated.

A weighted hybrid system generates such a recommended list which has a much higher personalization score and a much better diversity as compared to the cascaded system, it also fairly resolves the new item cold-start problem since the CBF has been given a slightly higher weight. These results indicate that the proposed weighted hybrid system performs better than the CBF and CF acting alone, as well as the cascaded hybrid system.

References

1. Wang, L.-C., Meng, X.-W., Zhang, Y.-J.: Context-aware recommender systems. J. Softw. **23**, 1–20 (2012). https://doi.org/10.3724/sp.j.1001.2012.04100
2. Ricci, F., Rokach, L., Shapira, B., Kantor, P.B. (eds.): Recommender Systems Handbook. Springer, Boston, MA (2011). https://doi.org/10.1007/978-0-387-85820-3
3. Schafer, J.B., Konstan, J., Riedi, J.: Recommender systems in e-commerce. In: Proceedings of the 1st ACM Conference on Electronic Commerce - EC 1999 (1999). https://doi.org/10.1145/336992.337035
4. Ratings, reviews, and where to watch the best movies & TV shows. In: IMDb. https://www.imdb.com/. Accessed 15 Aug 2021
5. Goldberg, D., Nichols, D., Oki, B.M., Terry, D.: Using collaborative filtering to weave an information tapestry. Commun. ACM **35**, 61–70 (1992). https://doi.org/10.1145/138859.138867
6. Resnick, P., Iacovou, N., Suchak, M., et al.: GroupLens. In: Proceedings of the 1994 ACM Conference on Computer Supported Cooperative Work - CSCW 1994 (1994). https://doi.org/10.1145/192844.192905

7. Hill, W., Stead, L., Rosenstein, M., Furnas. G.: Recommending and evaluating choices in a virtual community of use. In: Proceedings of the SIGCHI Conference on Human Factors in Computing Systems - CHI 1995 (1995). https://doi.org/10.1145/223904.223929
8. Balabanović, M., Shoham, Y.: Fab. Commun. ACM **40**, 66–72 (1997). https://doi.org/10.1145/245108.245124
9. Burke, R.: Hybrid recommender systems: survey and experiments. User Model. User-Adapt. Interact. **12** (2002). https://doi.org/10.1023/A:1021240730564
10. Chen, R., Hua, Q., Chang, Y.-S., et al.: A survey of collaborative filtering-based recommender systems: from traditional methods to hybrid methods based on social networks. IEEE Access **6**, 64301–64320 (2018). https://doi.org/10.1109/access.2018.2877208
11. Çano, E., Morisio, M.: Hybrid recommender systems: a systematic literature review. Intell. Data Anal. **21**, 1487–1524 (2017). https://doi.org/10.3233/ida-163209
12. de Campos, L.M., Fernández-Luna, J.M., Huete, J.F., Rueda-Morales, M.A.: Combining content-based and collaborative recommendations: a hybrid approach based on Bayesian networks. Int. J. Approx. Reason. **51**, 785–799 (2010). https://doi.org/10.1016/j.ijar.2010.04.001
13. Do, H.-Q, Le, T.-H, Yoon, B.: Dynamic weighted hybrid recommender systems. In: 2020 22nd International Conference on Advanced Communication Technology (ICACT). (2020). https://doi.org/10.23919/icact48636.2020.9061465
14. Suriati, S., Dwiastuti, M., Tulus, T.: Weighted hybrid technique for recommender system. J. Phys. Conf. Ser. **930**, 012050 (2017). https://doi.org/10.1088/1742-6596/930/1/012050
15. Hug N Home: In: Surprise. http://surpriselib.com/. Accessed 15 Aug 2021
16. Fatourechi, M.: The evolving landscape of recommendation systems. In: TechCrunch (2015). https://techcrunch.com/2015/09/28/the-evolving-landscape-of-recommendation-systems/. Accessed 15 Aug 2021
17. Statisticianinstilettos Statisticianinstilettos/Recmetrics: A library of metrics for evaluating recommender systems. In: GitHub. https://github.com/statisticianinstilettos/recmetrics.. Accessed 15 Aug 2021
18. Banik, R.: The movies dataset. In: Kaggle. https://www.kaggle.com/rounakbanik/the-movies-dataset. Accessed 15 Aug 2021
19. The movie database (TMDB). https://www.themoviedb.org/. Accessed 15 Aug 2021
20. GroupLens. https://grouplens.org/. Accessed 15 Aug 2021
21. Natural Language Toolkit. Natural Language Toolkit - NLTK 3.6.2 documentation. https://www.nltk.org/. Accessed 15 Aug 2021

Text Analysis and Classification for Preprocessing Phase of Automatic Text Summarization Systems

Vaishali P. Kadam[(✉)], Kalpana B. Khandale, and Namrata Mahender C.

Department of Computer Science and I.T, Dr. Babasaheb Ambedkar Marathwada University,
Aurangabad, Maharashtra, India
vaishu7817kadam@yahoo.in, kalpanakhandale1788@gmail.com,
cnamrata.csit@bamu.ac.in

Abstract. Stop word removal is a basic morphological analysis or a process used to remove those words that appear frequently or repeatedly within the document text. These words do not affect generating language expressions but appear many times in default language transformation and communication. This caused us to acquire more space in memory and the allocation of the same words caused problems with memory indexing. In the case of big data analysis and information management applications, identification of such particular words present within the document is an important task. In this paper, text analysis was carried out to identify various stop words within the text. It is used as a pre-processing operation for the development of various natural language text processing and data mining techniques. This is important in part-of-speech tagging, sentiment analysis, content classification, semantic analysis, search engines, text summarization, named entity recognition, etc. Our proposed system discusses the role of text analysis techniques and pre-processing of language by removing stop words from a given input document written in Marathi text using a linguistic and rule-based approach. The system is particularly discussed for Marathi. Very little work has been done on stop word removal using a supervised approach and evaluating its impact on the performance of information retrieval systems and summarizing the text. The result get accuracy at 32.6% to compress the text.

Keywords: Information retrieval · Marathi text · Natural language processing · Text analysis · Text mining

1 Introduction

Natural language processing is a sub-field of computer science and artificial intelligence. It deals study of various human communication or natural languages, which is useful for making a machine capable of understanding natural language and manipulating it for various language transformations for cultural and social standard development. It has been in existence since the 1958. It is also known as computational linguistics. Natural language research in Indian languages has mainly focused on the development of

© Springer Nature Switzerland AG 2022
K. K. Patel et al. (Eds.): icSoftComp 2021, CCIS 1572, pp. 382–396, 2022.
https://doi.org/10.1007/978-3-031-05767-0_30

rule based systems due to the lack of annotated corpora. The automatic summarization of text is a well- known task in the field of natural language processing. Significant achievements in text summarization have been obtained using text classification and text analysis [1]. NLP is useful for developing standards in human interaction by understanding various natural languages and knowledge exchange. Today's era is known for its information explosion and information overload. Unlimited information is available in the form of various language components through mobile, information technology and internet resources. Now it has become a challenge for researchers and technologies to access the specific required information using tools that can save time and money and provide it efficiently for the user. Most of this information is available in written text format. Sorting of all this information as per the needs of the user is a difficult task. Many times, information is not available in a good structured format or ready to use format that users require for processing. But for designing any natural language application, this information should have to be in a standard structured format. It is to make this information more specific for language understanding and language generation. Managing such a language component, it should go through some proper information analysis. This text analysis is based on the morphology of that specific language and the grammar of the language. It also depends on the intention of the user for which transformation of language is done. A language is actually a set of characters and standard notations using which meaningful language words are formed. These are used for daily life communication. But in the case of information systems, language is always used to generate the things that are expressed by language notation depending on the instance. Stop word removal is mostly used to improve system performance and it is achieved by increasing its standard in speed and accuracy. There may be various instances where the same word appears repeatedly within a document that is used in the system database, causing increased storage requirements. By removing stop words, system index files are reduced as the words present in the text are repeated in the documents. It is useful in handling memory management and time constraints efficiently. Stop words are common function words like articles, prepositions, conjunctions, prefix, suffix etc. that carry less important meaning than keywords. Hence should be eliminated.

1.1 Importance of Preprocessing in Natural Language Text

There are various operations that are used before processing of the data for any linguistic computational model designing or application development. This is important because most of the tremendous data is available through various digital and electronic resources in the form of written text, but it is not always available in a standard structured format. Preprocessing is generally done to maintain data in a standard format. The following are some important operations in preprocessing.

1.1.1 Text Analysis

In this, various morphological processes are performed on the text to make it useful for linguistic computational processing and application development.

1.1.2 Text Cleaning

In this, data undergoes various data mining techniques and methods to reduce the non-useful parts of the input text, like stop word removal or less important context removal, through using various document analysis levels or data validation.

1.1.3 Text Annotation

In this, various tools and techniques are used to illustrate the grammatical meaning and related details context to generate and understand the language, like part-of-speech (POS) tagging and some structural marking based on identification of gender, tense, plurals or singular, synonyms and antonyms etc.

1.1.4 Text Normalization

In this, mapping of the text is done or text undergoes for linguistic reduction using some techniques like stemming, lemmatization and other standard forms useful for computational linguistic analysis.

1.2 Importance of Stop Word Removal in Marathi Natural Language Text Processing

The Marathi language is an Indo-Aryan language. It has various forms of transformations used in communication and language generation. It is a free-order language consisting of a large set of stop words that repeatedly appear in the language text. These words do not affect the meaning generation. Many times, in the case of information retrieval, it creates confusion. As Marathi is a free-ordered language by its nature, it creates ambiguities when same words used in different statement formations. Less work has been done on Marathi language stop words. There are some standard stop words lists are available online by IIIT Hyderabad, TDIL, kaggle and Ranks NL where kaggle and Ranks NL provides stop words only for Hindi, Bengali, and Marathi languages. Some of the researchers also proposed methods for own list of stop words from corpus. No such standard Marathi stop word list available for computational processing that has cover list of all such words. Stop word removal is important to find relevant information and actual theme understanding and identification in quick information analysis and retrieval services.

2 Literature Survey

Stop words removal is difficult in the Marathi language. There are various stop word removal tools available for English and other foreign languages. Much work has been done for the foreign languages. In the initial days of NLP development, it was studied only for the English language. But studies related to Indic languages started after 2000. The below table shows the related studies for the various stop words removal systems available for various languages (Tables 1 and 2).

Table 1. Literature study of foreign languages

Author	Language	Technique and methods	Performance and features
Guembe Blessing et al. 2021 [2]	Multilingual	Machine learning and natural language processing techniques, stemming, POS tagging, lemmatization, stop words	This review performs study in the research domain of automatic text based assessment grading system for ODL courses
Manjit Jaiswal et al. 2021 [3]	English	Stop words removal, TF-IDF and stemming algorithm, Naïve Bayes algorithm	Accuracy of 99.67% for training data and 99.03% for testing data which is more accurate thanNaïve Bayes classifier
Serhad Sarica et al. 2020 [4]	English	Stop word list implantation technique based on the synthesis of alternative data driven approaches	Final stop words list implemented forNLTK and USPTO in NLP for text analysis tasks related to technology
Reda Elbarougy et al. 2019 [5]	Arabic	Graph-based summarization algorithm, morphological analyzer, sentence similarity measure	Study proved that performance of summarization is increased when stop words are removed
Olusola AbayomiAlli et al. 2018 [6]	English	Genetic algorithm, artificial immune system (AIS) algorithm	System improves feature selection for dimensionality reduction text classification for short text of SMS, tweets, WhatsApp, microblogs, email etc.
Jennifer P. et al. 2018 [7]	English	Overview of stop words, data mining, stemming	Efficiently studied the term of data mining applications
Alexandra Schofield et al. 2017 [8]	English	Methods for removing corpus specific stop words	Removing stop words after training is effective as removing them before

(continued)

Table 1. (*continued*)

Author	Language	Technique and methods	Performance and features
Stephen A. Adubi et al. 2016 [9]	Yoruba	Syllable extraction algorithm, statistical and dictionary-based methods	Among five syllable extraction algorithms of different languages.YSYLL records 100% accuracy for Yoruba language
Olugbenga Oluwagbemi et al. 2013 [10]	R, Python, Lua, Scala, Haskell, Ruby PHP, Perl, JavaScript, Erlang,	Techniques for comparison analysis of ten popular script languages according to the TIOBE language ranking	The study provides guidance to researchers to select a suitable script language in software engineering, bioinformatics, computational biology
Ibrahim Akman et al. 2011 [11]	Turkish	Text compression algorithm, set of 20 different texts varying sizes between 4.6 and 725 K bytes	Compression rate varying from 13.0% to 43.2%. It represents thathigher compression rates can be achieved with increasing text sizes
S. Kannan et al. 2011 [12]	English	Preprocessing methods, tokenization, stop word removal and stemming	Pre-processing techniques eliminate noisy text data and improve the performance of information retrieval
A N K Zaman et al. 2011 [13]	English	Latent semantic indexing (LSI) based information retrieval technique	For the large data set tailored stop word lists improves information retrieval

2.1 Challenges in Natural Language Processing Application Development

The development of any natural language processing application system requires proper language text with a standard format, lexical resources to understand the language and generate the meaning of the language. With this, NLP requires standard system tools to process the data and convert it into standard information based on the linguistic and structural analysis of the text. In the case of a language like Marathi, it has a free structure. There are various formats of word and sentence formations used to express the same thing and meaning. It is quite difficult to understand the language properly and present its meaning. So many times, researchers face difficulties with computational processing and text analysis.

Table 2. Literature study for Indian languages

Author	Language	Technique and methods	Performance and features
Deepali Kadam et al. 2021 [14]	Marathi	Stemming, stop word removal, sentence ranking, machine learning, tokenization	Study observed that comparatively less work has been done in Marathi language. Need to focus mainly on its preprocessing phase of summarization
Prafulla B. Bafna et al. 2021 [15]	Marathi	Clustering technique for incremental data set, TF-IDF, cosine-based document similarity measures	Entropy and precision used to evaluate results on different datasets to find the performance of proposed approach for the Marathi Corpus
Pradeepika Verma et al. 2020 [16]	English, Hindi, Tamil Marathi, Punjabi	Stop word removal, stemming, segmentation, feature extraction, sentence ranking methods	Results of precision, recall, and F1 measures for summarization methods compared for different languages to check performance of NLP tools
Prafulla B. Bafna et. al. 2019 [17]	Hindi	Unsupervised learning method, Multi-document clustering, word cloud technique, Fuzzy k-means	The entropy produced by Fuzzy K-means is improved by 40% for incremental data size, i.e. for 300 documents
Abhijit Barman et al. 2019 [18]	Bengali	Supervised machine learning decision tree, Naive Bayes multinomial, optimized SVM	System removes around 42% of total corpus or insignificant content using IDF and ICF measure
Rakib U. Haque et al. 2019 [19]	Bengali	Stop word detection and extraction techniques	Stop word elimination algorithm achieved an accuracy of 70 - 75%

(*continued*)

Table 2. (*continued*)

Author	Language	Technique and methods	Performance and features
Ruby Rani et al. 2018 [20]	Hindi	Term-Weight model and Information theory model	Stop word list is prepared for Hindi with 1475 words which is more general than domain specific lists and useful in generic information retrieval
Dhara J. Ladani et al. 2017 [21]	Hindi, Gujarathi, Multilingual	Static Approach Based stop word Identification	Stop words removal reduces text size and improves accuracy with space, time complexity reduction in searching and indexing
Pooja Bolaj et al. 2016 [22]	Marathi	Supervised machine learning methods, Naïve Bayes, modified K-nearest neighbor, SVM and ontology-based classification	Marathi document text classified using supervised and ontology-based methods for six class label like festival, sports, history, literature and tourism
Vandana Jha et al. 2016 [23]	Hindi	A stopword removal algorithm using DFA	The system has given an accuracy of 99%, which is tested on 200 documents
Jaideep singh K. Raulji et al. 2016 [24]	Sanskrit	The stop-word removal algorithm, dictionary-based approach	Effective indexing can be achieved by the removal of stop words for Sanskrit text
Jasleen Kaur et al. 2015 [25]	Language independent	Supervised learning algorithms Naive Bayes, SVM, Artificial Neural Network, N-gram technique	Text classifiers are analyzed for their performance among them supervised approaches have given good results for Indian languages
Ashish B. Tikarya et al. 2014 [26]	Gujarathi	Techniques for stemming, stop-word and elimination of duplicate sentence	Study discussed preprocessing of text summarization for reducing the size of text

(*continued*)

Table 2. (*continued*)

Author	Language	Technique and methods	Performance and features
Jayashree R. et al. 2013 [27]	Kannada	Naïve Bayesian algorithm for dimensionality reduction with TF, stop word removal	The system has increase the performance of the classifier by reduction of stop words
Jayashree R. et al. 2011 [28]	Kannada	Stop word removal method	Pre-categorized data and good classifiers provide classification with 55% accuracy
Amaresh Kumar Pandey et al. 2009 [29]	Hindi	Stemming and stop word removal techniques for Hindi using 3 stemmers	This study observed that stop word removal and stemming improves the performance of Information retrieval

2.2   Level of Scope for Text Document Summarization

Every text document or language text has four different types of scope in data analysis and information retrieval systems. The scope of the text document is considered as per the requirements of the application or the task for which data is considered to apply.

2.2.1   Document Level

The whole document is considered for morphological or linguistic processing.

2.2.2   Paragraph Level

A specific part or a single paragraph from a document is selected and applied to process the text and to obtain the result.

2.2.3   Sentence Level

A single sentence or a part of a paragraph is used for processing the text.

2.2.4   Word Level

Individual words from the sentence or sub-expression are used for processing.

3   Dataset

For the database, we created our own database corpus in Marathi text which is a Deonagri script. We have used 25 moral stories which have available online. From these different stories, we have collected a total of 3260 words. For our proposed work, we have trained our system with a total of 3260 words. The below table gives an idea about the database that we have created (Table 3).

Table 3. Sample of the text documents (Stories)

Name of sample document	Length of sample document in no. of words
Sd1	117
Sd2	136
Sd3	159
Sd4	146
Sd5	144
Sd6	202
Sd7	124
Sd8	136
Sd9	192
Sd10	145

3.1 Stop Word Identification Rules

The primary process for summary generation is to remove stop words from the document and identify main keywords. In this, sentence segmentation is done by its boundary detection. It is denoted with a period (.), exclamation mark (!), question mark (?), or a pipe (|). These four symbols are used to end sentences in the Indian language. Hindi, Bengali, and Punjabi languages use pipes to end the sentence, while other languages use a period to denote the sentence's end. In general stop words appeared with sentence ending or with a word used to join two sentences.

3.2 List of Marathi Stop Words

For our system, we have created a manual list of Marathi stop words that is maintained in the form of a dictionary. It is used to identify stop words in the document. It needs to identify these words during removal processing using text or string matching concepts. This list consists of a total of 300 stop words (Table 4).

Table 4. Sample list of manually created Marathi stop words

Marathi Stop word list			
अधिक	असा	असून	करण्यात
असल्याचे	असलेल्या	असे	करून
अनेक	अशी	आणि	काम
आता	आपल्या	आला	काय
आहे	एक	आली	की
आहेत	एका	आले	झाली
काही	किंवा	झाला	झाले
केला	घेऊन	जात	झालेल्या
केली	पुढचे	तिचे	काही
केले	च्याबद्दल	अधिक	देखील
गेल्या	स्वत:	गेले	आमचे
तर	एकदा	नंतर	एकदा
होता	खूप	आहेत	मला
तरी	आहे	आमच्या	पण
तसेच	ते	याचा	परंतु
ती	तो	यावर	म्हणून
तुला	मात्र	या	म्हणाले
ते	त्या	मी	म्हणजे
त्याची	त्याच्या	दिली	न
त्यांना	त्यांनी	त्यामुळे	नाही

4 Methodology

There are eight types of part-of-speech found in Marathi, namely noun, verb, adjective, adverb, etc. In the language, they appear in different forms with variations in their structure to express a specific meaning. The Marathi language is an ordered free language in its morphological structure. In general, the Marathi language follows the Subject + Object + Verb (SOV) format for language construction, which is more difficult. The proposed method follows rules for removing stop words accurately based on transformation grammar. Below, Fig. 1 shows the architectural view of the rule-based preprocessing tool for stop word removal in Marathi.

Proposed System follows steps for removal of stop words as shown below

4.1 Input Marathi Text

For input, we have selected an online available story document consisting of the Marathi language text. We have done manual validation of the text for its standard format. The documents are in Devanagari script and stored in UTF-8 format.

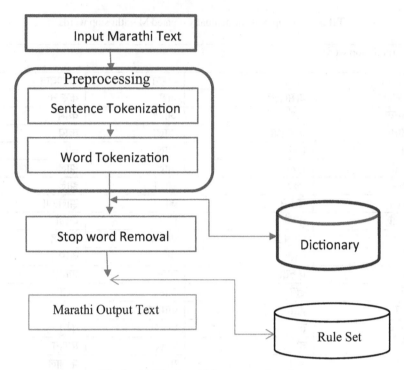

Fig. 1. Architecture of the proposed system

4.2 Preprocessing

In preprocessing, we have tokenized the sentences and the words individually from the input text. It has been done using the python language and the natural language toolkit (NLTK) library functions.

4.2.1 Sentence Tokenization

Sentence tokenization is the basic operation that divides the text in segments or splits text corpora. The whole text is divided into a number of individual sentences separated by a comma. In Marathi language, sentence is segmented by its end which is represented with a dot. The following example shows the sentence tokenization.

Input Sentence- गीता एक दहावीत शिकणारी मुलगी आहे. ती उषाची मोठी मुलगी आहे. ती खूप हुशार आणि गुणी आहे.

Tokenized Sentence - 'गीता एक दहावीत शिकणारी मुलगी आहे',' ती उषाची मोठी मुलगी आहे', ' ती खूप हुशार आणि गुणी आहे'

4.2.2 Word Tokenization

After sentence tokenization, we have to tokenize the sentence or split the words individually from the sentence. Sentences are split into the number of individual words or tokens from input text this process is called word tokenization. For example

Input Sentence-

गीता एक दहावीत शिकणारी मुलगी आहे. ती उषाची मोठी मुलगी आहे. ती खूप हुशार आणि गुणी आहे.

Tokenized Words-

'गीता',' एक',' दहावीत', 'शिकणारी',' मुलगी',' आहे',' ती',' उषाची',' मोठी',' मुलगी',' आहे',' ती ',' खूप',' हुशार',' आणि', ' गुणी', 'आहे'

4.3 Stop Word Removal

4.3.1 Select a word from the list of stop word

4.3.2 Check it with all the words from document by its similarity

4.3.3 If selected word is stop word then delete all the occurrences of that word in the text inputted.

4.3.4 If selected word is not match with the stop word then read next word from array for the processing.

4.3.5 Repeat the process till array completion. Delete all find stop words in the text inputted to avoid them in final output.

Input: 'गीता',' एक',' दहावीत', 'शिकणारी',' मुलगी',' आहे',' ती',' उषाची',' मोठी',' मुलगी',' आहे',' ती ',' खूप',' हुशार', 'आणि', 'गुणी', 'आहे'

Removed Stop words: एक, आहे, ती, मोठी, खूप, आणि

4.4 Marathi Output Text

Produce output Marathi Language text after preprocessing or stop word removal and by applying the language rules. Finally, it display text without the stop words.

Input Sentence- गीता एक दहावीत शिकणारी मुलगी आहे. ती उषाची मोठी मुलगी आहे. ती खूप हुशार आणि गुणी आहे.

Text after Stop Word Removal- गीता दहावीत शिकणारी उषाची हुशार गुणी मुलगी

5   Results

For the proposed system we train our system with the Marathi Language words. A list of stop word manually created using set of stop words collected from the Marathi language text corpora. We have collected a total of 3260 words corpora. From the overall corpora of text, 32.62% text is correctly removed by our Rule-based stop word removal system. Performance of the system is calculated in percentage using the below formula of accuracy (Table 5).

$$\text{Accuracy} = \frac{\text{Number of Correct Words} + \text{Number of Incorrect Words}}{\text{Total Number of Words}} \times 100$$

Table 5. Sample result of stop word removal

Name of sample document	Size of document in total no. of words	No. of stop words removed	Percentage of removal of stop words
Sd1	117	42	35.89
Sd2	136	80	58.82
Sd3	159	51	32.07
Sd4	146	43	29.86
Sd5	144	51	25.24
Sd6	202	39	31.45
Sd7	124	38	27.94
Sd8	136	53	27.60
Sd9	192	45	31.03
Sd10	145	43	26.54

6   Conclusion

Stop word removal is an important preprocessing phase in the development of various text analysis and information retrieval systems, such as text summarization. There are various words in Marathi language that have appeared frequently by its default nature. To remove such stop words and enhance system performance, our system will play an important role. There are other Indian languages that also follows the same Devnagari script, like Konkani, Hindi, Punjabi, and Gujarati, that have achieved more success by developing various language text preprocessing tools. But for Marathi, very little work has been done on stop word removal. To achieve better accuracy, we have designed a rule-based supervised machine learning approach for Marathi. Our system gives the average accuracy calculated at 32.62%. It is useful in language communication and language generation to efficiently handle text data and improve performance by reducing text for quick information retrieval. In the future, we have to concentrate on enhancing the database and increasing the size of the text with list of stop words.

Acknowledgement. Authors would like to acknowledge and thanks to CSRI DST Major Project sanctioned No. SR/CSRI/71/2015 (G), Computational and Psycholinguistic Research Lab Facility supporting to this work and Department of Computer Science and Information Technology, Dr. Babasaheb Ambedkar Marathwada University, Aurangabad, Maharashtra, India. Also thankful to the SARTHI organization for providing financial assistance and Ph. D. research fellowship. Sincere thanks to research guide C. Namrata Mahender (Asst. Professor), of the department for providing research facilities, constant technical and moral support.

References

1. Gaikwad, D.K., Mahender, C.N.: A review paper on text summarization. Int. J. Adv. Res. Comput. Commun. Eng. **5**(3), 2319–5940 (2016). ISSN (Online) 2278–1021 ISSN: (Print)
2. Blessing, G., Azeta, A., Misra, S., Chigozie, F., Ahuja, R.: A machine learning prediction of automatic text based assessment for open and distance learning: a review. In: Abraham, A., Panda, M., Pradhan, S., Garcia-Hernandez, L., Ma, Kn. (eds.) IBICA 2019. AISC, vol. 1180, pp. 369–380. Springer, Cham (2021). https://doi.org/10.1007/978-3-030-49339-4_38
3. Jaiswal, M., Das, S.,Khushboo: Detecting spam e-mails using stop word TF-IDF and stemming algorithm with Naïve Bayes classifier on the multicore GPU. Int. J. Electric. Comput. Eng. **11**(4), 3168–3175 (2021)
4. Sarica, S., Luo, J.: Stop words in Technical Language Processing. ResearchGate Publication (2020). https://www.researchgate.net/publication/341926808
5. Elbarougy, R., Behery, G., Khatib, A.E.: Extractive Arabic text summarization using modified Pagerank algorithm. Article Egypt. Inf. J. (2019). https://www.researchgate.net/journal/Egyptian-Informatics-Journal 1 110-8665
6. Abayomi-Alli, O., et al.: An improved feature selection method for short text classification. In: The 3rd International Conference on Computing and Applied Informatics 2018. IOP Conference Series: Journal of Physics: Conference Series, vol. 1235, pp. 012021. IOP Publishing (2019). https://doi.org/10.1088/1742-6596/1235/1/012021 (2021)
7. Jennifer, P., Muthukumaravel, A.: A study on stop words, stemming and Tet mining. Int. J. Pure Appl. Math. **118**(no. 20) 155–167 (2018)
8. Schofield, A., Magnusson, M., Mimno, D.: Pulling out the Stos: Rethinking Stop word Removal for Topic Models. ReserachGate Publication (2017). https://www.researchgate.net/publication/318741781
9. Oluwagbemi, O., Adewumi, A., Misra, S.: An Analysis of Scripting Languages for Research in Applied Computing (2013). https://www.researchgate.net/publication/259147772
10. Adubi, S.A., Misra, S.: Syllable-based text compression: a language case study. Arab. J. Sci. Eng. **41**(8), 3089–3097 (2016). https://doi.org/10.1007/s13369-016-2070-1
11. Akman, I., Bayindir, H., Ozleme, S., Akin, Z., Misra, S.: Lossless text compression technique using syllable based morphology. Int. Arab J. Inf. Technol. **8**(1) (2011)
12. Kannan, S., Gurusamy, V.: Preprocessing Techniques for Text mining. ResearchGate Publication (2011). https://www.researchgate.net/publication/273127322
13. Zaman, A.N.K., Matsakis, P.: Charles Brown: Evaluation of Stop Word Lists in Text Retrieval Using Latent Semantic Indexing. ResearchGate Publication (2011). https://www.researchgate.net/publication/221254145
14. Kadam, D.: A survey of extractive text summarization for regional language. Int. Res. J. Eng. Technol. **08**(06) (2021). e-ISSN: 2395–0056 www.irjet.net p-ISSN: 23950072
15. Bafna, P.B., Saini, J.R.: Marathi text analysis using unsupervised learning and word cloud. Int. J. Eng. Adv. Technol. **9**(3) (2020). ISSN: 2249 – 8958

16. Verma, P., Verma, A.: Accountability of NLP tools in text summarization for Indian languages. J. Sci. Res. **64**(1) (2020). http://dx.doi.org/https://doi.org/10.37398/JSR.640149.
17. Bafna, P.B., Saini, J.: Hindi Multi-document Word Cloud based Summarization through Unsupervised Learning. In: 9th International Conference on Emerging Trends in Engineering and Technology -Signal and Information Processing (ICETET-SIP-19) (2019). https://doi.org/10.1109/ICETET-SIP-1946815.2019.9092259. (2019)
18. Barman, A., Saha, D.: Algorithm for removal of semantically insignificant content words. JCSE Int. J. Comput. Sci. Eng. **7**(1) (2019). E-ISSN: 2347–2693
19. Haque, R.U., Mehera, P., Mridha, M.F., Hamid, Md.A.: A Complete Bengali Stop Word Detection Mechanism. ResearchGate Publication (2019). https://www.researchgate.net/publication/336327658
20. Rani, R., Lobiyal, D.K.: Automatic construction of generic stop words list for Hindi text. In: International Conference on Computational Intelligence and Data Science (ICCIDS 2018) (2018). https://www.researchgate.net/publication/325657435
21. Ladani, D.J., Desai, N.P.: Stop word identification and removal techniques on TC and IR applications: a survey. In: 6th International Conference on Advanced Computing and Communication Systems (ICACCS) (2020). 978-1-728151977/20/$31.00IEEE
22. Bolaj, P., Govilkar, S.: Text classification for Marathi documents using supervised learning methods. Int. J. Comput. Appl. **155**(no 8), 0975–8887 (2016)
23. Jha, V., Manjunath, N., Shenoy, P.D., Venugopal, K.R.: HSRA: Hindi Stopword Removal Algorithm. IEEE (2016). 978-14673-6621-2/16$31.00
24. Raulji, J.K., Saini, J.R.: Stop-word removal algorithm and its implementation for Sanskrit language. Int. J. Comput. Appl. **150**(2), 0975–8887 (2016)
25. Kaur, J., Saini, J.R.: A study of text classification natural language processing algorithms for Indian languages. VNSGU J. Sci. Technol. **4**(1), 162–167 (2015). ISSN: 0975-5446
26. Tikarya, A.B., Mayur, K., Patel, P.H.: Pre-processing phase of text summarization based on Gujarati language. Int. J. Innov. Res. Comput. Sci. Technol. **2**(4) (2014). ISSN: 23475552
27. Jayashree, R., Srikantamurthy, K., Anami, B.S.: Suitability of naïve Bayesian methods for paragraph level text classification in the Kannada language using dimensionality reduction technique. Int. J. Artific. Intell. Appl. **4**(5) (2013)
28. Jayashree, R., Srikanta Murthy, K., Sunny, K.: Document summarization In: Bracewell, D., et al. (eds.) Kannada Using Keyword Extraction. AIAA 2011, CS & IT 03, pp. 121–127. CS & IT-CSCP 2011 (2011). https://doi.org/10.5121/csit.2011.1311
29. Pandey, A.K., Siddiqui, T.J.: Evaluating effect of stemming and stop-word removal on Hindi text retrieval. In: Proceedings of the First International Conference on Intelligent Human Computer Interaction, pp. 316–326 (2009). https://doi.org/10.1007/978-81-8489-203-1_31

Arabic Cyberbullying Detection from Imbalanced Dataset Using Machine Learning

Meshari Essa AlFarah[✉], Ibrahim Kamel, Zaher Al Aghbari, and Djedjiga Mouheb

University of Sharjah, Sharjah, UAE
{U18200669,Kamel,Zaher,Dmouheb}@sharjah.ac.ae

Abstract. In recent years, the number of online social networks users is dramatically increased. Cyberbullying is a serious threat in online social networks especially toward children and teenagers. Victims are harassed by perpetrators even with no physical attendance. Cyberbullying causes psychological damage to victims resulting in anxiety, depression, and even suicide. Cyberbullying is language and culture sensitive. Many prior works in cyberbullying detection was conducted in English and only few papers studied Arabic cyberbullying detection. In this research, Arabic cyberbullying detection using five machine learning techniques were studied using real Arabic messages from Twitter and YouTube. Many performance techniques were used to evaluate the robustness of the model. Naïve Bayes (NB) achieves the highest Receiver Operating Characteristic (ROC) curve (AUC) of 89% while Support Vector Machine (SVM) and Logistic Regression (LR) both achieve 88%.

Keywords: Cyberbullying · Arabic messages · Machine learning (ML) · Online social networks

1 Introduction

With the rise of technology of communication and mobile devices, many teenagers tend to communicate through online social networks. This allows people to communicate anonymously using fake accounts. This behavior increases the tendency toward cyberbullying. Cyberbullying is the intentional and repeated harm committed through social media applications [1]. The main issue of cyberbullying is that the perpetrator can harass the victim without physical appearance and real identity. The committer usually posts aggressive messages with harsh comments or expose personal content to either insult or demean the victim. Cyberbullying can cause a lot of psychological and mental damage mostly for children and teenagers making them feel vulnerable and unsafe. Furthermore, it results in anxiety, depression and even suicide. In contrast with traditional bullying, cyberbullying can take place anywhere and anytime.

Arabic language is broadly spoken, and it is one of the five top popular languages in the world. The structure of Arabic language is complex in comparison with other languages with 30 different dialects nowadays [2]. Furthermore, the number of conducted

© Springer Nature Switzerland AG 2022
K. K. Patel et al. (Eds.): icSoftComp 2021, CCIS 1572, pp. 397–409, 2022.
https://doi.org/10.1007/978-3-031-05767-0_31

research in Arabic cyberbullying detection is few in comparison with English cyberbullying detection. Cyberbullying depends on language and culture. What is acceptable in one culture might not be acceptable in another and might be considered offensive. The contribution of cyberbullying detection in English language in the literature review is relatively different than the Arabic and Islamic cultural context.

In this research, five machine learning techniques were studied for cyberbullying detection in Arabic messages. Twitter and YouTube were used to construct the dataset which contains Arabic messages. Several preprocessing were performed to clean the data and simplify the dataset.

The reminder of the paper is organized as follows. Section 2 presents the related work. Section 3 presents the scheme of detecting cyberbullying in Arabic messages using machine learning. Section 4 describes several machine learning techniques with the evaluation metrics. Section 5 shows the experimental results. Section 6 concludes the paper.

2 Related Work

In this section, previous work in cyberbullying detection in both Arabic and English languages and oversampling and under sampling techniques are presented. The papers in the literature mainly use machine learning techniques to detect cyberbullying from messages.

Noviantho et al. constructed a classification model based on SVM with different kernels, Naïve Bayes, J48, and k-NN [4]. The dataset consists of 1600 conversations obtained from Formspring. Salawu and Lumsden presented a survey on automated cyberbullying approaches such as machine learning techniques and natural language processing [5]. Balakrishnan et al. presented cyberbullying detection based on psychological features and machine learning techniques such as J48, Naïve Bayes, and Random Forest [6]. The psychological features consist of personalities, sentiment, and emotions. The authors used a Twitter dataset that contain 5453 tweets collected from a hashtag #Gamergate.

Rosa et al. presented a systematic review on 22 studies on automatic cyberbullying detection methods [7]. The authors also used machine learning techniques such as Support Vector Machine, Logistic Regression, and Random Forest to detect cyberbullying. Two datasets were used which are 13160 texts from Formspring and 2999 tweets from Twitter. Al-garadi et al. present a model for cyberbullying detection in Twitter based on machine learning [8]. The dataset contains 10,606 tweets. The authors used several machine learning classifiers such as SVM, Naïve Bayes, KNN, and Random Forest.

Al-Hassan and Al-Dossari presented a survey on hate speech detection in social networks for multiple languages [9]. The survey consists of multiple languages including English, German, Turkish, Italian, Indonesian, and Arabic language. Mohaouchane et al. presents offensive language detection on social networks for Arabic messages using deep learning techniques [10]. The authors used 4 different neural network architectures which are CNN, Bi-LSTM, Bi-LSTM with attention mechanism, and a combination of LSTM and CNN. The dataset contains 15,050 comments in Arabic YouTube celebrities' content.

Mouheb et al. presents Arabic cyberbullying detection in social networks [11]. The authors used two datasets from Twitter and YouTube. The YouTube dataset consists of 50,000 comments from Arabic content videos with most of the comments are in Arabic language. Twitter dataset has Arabic 42,138 comments. The cyberbullying is detected using word spotting. Moreover, Mouheb et al. propose Arabic cyberbullying detection on social networks using machine learning [12]. The datasets are obtained from YouTube and Twitter APIs. Naïve Bayes technique is used along with word spotting.

Haidar et al. use ensembles machine learning to enhance the detection rate [13]. The data were collected from Twitter and the final dataset contains 31891 non bullying instances and 2999 bullying instances. The researchers used several classifiers such as KNN, Random Forest, and SVM. Also, boosting and bagging were used as ensemble methods. In addition, Haidar et al. [14] present a deep learning model for cyberbullying detection. The scholars use feed forward neural networks with multiple hidden layers. The results show that the one hidden layer produce output similar to 128 hidden layers. Thus, the performance does not improve by adding more layers.

Oversampling and under sampling techniques are also takes a significant part to balance imbalanced datasets. Synthetic Minority Over-sampling Technique (SMOTE) and its variations are famous oversampling and under sampling techniques. SMOTE is an oversampling technique which creates synthetic minority class examples by working in feature space [15]. The process of oversampling the minority class is to take each smaller class instances and producing synthetic instances along with the line segments joining any or all of the minority class K nearest neighbors. The number of KNNs is selected randomly based on the needed oversampling size. In addition, SMOTE can also be used with under sampling technique. First, under sample the major class then apply SMOTE technique.

SMOTE has many variations with more than 85 extensions of SMOTE have been proposed [16]. The extension Borderline-SMOTE is an oversampling techniques where the minority instances close to the borderline are oversampled [17]. Moreover, SMOTE-Tomek and SMOTEENN are combinations of oversampling and under sampling techniques [18]. In addition, ADASYN is an oversampling technique which uses a weighted distribution for unlike minor class instances regarding with the degree of difficulty in training. Complex and hard minority class examples tend to generate more synthetic data. On the other hand, simple and easy minority class tend to generate few synthetic data [19].

3   Proposed Scheme

The proposed scheme of this paper shows the flow of Arabic cyberbullying detection used in this paper and it is presented in Fig. 1. First, Arabic messages were collected to train the ML models for Arabic cyberbullying detection. First of all, the dataset is cleaned and preprocessed then annotated. Furthermore, TF-IDF feature extraction were applied for training and testing data. Then oversampling technique were applied to overcome the issue of imbalanced dataset.

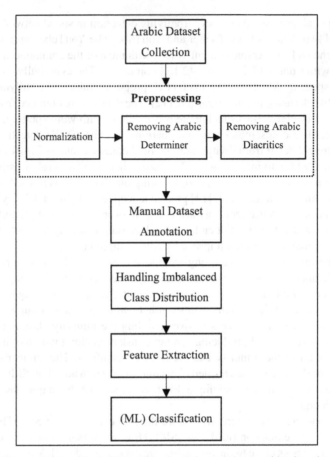

Fig. 1. Arabic cyberbullying detection scheme using machine learning

3.1 Data Collection

Twitter and YouTube social networks were selected as the main source of retrieving the data as most of the Arab teenagers use Twitter and YouTube platforms nowadays. The datasets of Twitter and YouTube were collected using Netlytic Social Network Analyzer (SNA) [20]. The SNA follows the Twitter API rules and limitations. Several parameters were used to obtain Arabic tweets by using the longitude and altitude with specific radius of the Arab geographical region along with Arabic text as the main language. Riyadh city was selected as the center point of a circle with a radius of 1400 km so that it covers all of the Arabian gulf countries and partially other near countries such as Egypt, Sudan, Yemen, and Jordan, etc. The collected Twitter dataset contains 50,000 tweets and retweets which are collected from the Twitter public stream. The dataset was collected between January 2020 and March 2020. The period of collecting the Twitter data is regular and does not follow an event or a trend.

In addition, YouTube is another source of our data because many young Arab and Arab teenagers use YouTube nowadays in their daily life. The YouTube dataset contains

4,860 comments. It has comments between May 2017 and July 2020. The collected YouTube comments were obtained from controversial videos related to sport, religion, etc. which relatively contain bullying messages. These comments might enhance the robustness of the classifier by getting the classifier exposed to different types of bullying so that it becomes more robust in real time classification.

3.2 Preprocessing

In this phase, the messages are cleaned and refined to simplify and reduce the size of the dataset and the sparse matrix of the TF-IDF that is discussed further in subsection E. In addition, each word in the messages goes through the preprocessing steps multiple times to refine it [12]. There are many steps of refinement. The order of the refinement steps matters so that no information or additional letters are missed. The refinement steps are as follows:

Normalization
In the normalization stage, words that contain misinterpretation letters are normalized. First of all, the letters 'أ','إ'are replaced with 'ا'to keep it in the standard form. Furthermore, the letter 'ة', is replaced with 'ه'at the end of the words. This step minimizes the sparse matrix size and simplify similar words to only one word.

Removing Arabic Determiner
In this step, the Arabic determiner 'ال'is removed. It is equivalent to the word 'the' in English. The step works when the number of letters in a word exceed three letters because words start with the Arabic determiner and has three letters can only be a proposition. This step does not change the meaning of the sentence but will affect the sparse matrix size and keep the meaning at the same time.

Removing Arabic Diacritics
Arabic diacritics are special characters the set above and below consonant letters. These precedes the last letter in the words, and it is also removed to simplify and reduce the size of sparse matrix of TF-IDF.

Manual Dataset Annotation
The collected dataset of Twitter and YouTube contain Arabic tweets and comments. These messages need to be annotated first before training the machine learning classifiers. The Arabic messages are categorized into two classes as follows:

- Cyberbullying: If a message contain bullying content.
- Non-cyberbullying: If a message does not contain bullying content.

The dataset was manually annotated, and it contains 14,717 annotated Arabic messages. (12,280) messages were labeled as non-cyberbullying and (2,437) messages were labeled as cyberbullying.

In this work, we did not outsource the process of annotation through for e.g., Mechanical Turk website (M. Turk). Instead, we manually annotated the dataset to assure the quality of the annotation and avoid online spam truckers who do not read the actual data when labeling [21]. Nevertheless, the process of manually annotation is requires more time in comparison with M. Turk.

3.3   Handling Imbalanced Class Distribution

In this research, the dataset is a binary class classification problem. Imbalanced datasets are a real issue when it comes to train machine learning classifiers. This is because the accuracy of the classifier will be similar to the distribution percentage of the majority class. For instance, if a dataset contains 95% positive instances, then the accuracy of the trained model will be about 95% as well. However, a classifier with lower percentage might be better when deployed as a real life application.

Many sampling techniques were proposed, and many of them were based on generating synthetic data such as SMOTE and its variations. Although it might work in text classification, it is difficult to work with these techniques especially when working with sparse matrix of TF-IDF that have high dimensional space. These oversampling and under sampling techniques works on the feature space, and our dataset has a high dimensional sparse vector which make it challenging for text classification. In addition, generating synthetic data of text does not make sense because the generated numbers are not corresponding to real words or meanings, and it might not help when it is deployed as a real time classification model.

The oversampling techniques oversample the minority class instances until the number of minority class instances matches the number of majority class instances such that the final dataset contains 50% positive instances and 50% negative instances. To tackle the issue of our imbalanced dataset, we iterate over the minority class instances and randomly oversample the minority instances until the size of minority class instances equal to the size of majority class instances. After oversampling the dataset, the dataset contains 24,560 annotated Arabic messages. (12,280) messages were labeled as non-cyberbullying and (12,280) messages were labeled as cyberbullying.

3.4   Feature Extraction

The Arabic messages is modeled by transforming into a vector space. These messages are represented within a vector of extracted features. The extracted features are words that forms a list of words. Then, the weight of the term frequency-inverse document frequency (*tf-idf*) is calculated for a term t of a document d in a document set as follows which is implemented in python scikit-learn library [22]:

$$tf - idf(t, d) = tf(t, d) * idf(t) \tag{1}$$

and the idf is computed as:

$$idf(t) = log\left[\frac{n}{df(t)}\right] + 1 \tag{2}$$

where n means whole number of documents in the document set. df(t) is the frequency of document t which is the number of documents in the document set where it includes phrase t.

4 Machine Learning Algorithms

This section discusses the five machine learning algorithms that are used to detect Arabic cyberbullying which are Support Vector Machine (SVM), Random Forest, Naïve Bayes (NB), Logistic Regression, and K-Nearest Neighbor (KNN). Each machine learning technique will be described in a subsection. Furthermore, the machine learning algorithms were implemented using python scikit-learn library.

Support Vector Machine (SVM): SVM is a popular machine learning technique that has been used in many fields nowadays. Given a training set of data $(x_1, x_2, ..., x_n)$, denoted as trajectories in a particular area $X \subseteq Rd$ and identified as $(y_1, y_2, ..., y_m)$, such that $y_i \in (-1, 1)$ [8]. SVMs can be expressed as hyperplanes which isolate the data of the training part by a maximum boundary where the negative category exists in one region of the hyperplane, and category is on the opposite region.

Random Forest: Random Forest is an ensemble machine learning classifier which fits a number of decision tree classifiers. It also utilizes averages techniques to increase the accuracy and to avoid over fitting [22].

Naïve Bayes (NB): Naïve Bayes is a machine learning technique which applies the Bayes theorem with naive independence assumptions given in the formulas below [8]. Given a class variable y and a dependent feature vector x_1 through x_n the theorem defines the expression in (3):

$$p(y|x_1, \ldots, x_n) = \frac{p(y)(x_1,...,x_n|y)}{p(x_1,...,x_n)}. \tag{3}$$

Based on naive independence assumption, the next acquired the formula is:

$$p(x_i|y, x_1, \ldots x_{i-1}, x_{i+1}, x_n) = p(x_i|y), \tag{4}$$

For all i, the association is simplified in (5):

$$p(y|x_1, \ldots, x_n) = \frac{p(y)\prod_{i=1}^{n} p(x_i|y)}{p(x_1,...,x_n)}. \tag{5}$$

Since p(y|) is constant because of the input, the next classification rule is applied:

$$p(y|x_1, \ldots, x_n) \propto p(y)\prod_{i=1}^{n} p(x_i|y). \tag{6}$$

Logistic Regression (LR): Logistic regression is a ML method which is very useful in text classification and natural language processing problems (NLP). The conditional probability of the Logistic Regression is as follows [23]:

$$P_w(y = \pm1|x) \equiv \frac{1}{1+e^{-yw^T x}} \tag{7}$$

where x is the data, y is the class label, and $w \in R^n$ is the weight vector. Given a binary classification problem training data $\{x_i, y_i\}_{i=1}^{l}$, $x_i \in R^n$, $yi \in \{1, -1\}$, LR reduces the regularized negative log likelihood [23]:

$$P^{LR}(w) = C \sum_{i}^{l} log\left(1 + e^{-yw^T x_i}\right) + \frac{1}{2}w^t w \tag{8}$$

where C > 0 is a penalty factor.

K-Nearest Neighbor (KNN): The K-Nearest Neighbor classifier is a popular machine learning technique. The KNN technique considers the k nearest examples $\{i_1, i_2, ..., i_k\}$ from instance (x), and the most frequent class in the set $\{c_1, c_2, ... c_k\}$. is expected to be the class of that example (x) [24]. To decide the nearest example, KNN method uses a metric based on distance that measures the proximity of example x to k of collected examples. Many distance functions can be used in KNN such as the Euclidean, Cosine or Jaccard [25].

Evaluation Metrics: The confusion matrix of the machine learning classifiers is the base of many different metrics where TP, TN, FP, FN are as follows:

- True positive (TP): The number of non-bullying instances where the model accurately predicts as non-bullying.
- True negative (TN): The number of bullying instances where the model accurately predicts as bullying.
- False positive (FP): The number of bullying instances where the model incorrectly predicts as non-bullying.
- False negative (FN): The number of non-bullying instances where the model incorrectly predicts as bullying.

The performance metrics of the classifiers are based on the accuracy, recall, precision, and F-measure which is the harmonic mean of precision and recall and are calculated as follows:

$$Precision = \frac{TP}{TP + FP} \tag{9}$$

$$Recall = \frac{TP}{TP + FN} \tag{10}$$

$$Accuracy = \frac{TP + TN}{TP + FP + TN + FN} \tag{11}$$

$$F - measure = \frac{2 * Precision * Recall}{Precision + Recall} \tag{12}$$

The area under Receiver Operating Characteristic (ROC) curve (AUC) is major measure as it is robust for imbalanced datasets [3]. The ROC curve is a two-dimensional diagram where the true positive rates are plotting on the Y axis and the false positive rates are plotting on the X axis. AUC has been used also in the medical decision making and nowadays in data mining and the machine learning fields. AUC is statistically important as the AUC of a ML model is the probability the ML technique will categorize a random positive instance greater than a random negative instance. Although AUC value can be in the range [0–1], no realistic ML model gets an AUC value lower than 0.5 [3].

5   Experimental Results

In this section, the results of Arabic cyberbullying detection study using machine learning techniques are presented. Five different machine learning classifiers were implemented and compared using several performance metrics.

In this work, after preprocessing the dataset, the dataset is split by 80% for the training part and 20% for testing part. The training set contains 19,676 messages after oversampling the dataset and the testing set contains 2,944 messages. Then, oversampling is applied after splitting the dataset. The dataset is oversampled to balance the instances of the negative class and the positive class. Oversampling before splitting the dataset could results in high performance and overfitting because the ML model has seen the instances in the training phase which will not work properly when deployed as a real time classifier.

The results of the performance metrics for the machine learning classifiers are shown in Table 1 and Table 2. All the machine learning methods are experimented with the same exact data. Table 1 shows the performance of different results between the trained ML techniques using precision, recall, and F-measure along with macro averaging and weighted averaging.

Macro averaging evaluate metrics for each label and obtain the unweighted average which avoid considering imbalanced examples. On the other hand, weighted averaging compute metrics for each label, and obtain the average weighted by support [22]. Many papers present their work using only one type of averaging method and they do not specify which type of averaging method they use. This misleads scholars because usually weighted average shows higher results in comparison with macro average especially when dealing with imbalanced dataset where the macro average emphasizes the minority class. Therefore, to avoid misconception, both averaging methods are presented in this work.

Table 1. The performance metrics for all machine learning classifiers

Classifier	Metrics	Bullying	Non-bullying	Macro average	Weighted average
SVM	Precision	33.86%	97.54%	65.70%	92.57%
	Recall	73.91%	87.77%	80.84%	86.68%
	F-measure	46.45%	92.40%	69.42%	88.81%
KNN	Precision	34.06%	83.91%	58.99%	74.36%
	Recall	30.32%	86.09%	58.21%	75.41%
	F-measure	32.08%	84.99%	58.53%	74.85%
Logistic Regression	Precision	69.32%	87.51%	78.42%	83.48%
	Recall	53.29%	93.28%	73.29%	84.41%
	F-measure	60.26%	90.30%	75.28%	83.64%
Naïve Bayes	Precision	83.67%	78.09%	80.88%	79.90%
	Recall	43.98%	95.88%	69.93%	79.04%
	F-measure	57.65%	86.08%	71.86%	76.86%
Random Forest	Precision	38.65%	96.07%	67.36%	90.41%
	Recall	66.90%	88.39%	77.65%	86.28%
	F-measure	48.99%	92.07%	70.53%	87.83%

In Table 1, the SVM achieves high performance in precision for the non-bullying class with 97.54%. However, the precision in the bulling class is noticeably low. SVM also has the highest recall in the bullying class reaching 73.91%. The KNN instead is the lowest performance in all performance metrics among other ML techniques.

On the other hand, the Logistic Regression model achieves good results in precision and recall in the minor class which is different than SVM and KNN. Naïve Bayes precision in the bullying class is the highest among other ML classifiers achieving 83.67%. The Random Forest classifier has similar results to the SVM but a bit lower. The Logistic Regression has the best macro averaging in comparison with other ML techniques. Thus, the best model is the Naïve Bayes. Although the minority class was oversampled, the machine learning classifiers are still biased toward the major class.

Table 2 shows the accuracy and AUC of the trained ML techniques. The results of accuracy are known to be not important when having imbalanced dataset. However, since the dataset is balanced by oversampling, it is provided in Table 2.The results in Table 2 shows that the Naïve Bayes classifier has the best AUC achieving 89%. SVM and Logistic Regression have slightly similar AUC with 88%. Random Forest shows good AUC achieving 85% which is slightly lower than SVM and Logistic Regression. The KNN has the lowest AUC with 64%. Figure 2 shows the ROC curves of true positive rates and false positive rates, and we see that SVM, Logistic Regression, and Naïve Bayes have the highest ROC curves achieving 88%, 88%, 89% respectively.

Table 2. A comparison of the performance between different machine learning techniques

Classifier	Accuracy	AUC
SVM	86.68%	88%
KNN	75.40%	64%
Logistic Regression	84.40%	88%
Naïve Bayes	79.04%	89%
Random Forest	85.97%	85%

The AUC of Random Forest is a little bit lower with 85% while KNN has the lowest ROC curve with 64%. The hyperparameters used for the trained models are shown in Table 3.

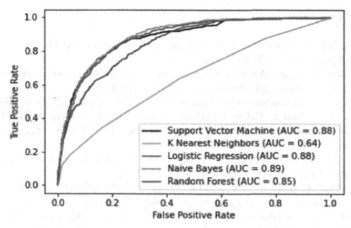

Fig. 2. ROC curves of different machine learning techniques

Table 3. Hyperparameters for each machine learning model

Classifier	Hyperparameters
SVM	C = 1.0
	Cache size = 200MB
	Kernel = RBF
	Degree = 3
KNN	Leaf size = 30
	Number of neighbors = 5
	Euclidean distance (L2)
Logistic Regression	C = 1.0
	Solver = lbfgs
Naïve Bayes	Alpha = 1.0
Random Forest	Number of estimators = 100

6 Conclusion

This research studies Arabic cyberbullying detection from online social networks. The dataset is obtained from Twitter and YouTube and it is manually annotated to ensure the quality of annotation. Since the dataset is imbalanced, the issue of imbalanced dataset was solved by oversampling the minority class. Five machine learning techniques were implemented to compare the performance of each classifier. An extensive results of all performance metrics were presented in details. The highest AUC achieved is 89% using Naïve Bayes where SVM and Logistic Regression achieve 88%.

References

1. Cyberbullying Research Center. What is cyberbullying? https://cyberbullying.org/what-is-cyberbullying. Accessed 27 Nov 2018
2. Asian Absolute Team, Arabic language dialects (2016). https://asianabsolute.co.uk/blog/2016/01/19/arabic-language-dialects/. Accessed 07 Apr 2020
3. Fawcett, T.: An introduction to ROC analysis. Pattern Recogn. Lett. **27**(8), 861–874 (2006)
4. Isa, N.S.M., Ashianti, L.: Cyberbullying classification using text mining. In: 2017 1st International Conference on Informatics and Computational Sciences (ICICoS), pp. 241–246 (2017). https://doi.org/10.1109/ICICOS.2017.8276369
5. Salawu, S., He, Y., Lumsden, J.: Approaches to automated detection of cyberbullying: a survey. IEEE Trans. Affect. Comput. **11**(1), 3–24 (2020)
6. Balakrishnan, V., Khan, S., Arabnia, H.R.: Improving cyberbullying detection using Twitter users' psychological features and machine learning. Comput. Secur. **90**, 101710 (2020)
7. Rosa, H., et al.: Automatic cyberbullying detection: a systematic review. Comput. Hum. Behav. **93**, 333–345 (2019)
8. Al-Garadi, M.A., Varathan, K.D., Ravana, S.D.: Cybercrime detection in online communications: the experimental case of cyberbullying detection in the Twitter network. Comput. Hum. Behav. **63**, 433–443 (2016)
9. Al-Hassan, A., Al-Dossari, H.: Detection of Hate Speech in Social Networks: A Survey on Multilingual Corpus, pp. 83–100 (2019)
10. Mohaouchane, H., Mourhir, A., Nikolov, N.S.: Detecting offensive language on Arabic social media using deep learning. In: 2019 Sixth International Conference on Social Networks Analysis, Management and Security (SNAMS), pp. 466–471 (2019). https://doi.org/10.1109/SNAMS.2019.8931839
11. Mouheb, D., Ismail, R., Qaraghuli, S.A., Aghbari, Z.A., Kamel, I.: Detection of offensive messages in arabic social media communications. Int. Conf. Innov. Inf. Technol. **2018**, 24–29 (2018). https://doi.org/10.1109/INNOVATIONS.2018.8606030
12. Mouheb, D., Albarghash, R., Mowakeh, M.F., Aghbari, Z.A., Kamel, I.: Detection of Arabic cyberbullying on social networks using machine learning. In: 2019 IEEE/ACS 16th International Conference on Computer Systems and Applications (AICCSA), pp. 1–5 (2019) https://doi.org/10.1109/AICCSA47632.2019.9035276
13. Haidar, B., Chamoun, M., Serhrouchni, A.: Arabic cyberbullying detection: enhancing performance by using ensemble machine learning. In: 2019 International Conference on Internet of Things (iThings) and IEEE Green Computing and Communications (GreenCom) and IEEE Cyber, Physical and Social Computing (CPSCom) and IEEE Smart Data (SmartData), pp. 323–327 (2019). https://doi.org/10.1109/iThings/GreenCom/CPSCom/SmartData.2019.00074
14. Haidar, B., Chamoun, M., Serhrouchni, A.: Arabic cyberbullying detection: using deep learning. In: 2018 7th International Conference on Computer and Communication Engineering (ICCCE), pp. 284–289 (2018) https://doi.org/10.1109/ICCCE.2018.8539303
15. Chawla, N.V., Bowyer, K.W., Hall, L.O., Kegelmeyer, W.P.: SMOTE: synthetic minority over-sampling technique. J. Artif. Intell. Res. **16**, 321–357 (2002)
16. Fernandez, A., Garcia, S., Herrera, F., Chawla, N.V.: SMOTE for learning from imbalanced data: progress and challenges, marking the 15-year anniversary. J. Artif. Intell. Res. **61**, 863–905 (2018)
17. Han, H., Wang, W.-Y., Mao, B.-H.: Borderline-SMOTE: A New Over-Sampling Method in Imbalanced Data Sets Learning, pp. 878–887 (2005)
18. Batista, G.E.A.P.A., Prati, R.C., Monard, M.C.: A study of the behavior of several methods for balancing machine learning training data. ACM SIGKDD Explor. Newsl. **6**(1), 20–29 (2004)

19. He, H., Bai, Y., Garcia, E.A., Li, S.: ADASYN: Adaptive synthetic sampling approach for imbalanced learning. In: 2008 IEEE International Joint Conference on Neural Networks (IEEE World Congress on Computational Intelligence), pp. 1322–1328 (2008). https://doi.org/10.1109/IJCNN.2008.4633969

20. Gruzd, A.: Netlytic: Software for Automated Text and Social Network Analysis (2016). http://netlytic.org

21. Ipeirotis, P.: Mechanical Turk: Now with 40.92% spam. Behind Enemy Lines blog (2010)

22. Pedregosa, F., et al.: Scikit-learn: machine learning in python. J. Mach. Learn. Res. **12**, 2825–2830 (2011)

23. Yu, H.F., Huang, F.L., Lin, C.J.: Dual coordinate descent methods for logistic regression and maximum entropy models. Mach. Learn. **85**, 41–75 (2011). https://doi.org/10.1007/s10994-010-5221-8

24. Aldayel, M.S.: K-Nearest Neighbor classification for glass identification problem. Int. Conf. Comput. Syst. Indust. Inf. **2012**, 1–5 (2012). https://doi.org/10.1109/ICCSII.2012.6454522

25. Jivani, A.G.: The novel k nearest neighbour algorithm. In: 2013 International Conference on Advances in Computing, Communications and Informatics, ICCCI 2013, pp. 4–7 (2013)

A CNN Based Air-Writing Recognition Framework for Linguistic Characters

Prabhat Kumar[✉], Abhishek Chaudhary, and Abhishek Sharma

DSPM, International Institute of Information Technology Naya Raipur, Naya Raipur, India
{prabhat19102,abhishek19102,abhishek}@iiitnr.edu.in

Abstract. Air writing is a practice of writing the linguistic characters in free space utilizing the six degrees of freedom of hand motion. We propose a system that uses a generic webcam to detect and recognize the virtually written characters by a user as per their will. This system performs detection using morphological operations for masking the tracker or writing object. This system gives the user the freedom to select a writing object of any color, shape, or material for tracking purposes. The trajectory of the object's mask is tracked and rendered on a virtual window. The air-written character is recognized using the Convolutional Neural Network (CNN). The CNN is trained on a dataset of English handwritten characters of 26 different classes (A-Z). The accuracy achieved by the proposed system for an isolated character is 99.75%.

Keywords: Air-writing · Computer vision · Convolutional Neural Network · Handwritten character recognition · Human-computer interaction

1 Introduction

Air-writing refers to a system in which a human interacts with the computer to draw something on a virtual window or environment. This system is of great use for enhancing the user interfaces for linguistic inputs. Unlike the traditional mechanical methods such as writing on paper with pen-up and pen-down motion, this system provides a means to write using hand moments with six degrees of freedom. There are no predefined start or endpoints in free space, neither any fixed reference position for orientation for the virtual plane to write on. The input for such a system becomes highly variable with every user; also, there is no standard protocol to guide the writer's hand to follow a particular path, in three-dimensional space, along the predefined axis. So, the task of object detection and tracking becomes a non-trivial task.

The state-of-the-art systems incorporate depth sensors like Time-of-flight (ToF) [1], Leap Motion [2], and Kinect [3], along with computer vision techniques for air-writing recognition. Since these sensors are not widely available for the general public, a generic camera and computer vision algorithms can produce comparable results and reduce the system's production cost.

The wide availability of the generic camera in computational devices makes the proposed system a good choice for deployment in a broad spectrum of real-world scenarios such as:

© Springer Nature Switzerland AG 2022
K. K. Patel et al. (Eds.): icSoftComp 2021, CCIS 1572, pp. 410–420, 2022.
https://doi.org/10.1007/978-3-031-05767-0_32

- To facilitate the Kūsho effect among younger and older adults [4].
- Developing and improving the cognitive perception skills among young kids at kindergarten and nursery level helps them write and remember every character's shape.
- To understand the letters or words scribbled in the air by a hyperlexic child [5].
- To allow doctors to take notes while performing a surgery where they cannot use the conventional writing methods.

The approach presented in this paper introduces flexibility for the user to utilize this system as a solution for multi-functional consumer devices. This framework is a multipurpose setup for diverse age groups ranging from preschoolers to older adults. The system comes with two modes, 'Detection' and 'Recognition.' which can be toggled according to the user's demand. Detection mode begins the human-computer interaction (see Fig. 1) by letting the user write in the air with active tracking and feedback with four ink colors which are red, blue, green, and yellow; this mode is suitable for scribbling, drawing, and taking notes in cases where conventional writing methods may or may not be limited. Recognition modes complete the human-computer-interaction cycle, which offers character recognition. This mode is suitable for determining the correctness of air-written characters, especially for preschoolers and those who practice the Kūsho [4].

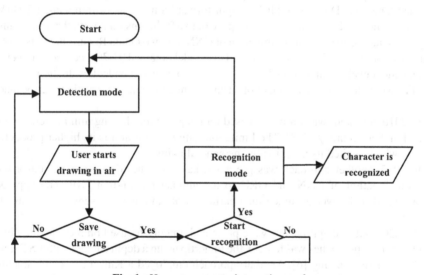

Fig. 1. Human-computer interaction cycle

The organization of this paper is as follows; some significant state-of-the-art systems are conferred in the next section; our proposed methodology is presented in section three. Section four comprises the experimentation result and its discussion, and finally, section five concludes this work along with some future scope.

2 Related Works

Most of the works proposed earlier to address air-writing relied on external depth sensors like Leap Motion, Kinect, or a wearable sensor-based recognition system. Though these methods achieve much higher recognition accuracy and effective object tracking, it incurs higher production cost and limits the deployment of these systems for the general consumers. These methods reduce the hardware redundancy as well.

Alam et al. [6] used a depth camera to collect the fingertip's trajectory in three-dimensional space for tracking; here, nearest neighbor and root point translation was used for feature selection and trajectory normalization. Finally, they deployed Long short-term memory (LSTM) and Convolutional Neural Network (CNN) as a recognizer. They achieved an accuracy of 99.32% on the 6D motion gesture (6DMG) alphanumeric dataset. Chen et al. [7, 8] utilized the Leap Motion for object tracking and Hidden Markov Model (HMM) for character recognition. They obtained an error rate of 1.9% for character-based and 0.8% for word-based recognition systems. Mohammadi et al. [9] utilized the color and depth images from the Kinect sensor for air-writing detection. Slope variations detection-based approach was employed for extracting the features from the trajectory of the written character, and then a novel analytical classifier was used for recognition. The recognition accuracy achieved was 98% for digits and numbers of the Persian language. Dash et al. [10] brought forward a novel algorithm named 2-DifViz that maps the hand movements in free space onto a 2-dimensional virtual canvas using a Myo armband sensor, and a combination of CNN and two Gated Recurrent Units (GRU) was used for recognizing the characters. This model obtained 91.7% accuracy in a person independent evaluation and 96.7% accuracy in a person dependent evaluation.

From the literature survey mentioned above, the following observations can be made:

- In 2016, a system was proposed based on Leap Motion, having minimal error count and higher accuracy [7, 8]. The limitation with this solution is its higher production cost which limited its adoption by the general public.
- In 2017, another approach was presented based on the Myo armband sensor with a combination of CNN and GRU to accomplish recognition [10]. The approach presented had lower accuracy than earlier work, and also the sensor cost was also high.
- In 2020, multiple approaches were presented, mainly focused on recognizing digits. One of the approaches was to track the fingertip using a depth camera with LSTM and CNN for recognizing [6]. Another approach employed a Kinect sensor for tracking and slope variations detection for recognition [9]. An OCR system was also proposed, which focused on improving the recognition ability of the CNN model [11]. All these systems had impressive comparable accuracies. The limitation was again the cost of sensors used and lower or no hardware redundancy.

The above case study suggests that all the proposed systems have two recurring problems. First, the higher cost of production, as an expensive dedicated sensor or depth camera is used for tracking, the second problem of lower hardware redundancy is a direct product of the first problem.

In this work, we propose a framework that is an optimum solution to these significant problems. This framework uses a generic camera for tracking purposes, which is factory embedded in widely available devices, thus cutting out the dependency on an expensive external sensor, making it a cost-effective approach. If the default camera fails, this system provides the flexibility to use any other device with a camera (such as a smartphone, external webcam, etc.) for tracking, thus giving it a higher hardware redundancy. A schematic workflow of the proposed system is shown in Fig. 2.

3   Proposed Work

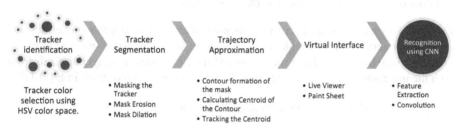

Fig. 2. Proposed system workflow

3.1   Color Selector for Tracker Identification

The proposed system employs a color selector to set the color of the object to be tracked. This feature lets the user set the color using the HSV color space. As shown in Fig. 3, the HSV color space consists of three descriptors, namely 'Hue,' 'Saturation,' and 'Value.' Hue describes the color; saturation describes the purity of the color; value describes the luminance of the color.

The color selector consists of 6 different sliders, namely "Upper hue," "Upper saturation," "Upper value," "Lower hue," "Lower saturation," and "Lower value." Upper hue and lower hue allow the user to set the color of the tracker from different shades of color. Upper and lower saturation are used to set the saturation value for the color (hue value) chosen above. The upper value and the lower value set the value for luminance or brightness of the color. Since the cameras installed on commonly available devices can capture approximately 16.78 million colors. Thus, this feature grant user the liberty to choose almost any object lying around them as a tracker. Once the system is configured for the tracker, this tracker shall act as an input method for the air-writing system.

3.2   Tracker Segmentation

Once the air-writing system identifies the tracking object (tracker). The next step is to extract out that object from the background. This system uses the masking technique for extracting the tracker, which is achieved using morphological operations. Morphological transformations are simple operations that are based on the shape of an image. It is

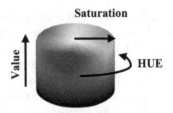

Fig. 3. HSV color space

typically applied to binary images (see Fig. 4). It requires two inputs: an original image and a structuring element or kernel that determines the nature of the operation. Erosion and dilation are two fundamental morphological operators:

Erosion: This morphological transformation calculates the local minima over the area of the given kernel and then replaces the minimum value calculated with the provided image or video frame. It reduces the size of an image by removing a layer of pixels from both the inner and outer edges of regions. The erosion operation is given in Eq. 1 [12].

$$dst(x, y) = min_{(x',y'):element(x',y') \neq 0} src(x + x', y + y') \tag{1}$$

Dilation: This operation convolves the pixels of the original image to the maximum value in the given kernel. It increases the size of an image by adding a layer of pixels to both the inner and outer edges of regions. The dilation operation is given in Eq. 2 [12].

$$dst(x, y) = max_{(x',y'):element(x',y') \neq 0} src(x + x', y + y') \tag{2}$$

Opening: This is a compound operation of the fundamental morphological transformation methods in which the operation of erosion is followed by dilation. This method is useful for removing noise. The opening operation is given in Eq. 3 [12].

$$dst(x, y) = open(src, element) = dilate(erode(src, element) \tag{3}$$

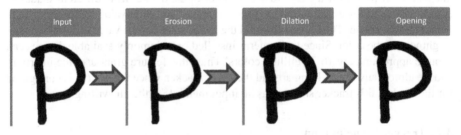

Fig. 4. Morphological operations on a test image

3.3 Trajectory Approximation

The output or the result of the tracker segmentation is the digital mask generated using morphological transformations. The mask produced (see Fig. 5) may be unique or multiple as there might be many objects of similar color in the surrounding. Thus, selecting the correct mask becomes a crucial step. It is done by identifying the contour of the mask, which has the highest dominance in terms of its area.

Once the contour of the correct mask is formed, the next step is to track this contour. For this, the centroid of the contour is calculated. This centroid is then used for tracking the motion of the tracker. The relation of the centroid is given in Eq. 4 and 5:

$$C_x = \frac{M_{10}}{M_{00}} \tag{4}$$

$$C_y = \frac{M_{01}}{M_{00}} \tag{5}$$

where, C_x and C_y are the coordinates of the centroid and M_{xy} is the moments of the contour.

For marking the contour on the user window, it is circumscribed in a circle such that the circle has the minimum area. This circle gives direct user feedback, marking the tracker's location on the virtual window.

3.4 Visual Interface

Live Viewer: Live viewer is a display window where all the controls of the air-writing system are present. This interactive window has five on-screen display buttons, each dedicated to a specific function like changing ink color or clearing the entire screen.

Paint Sheet: Paint sheet displays the live scribbling on a white background. Users can save the jottings produced on the paint sheet for future reference.

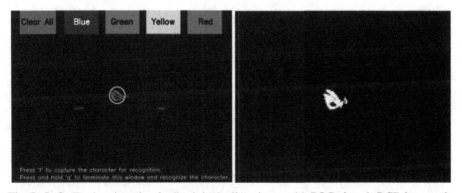

Fig. 5. Left: Tracker detection feedback in the live viewer with ROG phone's RGB logo as the tracker. **Right:** Mask formation of the detected tracker with suppressed noise.

3.5 Character Recognition

The approximated trajectory of the masked centroid is the projection of air-written characters from free space on a 2D feature space. We used CNN for predicting the written characters from these projected images. Preference of CNN based framework over the conventional algorithms like Support Vector Machine (SVM), Hidden Markov Model (HMM), etc. [13] is because of the following:

- CNN reduces the computational cost.
- CNN does not need manual feature extraction.

We trained the CNN model on A-Z handwritten data dataset. [14] This dataset was used primarily due to a large number of handwritten images, and also, during the development of this framework, any sizeable standard dataset was not available for English handwritten characters. Thus, this dataset was used. This dataset consists of 26 classes (A–Z) containing 372,450 handwritten grayscale images of 28 × 28 pixels, with each alphabet in the image center is fitted to 20 × 20 pixels.

In the network architecture (see Fig. 6), the extraction of features is followed by their categorization. The feature extraction block comprises three convolutional layers followed by a pooling layer. In this block, grayscale images of 28 × 28 pixels are fed as input. The first convolution layer uses 32 convolution filters and a convolutional kernel of 3 × 3 with valid padding. The second convolution layer uses 64 convolution filters and a convolution kernel 3 × 3 with the same padding. The third convolutional layer uses 128 convolutional filters and a convolutional kernel of 3 × 3 with valid padding. They all are followed by a maximum pooling layer of pool size 2 × 2 with strides of 2 × 2. Rectified Linear Unit (ReLU) is used as an activation function for all three layers. There are three fully connected layers for the classification block having 64, 128, and 26 neurons, respectively, for computation. The first two layers use ReLU for activation, and the output layer has Softmax for the same. For model optimization, we used Adam optimizer.

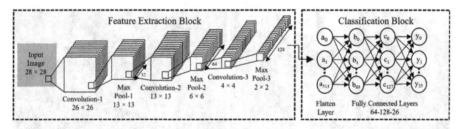

Fig. 6. CNN architecture

The output of the proposed CNN architecture on a test letter 'P' is shown in Fig. 7.

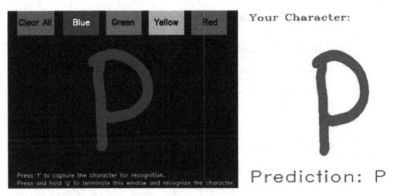

Fig. 7. Live viewer with test character 'P.' **Right:** 'P' predicted correctly in the prediction window.

4 Result and Discussion

4.1 Data Acquisition

The dataset used for the training purpose is A–Z handwritten data dataset. This dataset consists of 26 classes (A–Z) containing 372,450 handwritten grayscale images. The images are taken from NIST and NMIST large datasets, formatted in 28 × 28 pixels. Every alphabet in the image is center fitted to 20 × 20 pixels. Table 1 describes the dataset distribution for the experimental setup.

Table 1. Dataset distribution

Dataset	Type	Instances
train_x	Training	297,960
test_x	Testing	74,490
train_y	Training	297,960
test_y	Testing	74,490

4.2 Experimental Setup

The convolutional neural network was trained using two different optimizers, namely Adam (model_a) and RMSProp (model_r), in the following configuration.

- Training model_a with train_x and testing on test_x
- Training model_r with train_x and testing on test_x

4.3 Results

The resulting accuracy achieved by the CNN model for training followed by testing in the above-stated configuration is shown in Table 2. The comparison between the models' loss and accuracy during their entire runtime is visualized simultaneously in Fig. 8.

Table 2. Testing accuracy

Model	Train set	Test set	Accuracy
model_a	test_x	test_y	99.75%
model_r	test_x	test_y	98.17%

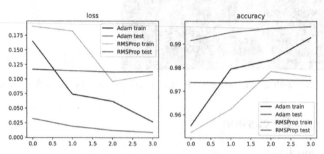

Fig. 8. Left: Comparison of loss function of Adam and RMSProp optimizers. **Right:** Comparision of accuracies of Adam and RMSProp optimizers.

4.4 Error Analysis

The confusion matrices for models model_a and model_r are shown in Fig. 9, Fig. 10, Fig. 11, and Fig. 12, respectively. Each model is trained and tested in the above-stated configuration. Upon observing these matrices, it is clear that the recognition rate is better in model_a than model_r for both training and testing.

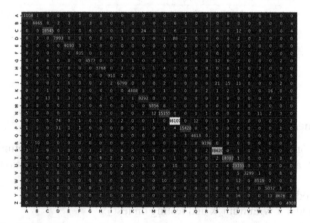

Fig. 9. Confusion matrix for the training model with Adam as an optimizer.

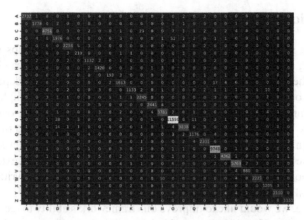

Fig. 10. Confusion matrix for the testing model with Adam as an optimizer.

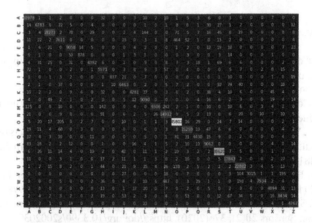

Fig. 11. Confusion matrix for the training model with RMSProp as an optimizer.

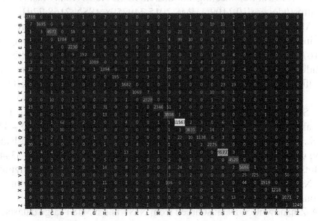

Fig. 12. Confusion matrix for the testing model with RMSProp as an optimizer.

5   Conclusion

In this work, a steady multipurpose framework for diverse age groups was proposed for recognizing air-written English language alphabets. This system can be configured according to the user's color of the tracker object and is entirely independent of any external depth or motion sensors like Leap Motion, Kinect, Time-of-flight (ToF), Myo, etc. These features make this framework ergonomic and cost-effective over other existing approaches. The recognition was performed using CNN, followed by performance optimization. In the experimental setup, the proposed framework achieved an accuracy of 99.75%. The results of this work suggest that this method can be extended over to recognizing short words, i.e., on a motion-based user interface, air-writing is suitable for short and infrequent texts.

In the future, we would like to improve this framework's ability to recognize diverse language characters. The objective would also be to improve the system's performance by training on larger datasets and making its interface more user-friendly.

References

1. Microsoft Corporation: Azure Depth Platform. Time-of-flight camera. https://devblo-gs.microsoft.com/azure-depth-platform/time-of-flight-camera-system-overview
2. Leap Motion Inc. Leap Motion. https://www.ultraleap.com
3. Microsoft Corporation. Azure Kinect Developer Kit. (2019). https://docs.microsoft.com/en-us/azure/kinect-dk
4. Itaguchi, Y., Yamada, C., Fukuzawa, K.: Writing in the air: facilitative effects of finger writing in older adults. PLoS ONE 14(12), e0226832. https://doi.org/10.1371/journal.pone.0226832. PMID:31881067; PMCID: PMC6934278
5. Rispens, J., Berckelaer, I.A., Hyperlexia: Definition and Criterion, Written Language Disorders, vol. 2 (1991). https://doi.org/10.1007/978-94-011-3732-4_8.ISBN: 978-94-010-5659-5
6. Alam, Md. S., et al.: Trajectory-based air-writing recognition using deep neural network and depth sensor. Sensors 20(2), 376(2020). https://doi.org/10.3390/s20020376
7. Chen, M., AlRegib, G., Juang. B.H.: Air-writing recognition–Part I: modeling and recognition of characters, words, and connecting motions. In: IEEE Trans. Hum. Mach. Syst. 46(3) (2016), 403–413. https://doi.org/10.1109/THMS.2015.2492598. ISSN:2168-2291.
8. Chen, M., AlRegib, G., Juang. B.H.: Air-Writing recognition–Part II: detection and recognition of writing activity in continuous stream of motion data. In: IEEE Transa. Hum. Mach. Syst. 46(3) (2016), 436–444. https://doi.org/10.1109/THMS.2015.2492599, ISSN: 2168-2291
9. Mohammadi, S., Maleki, R.: Air-writing recognition system for Persian numbers with a novel classifier. Vis. Comput. 36(5), 1001–1015 (2019). https://doi.org/10.1007/s00371-019-01717-3
10. Dash, A., Sahu, A., Shringi, R., et al: AirScript–creating documents in air. In: 14th International Conference on Document Analysis and Recognition, pp. 908– 913 (2017)
11. Ahlawat, S., Choudhary, A., Nayyar, A., Singh, S., Yoon, B.: Improved handwritten digit recognition using convolutional neural networks (CNN). Sensors 20(12), 3344 (2020)
12. OpenCV. Image Filtering. https://docs.opencv.org/4.5.2/d4/d86/group__imgproc__filter.html
13. Alzubi, J., Nayyar, A., Kumar, A.: Machine learning from theory to algorithms: an overview. J. Phys. Conf. Ser. 1142(1), 012012). (2018)
14. Patel, S.: A-Z handwritten Data: sourced from NIST and NMIST large datasets for handwritten letters (2016). https://www.kaggle.com/sachinpatel21/az-handwritten-alphabets-in-csv-format

Intraday Stock Trading Performance of Traditional Machine Learning Algorithms: Comparing Performance with and Without Consideration of Trading Costs

Kashyap D. Soni[✉] [ID]

Computer Engineering Department, Sardar Vallabhbhai National Institute of Technology, Surat, India
kashsvnit04@gmail.com

Abstract. This paper analyzes the performance of ML algorithm for trading into the U.S. stock market while considering the transaction costs and without considering the transaction costs. Investors trade stocks based on forecasts of stock price movements. Many academics have been focusing on using machine learning (ML) algorithms to forecast stock price movements in recent years. Their research, however, was confined to tiny stock datasets with few characteristics, a short back testing time, and no consideration of transaction costs. Furthermore, their experimental data were not subjected to a statistical significance test. In this article, we synthetically assess several ML algorithms on large-scale stock datasets and examine performance of the algorithms in intraday trading on stocks with and without consideration of transaction costs. We evaluate six standard machine learning methods on 8 years of data from 2010 of 425 S&P 500 stocks (SPICS). The result of this study shows that conventional ML algorithms outperform traditional DNN algorithms in most directional evaluation indicators. Furthermore, all machine learning models' trading performance is affected by changes in transaction cost. The trading performance of ML models is drastically reduced while considering transaction costs. The influence of direct and indirect costs on the performance of these algorithms varies. From this study it is observed that the performance of ML algorithm Extended Gradient Boosting is the best for trading into the U.S. stock market while considering the transaction costs and without considering the transaction costs.

Keywords: Algorithmic trading · Machine learning models · Intraday stock trading · Trading costs

1 Introduction

In everyday economic and social life, the stock market is extremely significant. By investing in the shares of a publicly traded firm with greater projected earnings, investors hope to grow their investments. Issuing stocks by publicly traded businesses is a key instrument for raising cash from the general public and expanding the industry's scope.

© Springer Nature Switzerland AG 2022
K. K. Patel et al. (Eds.): icSoftComp 2021, CCIS 1572, pp. 421–432, 2022.
https://doi.org/10.1007/978-3-031-05767-0_33

Investors make stock investing selections by predicting the up or down trend of equities in the future. For many decades, academics have primarily used statistics to explain timeseries stock price data in order to anticipate future stock return patterns [1–3]. It's interesting to note that, with the advancement of artificial intelligence technology, the modern computing methods presented by machine learning algorithms are also showing a strong increase in interest in stock market prediction [4].

The following are the primary causes.

(1) HFT trading information, diversified technical analysis indicators data, economic and financial are all readily available multi source diverse financial data.
(2) Intelligent algorithm research has been expanded. Image recognition and text analysis are two sectors where they've been put to good use.
(3) Due to advancements in computing technologies, we now have access to high computational and storage capabilities along with services like cloud hosting which allows users to have exceptionally high computing capacity within reach without actually having the computers to perform computing.

This has rapidly changed the way finance works in modern days [5]. Traditional machine learning algorithms have demonstrated a great capacity to forecast stock price trends over time. However, most prior research has concentrated on predicting stock indexes of key economies across the world or choosing a few stocks with restricted attributes based on their personal tastes, or not taking transaction costs into account, or using a very short backtesting time. Furthermore, there is no statistically significant test that compares different stock trading algorithms. Which is, large-scale stock datasets for comparison and assessment of alternative trading algorithms, as well as transaction costs and statistical significance tests, are lacking [6–8]. As a result, backtesting's results may be too optimistic [9–11]. In this context, two problems based on a large-scale stock dataset must be addressed:

(1) Can trading strategies based on ML models produce statistically significant outcomes when compared to the other standard ML algorithms that do not include transaction costs?
(2) Can trading strategies based on ML models produce statistically significant outcomes when compared to the other standard ML algorithms that include transaction costs?

The major goal for this research is to address these issues, which are critical for quantitative investing professionals and investment managers. The solutions to these challenges are quite useful for stock traders. The research object in this study is 425 SPICS from 2010 to 2017. The SPICS symbolises the world's top economy's industrial development and is attractive to investors from all around the globe. We create 44 technical indicators for each stock in SPICS using various moving averages and stochastic oscillators. The symbol representing the gain of the N + 1-th trading day compared to the N-th trading day appears on the N-th trading day's label. If the gain is positive, the label value is set to one; otherwise, it is put to zero. To create a stock dataset, we used 44 technical analysis indicators from 2000 trading days from January 1, 2010.

Training of the ML models is performed on the dataset by Walk Forward Analysis (WFA) methodology. To train and anticipate stock price trends based on technical indicators, we employ six conventional machine learning methods: Extreme Gradient Boosting (EGB) algorithm, Classification and Regression Tree (CRT) algorithm, Random Forest (RF) algorithm, Naive Bayes model (NB) algorithm, Logistic Regression (LR) algorithm, and Support Vector Machine (SVM) algorithm. Finally, we assess the performance outcomes of these machine learning algorithms using the method of directional evaluation indicators such as Area under the Curve (AUC), F1-Score (FS), Recall Rate (RR), Precision Rate (PR), Accuracy Rate (AR). Performance of these ML models is assessed by the performance evaluation indicators such as Maximum Drawdown (MD), Annualised Sharpe Ratio (ASR), Annualised Return Rate (ARR), and Winning Rate (WR).

Transparent transaction costs have a lot more influence on SPICS than slippage. Trading algorithm performance is substantially lower than that without transaction cost, as evidenced by many comparative analyses of various transaction cost architectures, indicating the sensitivity of trading performance to the transaction cost [12, 13]. The significance of this work is that it compares differences in trading results of various machine learning algorithms in both the cases, with and without trading transaction cost scenarios using non - parametric statistical test techniques. As a result, we may choose the best appropriate algorithm for intraday trading in the US stock market from among various machine learning algorithms.

The following is how the rest of the paper is organised: The architecture of this work is described in next Section. The settings for the parameters of the ML models and the procedure to obtain trading signals that are further discussed in this article, are presented in Sect. 3. The performance evaluation indicators, directional assessment indicators, and backtesting are presented in Sect. 4. Section 5 analyses and evaluates the intraday trading performance of these alternative algorithms in the US markets using nonparameter statistical test techniques.

Section 6 examines the influence of trading transaction costs on the intraday trading performance of machine learning algorithms. In Sect. 7, trading signals, transaction costs and the overall performance of the ML models is discussed. The Sect. 8 concludes the paper with a detailed conclusion and recommendations for further research.

2   Architecture of the Work

Figure 1 depicts the overall architecture for forecasting future stock price movements, trading, and backtesting using machine learning algorithms. The data collection, data cleaning and preparing, machine learning algorithms, and intraday evaluation of the trading performance of the 6 machine learning algorithms are all covered in this article. The initial stage in this investigation is to collect data. We need to think about where we should obtain data and what tools we should utilise to get it fast and accurately. All computational processes that are carried out in this article are using the Python programming language. In the meantime, we collect SPICS data from Yahoo Finance. Next, data preparation comprises the creation of a large number of widely known technical trading indicators, and dealing with the attributes, so that this data can be fed into the ML models. Finally, ML algorithms are used to create stock trading signals.

In this section, we use the WFA approach to train standard ML algorithms; these models will forecast the trend of the stock after training, this forecasting is referred to as a trading signal. We present a set of commonly used directional and performance assessment indicators, as well as a back testing technique for calculation of the indicators. Then we perform the back testing of the ML models by utilising the trading signals generated by these models, and we use the statistical method to see if the outcomes of these machine leaning algorithms have any statistical significance or not, while considering and not considering the trading transaction.

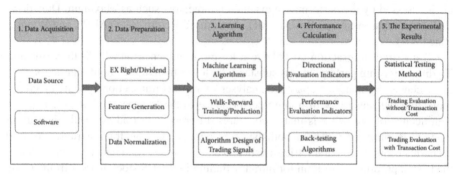

Fig. 1. Framework to predict stock price trend using machine learning models

3   Machine Learning Algorithms

3.1   Parameterization of the Machine Learning Algorithms

The objective of the machine learning algorithms is to accurately categorise the labels of the classes that are given to a training dataset. We will utilise six standard machine learning models (LR, SVM, CART, RF, BN, and XGB) as classifiers to forecast up or down movements of intrada prices of the stocks in this article. The key parameters of these ML models along with the parameters for their training are shown in Table 1.

Table 1 shows presents the way to set class labels and the features on the input of several machine learning algorithms Python programming language. In the paper, $m = 250$ denotes that in each round of WFA, we utilise the data of the previous $d = 250$ trading days for training the models; $n = 44$ denotes that each day's data includes 44 characteristics.

Table 1. Primary parameters of the machine learning algorithms

	Input	Label	Primary parameters
XGB	M(250,44)	M(250,1)	A tree can have a maximum depth of 10 levels, a maximum number of repetitions of 15, and set the rate of learning to 0.30
BN	M(250,44)	M(250,1)	The training set's class proportions are the prior probability of class membership
RF	M(250,44)	M(250,1)	There are 500 trees in total, with seven variables that are selected at random as candidates for each split
CART	M(250,44)	M(250,1)	Any node in the final tree can have a maximum depth of 21; Gini coefficient is taken for splitting index
SVM	M(250,44)	M(250,1)	Radial Basis Kernel is the kernel function utilised, and the cost of constraint violation is 1
LR	M(250,44)	M(250,1)	Logit is used for the model link function

3.2 Walk Forward Analysis

WFA [14] is the technique to train the models in rolling manner. We train the model with the most recent data rather than all previous data, and then use that model to make predictions for out-of-sample dataset. The next round's training is carried out using the next training dataset. This training dataset is one walk ahead of the previous training dataset. In real-time trading, the credibility and reliability of the trading strategy is bolstered by walk forward analysis method. In each phase of WFA, we take the data of 250 days to train the model and then the data of next 5 days is used to test the trained ML models. Because each stock has 2000 trading days of data, it takes 350 training sessions to generate 1750 forecasts, these are intraday trading signals that are generated for trading. Figure 2 depicts the WFA methodology.

3.3 Designing Algorithm for Generating the Trading Signal

In this section, machine learning algorithms perform the work of classifiers in order to forecast the future trend of the SPICS stocks to generate trading signals for each trading day, and then these generated daily trading signals from the forecasted findings are used to execute trades. Each ML algorithm is trained using the WFA technique. As per Fig. 2, we offer the producing algorithm of trading signals.

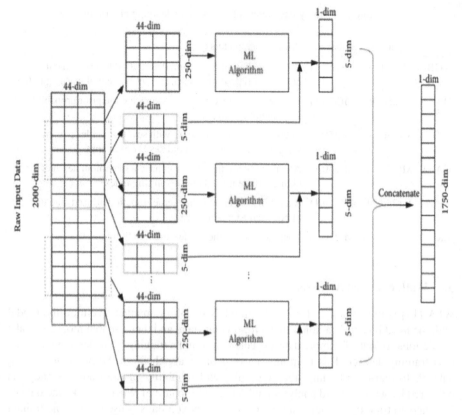

Fig. 2. WFA training and testing schematic diagram

4 Backtesting Algorithm and Evaluation Indicators

4.1 Indicators for Directional Evaluation

The major purpose of the machine learning algorithms in this article is to categorise the intraday returns of the stocks, therefore the main task of the machine learning algorithms is to forecast the trend of stocks. As a result, we must analyze the classification capabilities of these algorithms using directed evaluation markers. The dataset's real label values are sequences of sets of DN, UP. As a result, there are four types of anticipated and actual label values, denoted by the letters TU, FU, FD, and TD. TU is number of times when trading signal generated was uptrend and the actual movement of the stock was also an uptrend (UP). FU is the number of times when the trading signal was uptrend and the actual movement was a downtrend (DN). As indicated in Table 2, TD is the number of times when the trading signal was downtrend and the actual movement was also downtrend, whereas FD is the number of times when the trading signal was downtrend and the actual movement was uptrend. Table 2 is a confusion matrix, which is a 2 dimension table which categorises projected value of the label based on whether or not they match actual value of the label. The table's columns represents all potential predicted values of the labels, while the rows reflects all actual values of the labels. Correct classifications are those in which projected label values match true label values. The right predicted values of the labels are along the confusion matrix's diagonal line. What we are interested about in this article is that we purchase the stock at current day's closing price if the stock price direction is projected to be UP for next day and sell it at closing price of the next day; we take no action if the direction of the price of the stock is predicted as DN for next day. As a result, UP is a "positive" term for our problem.

Table 2. Classification result matrix of machine learning algorithms

		Predicted label values	
		UP	DN
Actual label values	UP	TU	FD
	DN	FU	TD

Accuracy Rate is used commonly to evaluate classifier performance in most classification applications. The number of right predictions divided by the total number of forecasts is known as AR. Here's how it works:

$$AR = \frac{TU + TD}{TU + FD + FU + TD} \tag{1}$$

"UP" is the profit source of the techniques in this article. The classification capacity of ML algorithms is measured by their capability to recognise "UP". As a result, RR and PR must be used to assess categorization results. These two assessment indicators were first used to assess the relevance of obtained results.

The ratio of properly predicted UP to all anticipated UP is known as PR. This is given as follows:

$$PR = \frac{TU}{TU + FU} \tag{2}$$

With a high PR, Machine Learning algorithms may concentrate on "UP" instead of "DN".

The ratio of properly anticipated "UP" to truly labelled "UP" is known as the "RR". This is how it goes:

$$RR = \frac{TU}{TU + FD} \tag{3}$$

High Recall Rate can efficiently identify huge number of true "UP" and catch them. In reality, presenting an algorithm with high RR and PR simultaneously is quite challenging. As a result, it is important to assess the ML algorithm's ability to perform classification by using various assessment indicators that mix PR and RR. F_1 is given by:

$$F_1 = 2 * PR * \frac{AR}{AR + PR} \tag{4}$$

When computing F1, it is presumed that the ratios of RR and PR are identical, however this presumption is not correct for all cases. F1 may be calculated with varying weightage for RR and PR, but establishing weightage is a difficult task. The area under the curve of ROC is referred to as AUC. The ROC equation is frequently used to evaluate tradeoffs between discovering TU and minimizing FU. It has a horizontal X-axis which is FU, while the vertical Y-axis is the TU. Every point on this graph depicts the percentage of TU under various thresholds for FU.

AUC represents how good is the classifier's capability to classify. The higher it is, the more accurate the categorization. It is important to mention that two distinct ROC graphs may result in identical value of AUC, thus when utilizing AUC value, it is important to perform qualitative analysis in conjunction with the ROC curve.

4.2 Performance Evaluation Indicators

The profit and risk control capabilities of the ML models are evaluated using performance assessment indicators. Backtesting has been carried out in this article using trading signals generated by ML systems, and trading performance is evaluated using the MDD, ASR, ARR, and WR. The values for these evaluation indicators are provided in Table 3. Here,

MDD stands for Maximum Drawdown.
ASR stands for Annualised Sharpe Ratio.
ARR stands for Annualised Rate of Return.
WR stands for Winning Rate.

4.3 Backtesting Algorithm

Backtesting is the process of implementing a trading strategy using previous data. Backtesting is generally done with a new set of data during the research and development phase of a proposed idea of trading. Furthermore, the backtesting time should be sufficient enough to make sure that the machine learning trading model can minimise data sampling bias by using the significant amount of the previous data. Backtesting allows us to obtain empirical performance of trading models. We obtain 1750 trading indications for each stock in this study. If the trading signal generated for the next day is 1, we purchase the stock at close of the market today and sell it at the next day's close; otherwise, there will be no trade executed.

5   Comparative Analysis of the Machine Learning Models

5.1 Non-parametric Statistical Tests

We utilise the backtesting method to construct the indicators to evaluate the various trading algorithms in this section. It is important to perform multiple comparisons and use variance analysis to determine any substantial variations among the assessment indicators of the machine learning algorithms and the benchmark index. Therefore, the following nine hypothesis are proposed to perform significance tests which has the null hypothesis HY_{aj} (j = 1, 2, 3, 4, 5, 6, 7, 8, 9) and the assumptions that correspond to these null hypotheses are HY_{bj} (j = 1, 2, 3, 4, 5, 6, 7, 8, 9).

For the evaluation indicators, $j \in$ {MDD, ASR, ARR, WR, AUC, F1, RR, PR, AR} and $i \in$ {BAH, XGB, RF, CART, SVM, SVM, LR}.

5.2 Performance of the Machine Learning Algorithms in SPICS: A Comparative Analysis

In MDD, ASR, ARR, WR, AUC, F1, RR, PR, and AR, Table 3 displays the average value of several trading algorithms. We can observe that XGB has the best values in AUC, F1, RR, and AR of all trading algorithms. In all trading methods, the WR of NB is the highest as clearly seen from Table 3. In all trading methods, including the benchmark portfolio (S&P 500 index), CART has the highest ARR as evident from the Table 3. In all trading methods, the ASR of RF is the highest. In all trading methods, the benchmark index's (BAH) MDD is the minimum.

Table 3. Trading performance of different trading strategies in the SPICS. All trading strategies with the best result are highlighted in boldface.

	Index	CART	NB	RF	LR	SVM	XGB
MDD	0.1938	0.3411	0.3429	**0.3283**	0.3445	0.3427	0.3339
ASR	0.8357	1.3930	1.6242	**1.6769**	1.5823	1.6021	1.6301
ARR	0.1226	**0.3318**	0.2975	0.3132	0.2945	0.3069	0.3044
WR	0.5448	0.5265	**0.5932**	0.5911	0.5856	0.5833	0.5892
AUC	-	0.6294	0.5488	0.6419	0.6492	0.6202	**0.6751**
F1	-	0.6492	0.5480	0.6596	0.6592	0.6517	**0.6752**
RR	-	0.6471	0.5761	0.6598	0.6722	0.6324	**0.6768**
PR	-	0.6515	0.5275	0.6594	0.6472	0.6732	**0.6783**
AR	-	0.6308	0.5467	0.6413	0.6447	0.6238	**0.6610**

6 Performance of Machine Learning Algorithms with Transaction Cost

The cost of trading can have an impact on the profitability of a stock trading strategy. In day trading, transaction costs that might be overlooked in long-term methods are often compounded. Many algorithmic trading research, on the other hand, assume that transaction costs are nonexistent. Frictions such as transaction costs can skew the market from the textbook ideal model. The exchange fees and commissions and taxes are examples of transparent costs that are known prior making trades. Implicit costs are those that must be assessed and include bid-ask spread, delay or losses due to slippage, and related market effect. In this section, our focus is the impact of implicit and transparent costs.

In the literature, it is assumed that transparent transaction cost is estimated as a percentage of turnover, such as less than half percentage, 0.2 percent, and half percentage. The assessment of slippage is varied. Slippage is set at 0.02 in several quant trading software. When purchasing and selling, the visible transaction fee and implicit transaction cost are levied in both directions.

The implicit costs of transactions are same, but the transparent costs of the transactions differ from broker to broker. One of the most significant variables influencing trading success is transaction cost. Transparent transaction costs in US stock trading can be calculated as a flat fee for each transaction, or as a variable fee calculated on the trading volume and turnover of each transaction. Traders often can bargain with brokers to negotiate transaction costs. Various brokers charge different fees for transactions. Implicit transaction costs are unknown in advance and estimating them is extremely difficult. As a result, for convenience of computation, we assume that the percentage of turnover equals the visible transaction cost.

For any trading algorithm, the WR decreases as the transaction cost rises, which is obvious. The WR is considerably lower than the WR without transaction costs in all transaction cost architectures. For every trading algorithm, the ARR decreases as the transaction cost rises. The ARR is considerably lower than the ARR without transaction costs in all transaction cost models. For every trading algorithm, the ASR decreases as the transaction cost rises. The ASR is considerably lower than the ASR without transaction costs in all transaction cost models. Similarly, for all the other performance metrics, the performance with transaction costs is significantly lower than the performance without considering the transaction costs.

7   Results and Discussion

Predicting future stock price movements and making trades is always difficult undertakings. But, the stock market's high rate of return attracts an increasing number of participants, and the investors are encouraged by high risks to give their best to develop lucrative trading techniques. The rapidly changing nature of financial markets, the exponential rise of large financial data, and the rapid acquisition of trading opportunities present academic circles with an expanding number of study subjects. In this work, we employ various well-known and frequently used machine learning techniques to trade stocks. Our goal is to see if there are any substantial variations in stock trading performance across different machine learning algorithms. Furthermore, we investigate whether highly successful trading algorithms may be found with consideration of transaction costs. High dimensionality, non-linearity, poor signal - noise ratio, and high randomness are the characteristics of rapidly evolving financial markets. As a result, employing computers to identify underlying patterns in financial large data is tough. This idea is also demonstrated in this study. Traditional machine learning methods, such as XGB, are better at predicting stock price trend. As a result, when collecting inherent patterns in financial data, simple models are less prone to overfitting and can generate better forecasts regarding the direction of stock price movements. When using the ML method to categorise jobs, we really assume that sample data is independent and uniformly distributed. The learning model has a small number of parameters. As a result, with fewer data, the learning aim may be achieved more effectively. We split transaction cost into two categories in this paper: Implicit costs and transparent costs. The influence of the two transaction costs on performance varies depending on the market selected. All the aspects of real-world trading have been taken into consideration; the transaction cost assumption used in this study is rather basic. As a result, in future study, we may evaluate the influence of market impact cost and opportunity cost on performance of the machine learning algorithms.

8   Conclusions

In this paper, we have used 425 SPICS stocks of US stock market as study object where we have picked up data of past 2000 days upto December 31, 2017. 44 indicators were built on this dataset and were fed to the 6 machine learning algorithms as input features. We create trading strategy based on the trading signals generated by these machine learning algorithms. We analyse the cases where we account transaction costs and the case where we ignore the transaction costs. The contribution of this paper is that we present the result of this study where we safely conclude from the results of the study that raw trading performance is not the accurate measurement of a trading strategy. We need to account for the trading costs in order to actually vouch for the performance of the trading strategy. The performance of all 6 machine learning models went down significantly after factoring in the transaction costs. From comparison of the results of the backtesting of the 6 machine learning models without considering transaction costs and with considering transaction costs, we conclude that a high return yielding strategy tested without transaction costs can be disastrous when executed in real world trading with transaction costs. Therefore it is very important to analyse the performance of the trading strategy with transaction costs to understand true potential of the strategy.

The advancements in the field of machine learning are occurring at phenomenal rate due to increasing access to the financial data which was once not accessible to everyone. Due to these advancements, we hope the future work can be carried out in these areas:

1. Analyse the markets for statistical arbitrage using machine learning models to perform high frequency trading.
2. More complex implicit costs can be studied and the impact they have on the machine learning models and how to mitigate these costs and optimize the performance of the machine learning models.
3. Machine learning can be used to create an optimal and dynamic portfolio and how risk can be minimized by using such a portfolio.

The solution to these problems shall help us to create a highly profitable trading strategy that can be deployed in real world stock markets which can manage risks on it's own while maximizing returns and adapt to the ever changing financial markets. This shall also help us to scale the trading strategy in terms of number of stocks which it can trade to the number of trades it can make in each of the stocks.

Compliance with Ethical Standards.   The Author declares that he has no conflict of interest.

The research does not involve any human or animal participants as test subjects or in any other capacity.

The Author has not received any funding for this research.

References

1. Cavalcantea, R.C., Brasileirob, R.C., Souzab, V.F., Nobregab, J.P., Oliveir, A.I.: Computational intelligence and financial markets: a survey and future directions. Expert Syst. App. **55**, 194–211 (2016)
2. Huang, W., Nakamori, Y., Wang, S.-Y.: Forecasting stock market movement direction with support vector machine. Comput. Oper. Res. **32**(10), 2513–2522 (2005)
3. Chen, J.: SVM application of financial time series forecasting using empirical technical indicators. In: Proceedings of the International Conference on Information Networking and Automation, ICINA 2010, pp. 1–77, China, October 2010
4. Xie, C.Q.: The optimization of share price prediction model based on support vector machine. In: Proceedings of the International Conference on Control, Automation and Systems Engineering, CASE 2011, pp. 1–4, Singapore, July 2011
5. Ładyżyński, P., Żbikowski, K., Grzegorzewski, P.: Stock trading with random forests, trend detection tests and force index volume indicators. In: Rutkowski, L., Korytkowski, M., Scherer, R., Tadeusiewicz, R., Zadeh, L.A., Zurada, J.M. (eds.) ICAISC 2013. LNCS (LNAI), vol. 7895, pp. 441–452. Springer, Heidelberg (2013). https://doi.org/10.1007/978-3-642-38610-7_41
6. Zhang, J., Cui, S., Xu, Y., Li, Q., Li, T.: A novel data-driven stock price trend prediction system. Expert Syst. Appl. **97**, 60–69 (2018)
7. Ruta, D.: Automated trading with machine learning on big data. In: Proceedings of the 3rd IEEE International Congress on Big Data, BigData Congress 2014, pp. 824–830, USA, July 2014
8. Patel, J., Shah, S., Thakkar, P., Kotecha, K.: Predicting stock and stock price index movement using trend deterministic data preparation and machine learning techniques. Expert Syst. Appl. **42**(1), 259–268 (2015)
9. Luo, L., Chen, X.: Integrating piecewise linear representation and weighted support vector machine for stock trading signal prediction. Appl. Soft Comput. **13**(2), 806–816 (2013)
10. Zbikowski, K.: Using volume weighted support vector machines with walk forward testing and feature selection for the purpose of creating stock trading strategy. Expert Syst. Appl. **42**(4), 1797–1805 (2015)
11. Dash, R., Dash, P.K.: A hybrid stock trading framework integrating technical analysis with machine learning techniques. J. Fin. Data Sci. **2**(1), 42–57 (2016)
12. GorencNovak, M., Velušček, D.: Prediction of stock price movement based on daily high prices. Quant. Fin. **16**(5), 793–826 (2015)
13. Cervello-Royo, R.: Stock market trading rule based on pattern recognition and technical analysis: forecasting the DJIA index with intraday data. Expert Syst. Appl. **42**(14), 5963–5975 (2015)
14. Chong, E., Han, C., Park, F.C.: Deep learning networks for stock market analysis and prediction: methodology, data representations, and case studies. Expert Syst. Appl. **83**, 187–205 (2017)
15. Lv, D., Yuan, S., Li, M., Yang, X.: An empirical study of machine learning algorithms for stock daily trading strategy. Math. Probl. Eng. **2019** (2019). (Figure 1 and Figure 2 Article ID 7816154 2019)

Author Index

Abudayor, Artee 51
Al Aghbari, Zaher 397
AlFarah, Meshari Essa 397
Angel Arul Jothi, J. 235
Arunnehru, J. 249

Bajaj, Rakesh Kumar 78
Bandyopadhyay, Aritra 65
Bhattacharya, Indradeep 90

C, Namrata Mahender 382
Chaudhary, Abhishek 410
Chlebiej, Michal 28
Chuma, Euclides Lourenço 302
Chuttur, Yasser 312

Das, Devadatta 65
de Oliveira, Gabriel Gomes 302
Deepak, Gerard 277, 356
Deva Prasad, A. M. 235
Doctor, Gayatri 181
Dosi, Saurav 15

Furqan, Ghazala 116

Gaba, Vishal 140
Gadebe, Moses L. 167
Gaikwad, Prajwal 290
Gandhi, Nilay 367
Gielecki, Jerzy 28
Gupta, Shibakali 90
Gurijala, Sudhamsu 341

Haus, Benedikt 41
Hemalatha, K. 264

Iano, Yuzo 302

Jaiswal, Arunima 116
JaswanthReddy, D. 153
Joshi, Hardik 209
Joshi, Hiren 209

Kadam, Vaishali P. 382
Kamel, Ibrahim 397
Karthik Upendra, P. V. 264
Khandale, Kalpana B. 382
Kogeda, Okuthe P. 167
Krishnan, Akhil S. 277
Krishnan, N. 356
Kumar, Prabhat 410

Mamode Ally, Nooreen 312
Mercorelli, Paolo 41
Mittal, Shruti 194
Mouheb, Djedjiga 397

Naga Sahithya, N. 264
Nagpal, C. K. 194
Nalbantoğlu, Özkan Ufuk 51
Naqvi, Najme Zehra 116
Narasimha, D. 15
Nazurally, Azina 312
Neetoo, Hudaa 312
Negrete, Juan Carlos Minango 302
Negrete, Pablo David Minango 302

Ojo, Sunday O. 167
Oommen, Deepthi 249

Pardeshi, Sanskruti 290
Pathi, Preetham Reddy 222
Pendyala, Vishnu S. 127
Pershin, Ilya 103
Poonia, Mahima 78

RadhaKrishna, P. 153
Raisinghani, Yogesh 367
Rasagna, K. 264
Raut, Samarth S. 15
Richa 3
Ritu 327
Rudrawar, Shrushti 290

Sabnis, Hrishikesh   235
Sai Charan, B.   264
Sai Chitti Subrahmanyam, V.   341
Saleti, Sumalatha   153, 264
Salvi, Nishani   181
Sangwan, O. P.   327
Sawant, Shreya   290
Schäfer, Lennart   41
Shah, Khushboo   209
Shah, Sanket   367
Sharma, Abhishek   410
Shimgekar, Soorya Ram   222
Soni, Kashyap D.   421
Srinivasan, Kavitha   341
Swetha, B.   341

Takooree, Sandhya   312
Tumakov, Dmitrii   103

Upare, Ketaki   290

Vamsi, Bala   15
Vaz, Gabriel Caumo   302
Verma, Vijay   140
Vijayarajan, V.   222

Yadav, Jyotsna   3
Yap, Jin Siang   41

Zurada, Anna   28

Printed in the United States
by Baker & Taylor Publisher Services